PROGRESS IN LASERS AND LASER FUSION

Studies in the Natural Sciences

A Series from the Center for Theoretical Studies
University of Miami, Coral Gables, Florida

A Continuation Order Plan is available for this series. A continuation order will bring delivery of each new volume immediately upon publication. Volumes are billed only upon actual shipment. For further information please contact the publisher.

ORBIS SCIENTIAE

PROGRESS IN LASERS AND LASER FUSION

Chairman
Behram Kursunoglu

Editors
Arnold Perlmutter
Susan M. Widmayer

Scientific Secretaries
Uri Bernstein
Joseph Hubbard
Christian Le Monnier de Gouville
Laurence Mittag
Donald Pettengill
George Soukup
M. Y. Wang

Center for Theoretical Studies
University of Miami
Coral Gables, Florida

PLENUM PRESS • NEW YORK AND LONDON

Library of Congress Cataloging in Publication Data

Orbis Scientiae, University of Miami, 1975.
   Progress in lasers and laser fusion.

   (Studies in the natural sciences; v. 8)
   "Part of the proceedings of Orbis Scientiae held by the Center for Theoretical
Studies, University of Miami, January 20-24, 1975."
   Includes bibliographical references and index.
   1. Laser fusion—Congresses. 2. Lasers—Congresses. I. Kursunoglu, Behram,
1922-   II. Perlmutter, Arnold, 1928-   III. Widmayer, Susan M.
IV. Miami, University of, Coral Gables, Fla. Center for Theoretical Studies. V.
Title. VI. Series.
QC791.7.072 1975   535.5'8   75-16375
ISBN-13: 978-1-4684-2921-3   e-ISBN-13: 978-1-4684-2919-0
DOI: 10.1007/978-1-4684-2919-0

Part of the Proceedings of Orbis Scientiae held by the Center for Theoretical
Studies, University of Miami, January 20-24, 1975

© 1975 Plenum Press, New York
Softcover reprint of the hardcover 1st edition 1975

A Division of Plenum Publishing Corporation
227 West 17th Street, New York, N.Y. 10011

United Kingdom edition published by Plenum Press, London
A Division of Plenum Publishing Company, Ltd.
Davis House (4th Floor), 8 Scrubs Lane, Harlesden, London, NW10 6SE, England

PREFACE

This volume contains a portion of the presentations
given at the session on Laser-Fusion and Laser Development of Orbis Scientiae II, held at the Center for
Theoretical Studies, University of Miami, from January
20 through January 24, 1975. This second in the new
series of meetings held at the CTS strove to implement
the goals professed in the organization of Orbis Scientiae
in 1974, namely to encourage scientists in several disci-
plines to exchange views, not only with colleagues who
share similar research interests, but also to acquaint
scientists in other fields with the leading ideas and
current results in each area represented. Thus, an
effort has been made to include papers in each session
that discuss fundamental issues in a way which is com-
prehensible to scientists who are specialists in other
areas. Also in keeping with the philosophy of Orbis
Scientiae, the major topics each year are to be varied,
with the invariant being the inclusion of developments
in fundamental physics.

The discussions of the current state of the art in
lasers and fusion represented in this volume are not
only of interest because they deal with newly unfolding
branches of physics, but also because of their potential
technological and societal significance. The paper by
V. N. Lugovoi and A. M. Prokhorov was not presented at
Orbis Scientiae II, but is included because of its
relevance to the topics in this volume.

Special gratitude is due to the following for
their contributions as organizers and moderators of the

sessions on lasers and laser fusion: Edward Teller,
Richard Morse, Arthur Kantrowitz, Marlan Scully and
Willis Lamb, Jr. The editors wish to express their
appreciation to Mrs. Helga Billings and Mrs. Jacquelyn
Zagursky for their dedication in the preparation of the
manuscripts for publication and for their capable
assistance during the meetings.

A companion volume, entitled Theories and Experiments in High Energy Physics, incorporates the papers
delivered at Orbis Scientiae II complementary to those
included in the present one.

The Orbis Scientiae II was supported in part by
the United States Energy Research and Development Administration, High Energy Physics Division.

                        The Editors

CONTENTS

Some of the participants of the Orbis Scientiae II in attendance during the Laser and Laser Fusion Session

OPENING REMARKS

Edward Teller

Lawrence Livermore Laboratory

University of California

Livermore, California 94550

From the time that the first hydrogen bomb was
exploded, administrators and politicians have urged the
rapid development of controlled fusion. It is indeed
a wonderful prospect to harness fusion energy since the
fusion fuel is abundant, fusion reactors would be ex-
ceptionally safe, and the storage of waste products
would present practically no problem. Some of us fore-
saw that to achieve controlled fusion would take research
that would not only take a long time, but would also
be extremely interesting, and yield many scientific by-
products.

This prediction was justified. Today throughout
the world plasma physics is being pursued not only for
the sake of producing controlled fusion, but also for
the sake of understanding the stars and for the sake of
many practical applications. Incidentally, plasma
physics in its own right is one of the most exciting
branches of applied research. In all this work we have

been considering plasmas which were to burn in a quiet
and more or less continuous fashion.

In recent years interest has turned to a new
principle: the micro-explosion. If one compresses
thermonuclear fuel to a thousand times its liquid density, simple similarity considerations (based on the
fact that the most important processes are binary collisions) show that the scale of a hydrogen bomb explosion can be reduced a million fold. The similarity
consideration is not exact, but is a good approximation.
In actual fact, a reduction by more than one million
can be accomplished.

This opens the possibility of a nuclear-fusion
"internal combustion" engine. Tiny droplets of thermonuclear fuel may be exploded and the process may be
repeated billions of times. This could in the end lead
to a new practical energy source.

There are, of course, a few difficulties. First,
we have to concentrate energy into a volume of about a
cubic millimeter in a time shorter than a nanosecond.
This might be done in a variety of ways. Today the
most popular scheme employs lasers whose energy is
focussed on the small droplet. The lasers of sufficient
energy and hopefully short wavelength are not yet available. The short wavelength is a great advantage because
CO_2 laser light gets reflected and absorbed in exceedingly dilute plasma which always will surround the droplet that is to be imploded.

The mechanism of the implosion can be easily
understood in a crude way. Lasers of the requisite
energy carry electric fields greater than the fields
that hold outer electrons in their atomic orbits. Thus
absorption is connected with instant generation of

of plasma.   The plasma will evaporate and, by recoil,
compress the remaining part of the droplet.

But at this point our difficulties have merely
started.   A thousand fold compression (an even higher
compression would be preferable) requires great symmetry.   Otherwise the droplet will disintegrate into a
spray of tiny fragments rather than be compressed to an
exceedingly high density.   Even so, given enough time
and work, I believe the experiment will succeed.

But after this is done we will have to face problems of engineering.   The individual explosions will
not be small.   To create an economically viable system
several explosions per second will be needed and the processes will have to continue--in a somewhat radioactive
surrounding--for many years.   How to make such a system
survive, how to keep it adjusted, and above all, how to
produce it for a moderate amount of money seem to be
tremendous problems.   I believe that laser fusion can
in no sense be the short-term answer to the energy
crisis.   Unfortunately, this crisis does demand short-
term answers.

It has been argued that in a few years laser
fusion has made great progress and is going to catch up
and surpass the older procedure of burning plasmas at
a low density.   To a superficial observer this prediction may seem justified.   I want to quote Niels
Bohr's definition of an expert: "A person who through
his own painful experience has found out all the mistakes which one can commit in a narrow field".   In controlled fusion there are not experts as yet.   But in
the burning of dilute plasmas (within confining magnetic
fields) we are approaching the stage of expertise.   In
the field of laser fusion we have the enjoyable experience ahead of us to commit many interesting mistakes.

The commission of these mistakes will mean physics research in the truest sense of the word. We are already beginning to compress small pieces of matter to high densities and these pieces of matter are out in the open, except that they are surrounded by a dilute envelope of plasma. This makes it possible to explore the state of matter at high densities in a direct and novel way.

Furthermore, x-ray bursts and neutron bursts derived from compressed matter and from initial thermonuclear burning will be research tools of real interest. As in all cases, research in a new field does not stand by itself, but produces stimulation in many neighboring areas.

There is one statement one can make about laser fusion which is certainly true. Laser fusion is a challenge. As physicists, we should not be deterred by the fact that the pay-off in the foreseeable future will be in physics, rather than in the production of cheap energy.

THEORETICAL INTERPRETATIONS OF ENHANCED LASER LIGHT ABSORPTION*

W. L. Kruer

Lawrence Livermore Laboratory

University of California, Livermore, California

I.   INTRODUCTION

The absorption of intense laser light is obviously
one of the very important questions for laser fusion
applications. In experiments this absorption has been
observed to be substantially more efficient than expected on the basis of classical inverse Bremsstrahlung.
An absorption efficiency of $\sim$ 70% has been typically
observed in experiments with slab targets[1-4]--even using
laser light intensities exceeding 10^{16} W/cm^2 (Nd). It
should be noted that a number of experiments with curved
targets such as spheres or cylinders have shown a somewhat lower absorption efficiency of $\sim$ 30%, with about
half the energy lost to refraction around the target.
But even in these experiments the absorption is usually
found to be greater than expected classically at high
intensities.

---

*All research performed under the auspices of the U. S.
 Energy Resource and Development Agency.

5

We can theoretically understand this enhanced
absorption on the basis of collective processes in the
plasma; i.e., the conversion of laser light into plasma
waves. An overview of our present understanding of
laser light absorption will be presented. The aim is to
convey the physical ideas using very simple estimates
rather than elaborate on the latest technical detail.
We will first discuss classical inverse Bremsstrahlung,
showing why it becomes inefficient even at moderate
intensities (10^{13} - 10^{14} W/cm^2, Nd), and then show
how a plasma can be heated collectively. The discussion
is made more concrete by applying our estimates to some
recent laser plasma experiments. We conclude with a
brief discussion of light absorption at high intensity
($I \gtrsim 10^{15}$ W/cm^2, Nd), emphasizing the importance of
density profile modifications and the possibility of
stimulated scattering of light from the plasma.

II. CLASSICAL INVERSE BREMSSTRAHLUNG

First let's look at some simple estimates in order
to see what to expect from classical inverse Bremsstrahlung. In more physical terms, this is simply
Joule heating of the plasma by the high frequency laser
light. The rise in the kinetic energy of the plasma
is

$$\frac{dkE}{dt} = \nu_{ei} \frac{E_L^2}{8\pi} ,$$

where ν_{ei} is the electron-ion collision frequency and
E_L is the electric field of the laser light. As is
well known, $\nu_{ei} \propto 1/\Theta_e^{3/2}$, where Θ_e is the electron

temperature.   The decrease with temperature follows
from general properties of the Coulomb force law.   And
so the problem becomes obvious.   A hot plasma becomes
collisionless, meaning that Joule heating becomes in-
effective.

We can illustrate the numbers involved by a very
simple "back-of-the-envelope" calculation.   Consider the
propagation of light into an inhomogeneous plasma slab,
assuming for simplicity a linear rise in density from
zero to the critical density in a distance L.   Then a
simple integration shows how the light is classically
attenuated as it traverses the plasma (in and out).

$$I_{ABS} = I_o \left[1 - \exp\left(-\frac{32}{15} k_o L \frac{\nu_{CR}}{\omega_o}\right)\right], \qquad (1)$$

where I_{ABS} is the absorbed intensity, I_o the incident
intensity and ν_{CR} the collision frequency evaluated at
the critical density.   The collision frequency depends
on temperature.   For our purposes we will crudely estimate a temperature by using the flux limit, which
essentially determines the minimum temperature the plasma
must reach in order to carry off the absorbed energy.
So, if anything, we are giving an over-estimate of classical absorption.   With a little algebra, we then obtain

$$\frac{I_{ABS}}{I_o} \; \ell n \; \left[1 - \left(\frac{I_{ABS}}{I_o}\right)\right]^{-1} = 10^{11} \, k_o L \qquad (2)$$

This estimate of the fractional absorption versus
incident intensity is plotted in Figure 1.   A scale
length of 100 λ_o (free space wavelengths) has been
assumed, a value estimated for some experiments to be
discussed shortly.   The absorption efficiency is quite

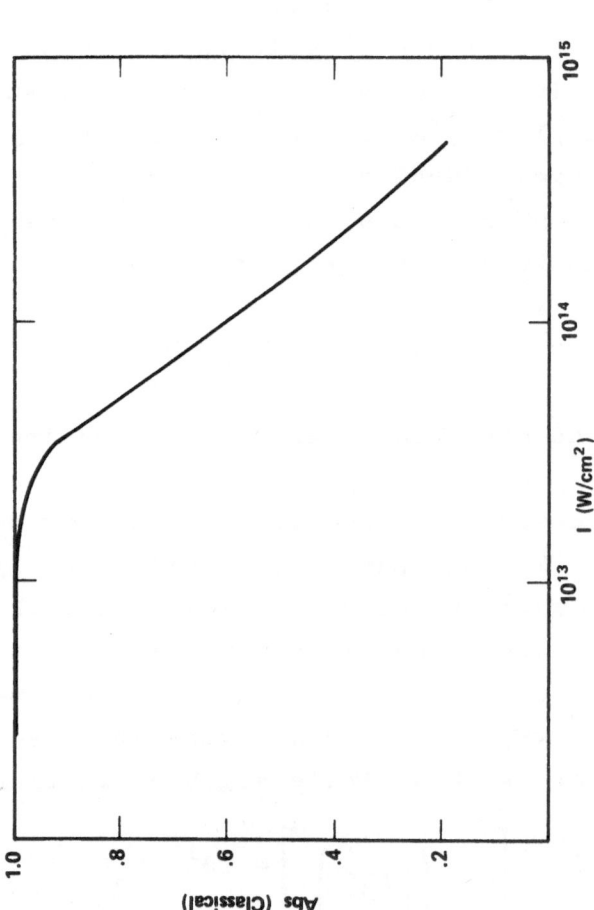

FIG. 1 An estimate of the fractional absorption due to inverse Bremsstrahlung as a function of laser light intensity.

good for intensities up to $\sim 10^{13}$ W/cm^2 but then falls
off abruptly. This decrease reflects the "bleaching
out" of the plasma-the fact that as it becomes hot, it
becomes relatively collisionless. Even sacrificing
efficiency, one can only classically absorb intensities
of $\sim 10^{14}$ W/cm^2, which corresponds to heating the plasma
to temperatures of $\sim$ .5 - 1 keV. Now obviously one can
change these numbers somewhat by using high Z targets or
different scale lengths, and one can do much more sophisticated calculations, but this calculation is indicative of what to expect from classical heating.

III. PLASMA HEATING BY COLLECTIVE PROCESSES

However, there is a way to efficiently heat even
a collisionless plasma. The laser light can convert
its energy into electron plasma waves. These electron
plasma waves, which are simply high frequency oscillations
of charge density, in turn accelerate and heat the electrons. The generation of electron plasma waves is most
efficient near the critical density surface, where
the laser light frequency equals the local electron
plasma frequency. Here virtually any spatial variation
in ion density will couple laser light into electron
plasma waves.[7] Physically this is very simple to understand. As illustrated in Figure 2, the electric field
of the laser light rapidly oscillates electrons from
regions of higher density to regions of lower density,
and vice versa, at a frequency near that at which the
plasma naturally responds.

The spatial variations in ion density are produced
in many different ways.

1. They can be spontaneously generated from the
noise by parametric instabilities.[8] In the

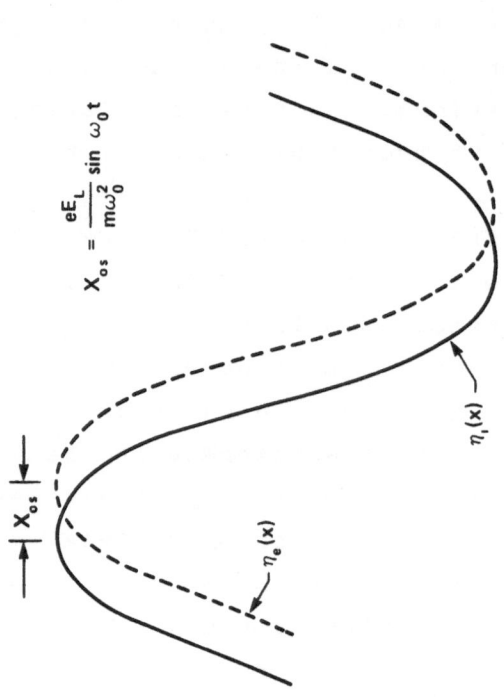

FIG. 2 A schematic which illustrates how laser light is converted into electron plasma waves near the critical density.

simplest case, such instabilities can be
thought of as the resonant decay of a light
wave into an electron plasma wave and an ion
wave.

2. The gradient in ion density can result from
the plasma expansion into the vacuum, provided
the light is obliquely incident. This is the
well-known phenomenon of resonant absorption,
discussed in standard textbooks.[9,10]

3. Particularly at high intensities ($\sim 10^{16}$ W/cm^2),
the variations in ion density can be generated
by many additional processes, such as laser
light filamentation and streaming instabilities
due to heat transport.[11] Such processes are
especially important since large density fluctuations efficiently couple laser light into
electron plasma waves even when the coupling
is quite off-resonant.[12] Roughly the frequency mismatch normalized to the laser light
frequency can be as large as the relative
magnitude of the density fluctuation, i.e.,
$\frac{\Delta\omega}{\omega_o} \sim \frac{\delta n}{n_o}$. Hence the absorption is not so finely
tuned to the critical density surface and
not so sensitive to density profile changes
which can be quite substantial at high intensities.

So there are many different ways to couple the
light into electron plasma waves. It's worthwhile to
quickly review how the plasma waves in turn heat the
particles. The charge density fluctuations have an
associated high frequency electric field which can be
described in terms of waves of the form $E_k \sin(kx - \omega t)$.

Here E_k is the Fourier amplitude, k the wave number and
ω the frequency ($\omega \sim \omega_{pe}$). Very slow electrons see
only a rapidly oscillating field and are not heated.
On the other hand, those electrons with velocity $\sim \omega/k$
experience an essentially constant field and are efficiently accelerated. The details of the energy transfer depend on the velocity distribution of the particles.
But the important point is that both instabilities and
resonant absorption tend to produce high phase velocity
electron plasma waves, ($\omega/k \gtrsim 3 \nu_{te}$ where ν_{te} is the
electron thermal velocity), and so a quite energetic
heated electron spectrum results. In contrast, processes such as electron-ion streaming instabilities
due to heat transport generate short wavelength ion
fluctuations which then (off-resonantly) couple laser
light into much slower electron plasma waves. A less
energetic spectrum of heated electrons then results,
as confirmed in computer simulations.

IV.   AN ESTIMATE OF INSTABILITY HEATING

Let us again consider a very simple example in
order to more quantitatively illustrate heating via
collective processes. In particular, consider rather
moderate laser light intensities near those at which
the classical heating strongly diminishes. We will see
that /the description becomes more complex for higher
laser light intensities like 10^{15} - 10^{16} W/cm^2, since
a larger number of processes can enter and profile
modifications become quite large. Hence this latter
regime is less amenable to "back-of-the-envelope" estimates.

With these restrictions, we again consider laser
light as normally incident on a inhomogeneous plasma

slab and now estimate the effect of instability heating.
Of course, if light is obliquely incident with the
proper polarization, it can be partially absorbed by
resonant absorption. But at moderate intensities this
is an additive effect; that is, we are probably under-
estimating the collective heating which is fine for this
discussion.

Simple models of instability heating near the
critical density have been investigated in some detail
both analytically[13] and in computer simulations.[8] For
our purposes, two points suffice. First, there is an
efficient plasma heating which can be estimated in terms
of an effective collision frequency (ν^*) which is
$\nu^* \sim 2\gamma$. Here γ is the linear growth rate of the instability. Secondly, there is a production of suprathermal tails on the electron distribution function.
Both these features have been observed in microwave
experiments[14] in which the turbulence can be directly
probed.

It is straight-forward to apply these estimates to
compute the energy absorption as a function of intensity
for the same case in which we have estimated the classical absorption. The result is shown in Figure 3, where
the fractional absorption is plotted versus intensity.
The absorption due to instabilities starts to be appreciable for an intensity of $\sim 10^{13}$ W/cm^2 and steadily
increases with intensity as the plasma is driven more
and more turbulent. Comparing with the results in
Figure 1, we note that 10^{13} W/cm^2 is roughly the intensity at which the classical absorption begins to
rapidly diminish. This complementary behavior is not
accidental but reflects some very simple physics. As
the plasma becomes hot and relatively collisionless, the

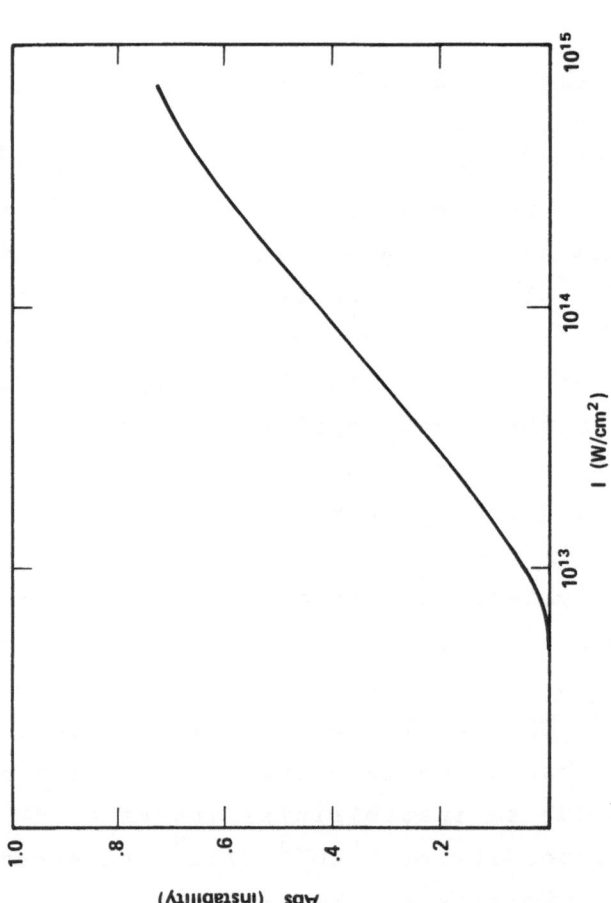

FIG. 3   An estimate of the fractional absorption due to instabilities near the critical density as a function of laser light intensity.

Joule heating goes away but plasma waves are then
easily excited. As already mentioned, we shouldn't
extrapolate our estimate for instability heating to
very large intensities because profile modifications can
limit the efficiency of instability heating vis-a-vis
other mechanisms.

V.   A COMPARISON WITH EXPERIMENTS.

It is instructive to apply these estimates to some
recent laser plasma experiments. These experiments[1]
were carried out by Professor Yamanaka and colleagues
in Japan. They focused a 2 ns pulse length Nd laser
on a D_2 slab. In these experiments the intensity
ranged from $\sim 10^{13}$ W/cm^2 to $\sim 3 \times 10^{14}$ W/cm^2. Figure
4 shows their measurements of electron temperature
versus intensity as deduced from x-ray measurements.
In the lower part of this figure, we have replotted our
estimates of the absorption efficiency due to both inverse Bremsstrahlung and instability heating.

The observed absorption remained $\gtrsim 80\%$ for all
intensities. We note that this absorption is larger
than estimated classically at the higher intensities
but is readily accounted for by instability heating.
More convincingly, above an intensity of $\sim 2 \times 10^{13}$ w/cm^2,
they deduce two distinct electron temperatures from the
x-ray measurements. This represents the onset of non-
Maxwellian velocity distributions, a signature of heating
via plasma waves. Indeed this onset occurs at an intensity near that for which we have estimated the onset
of substantial instability heating, as shown by the
lower curve. This represents reasonable evidence that a
plasma can in fact be efficiently heated by the conversion
of laser light into electron plasma waves.

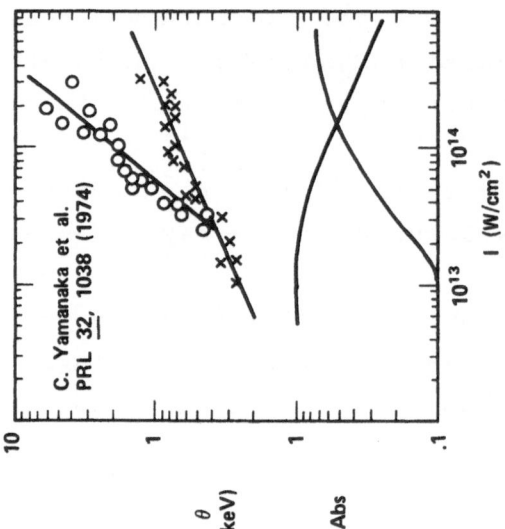

FIG. 4   Measurements of electron temperature versus laser light intensity (from Ref. 1). The lower curves are the estimates of light absorption as a function of intensity given in Figs. 1 and 3.

VI.   LIGHT ABSORPTION WITH DENSITY PROFILE MODIFICATION

We have concentrated on presenting the basic ideas with simple estimates. Lest we oversimplify the complexity of quantifying the light plasma interaction, let us conclude with a brief discussion of several important complications which enter for high intensity light. First it becomes essential to allow for the reaction back of the strong turbulence on the plasma density profile. This is particularly true since the electron plasma wave generation most efficiently occurs at preferred values of the density; i.e., when the density is within ∿20% of the critical density.

Strong density profile modification[15] has been clearly established in recent computer simulation studies of laser light absorption near the critical density. These studies use a 2-D particle code[16] which solves the complete set of Maxwell's equations and allows for fully relativistic particle dynamics. Hence the simulations self-consistently treat the laser light, the plasma turbulence both parallel and orthogonal to the direction of the density gradient, and the reaction of the turbulence back on the density profile. Their principal limitation lies in the somewhat limited region of plasma which can be simulated on the fine time and space scales characteristic of the turbulence. However, this is not too severe a restriction for a study of processes near the critical density, since the density profile becomes quite steep nonlinearly.

A typical nonlinearly steepened density profile is shown in Figure 5. This is a plot of the spatial profile of ion density from a 2-D simulation in which the laser light is obliquely incident on a plasma slab. The light intensity is 5×10^{15} W/cm^2, the background

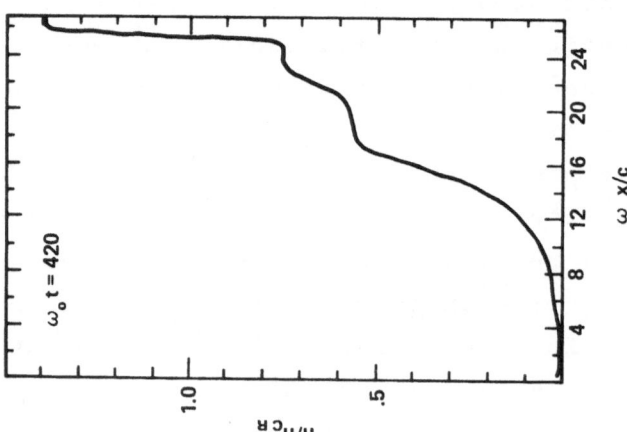

FIG. 5   A plasma density profile from a two-dimensional computer simulation of laser light absorption.

plasma temperature is 4 keV, and the angle of incidence is 24°. The laser light propagates into the
plasma and both resonantly absorbs and instability heats
near the critical density. However, the large plasma
wave fields then eject plasma by their ponderomotive
force and by the strong localized heating. As shown
in Figure 5, there results a sharp step from sub to
super-critical density. The effective gradient in
density becomes quite steep, for example, a jump in
density of .6 n_{CR} in a distance of $1/3 \ \lambda_0$.

This profile steepening is easy to understand physically. Basically one has a freely expanding plasma to
which a pressure is being applied at some preferred
densities; i.e., near the critical density. The density
locally piles up behind this region, forming a sharp
step -- an effect first pointed out by Ray Kidder.[17]
Such a profile modification has been recently observed
in microwave experiments[18] and has also been inferred
from laser plasma experiments.[4]

The nonlinear steepening has a number of important
consequences for the light absorption. For one thing,
instability heating at the critical surface is strongly
reduced since there is only a small region of plasma
then accessible to the instability generation. Instability heating can still take place in the lower
density region before the step, but simulations indicate that this plateau is at too low a density for
efficient instability generation for light intensities
greater than $\sim 3 \times 10^{14}$ W/cm^2. (This number can be
substantially increased if the electron transport is
inhibited due to magnetic fields, for example.) On the
other hand, the steepening favors resonant absorption
since the critical density surface becomes more acces-

sible to the obliquely incident light. The absorption
in this simulation is 60%, with over 80% of this due
to resonant absorption. The profile modification also
strongly affects resonant absorption. For example,
it much reduces the heated electron energies, since the
plasma oscillations then have a shorter wavelength and
thus effectively a lower phase velocity.

It should be emphasized that more complicated
models are needed to quantitatively predict the light
absorption at high intensity. These complications include the generation of short wavelength ion fluctuations
by heat flow into the overdense plasma, the production
of dc magnetic fields, a potential hydrodynamic break-
up of the critical density surface,[19] and the allowance for spatial variations in the intensity profile of
the incident light. Furthermore, some absorption also
occurs in the underdense plasma where the density is
less than the critical density due to such processes
as the $2\,\omega_{pe}$ instability, in which the light decays
into two electron plasma waves.

VII. STIMULATED SCATTERING OF LASER LIGHT

Collective processes in the underdense plasma are
especially important for plasmas with long density
gradients, since part of the light can be reflected
before it reaches the critical density. This induced
scattering[20,21] of the light occurs due to instabilities
in which the light decays into another (scattered) light
wave plus either an ion or electron plasma wave. The
strongest reflection occurs due to the instability
which involves ion waves (the Brillouin instability).
This instability occurs throughout the underdense plasma,
although it becomes quite weak in the very low density

regions since the ion waves are then heavily damped.

The intensity thresholds for these instabilities are usually determined by plasma inhomogeneity[22] which limits the region over which the interaction can take place. Sidescatter of the light is often preferred, since the light wave generated in this direction experiences the least inhomogeneity. However, in many experiments the focal spot is small and has even finer scale structure (hot spots), so that sidescatter is not really preferred over backscatter. Hence let us here consider the simpler and more tractable case of backscatter.

Figure 6 shows a plot of back reflection as a function of intensity as computed in a one-dimensional particle simulation code.[21] The points denote the simulation results, and the line shows a theoretical prediction of the reflection using a fluid model of the plasma (neglecting wave particle interactions). For these results, the plasma has an initial linear rise in density from 0 to .6 n_{CR} with an effective density gradient of 16 μ, and the initial plasma temperature is 1 keV. Appreciable reflection onsets for an intensity of ∿ 2 X 10^{15} W/cm^2 and rapidly increases to a value of 50% for an intensity of 10^{16} W/cm^2. It should be noted that these results are averaged over a time which is many instability growth times but which is still short on hydrodynamic time scales. For much longer density gradients the short-term reflection becomes even larger and approaches the theoretical limit of over 90%.

Experiments to date have not shown a large net stimulated scattering. The total energy reflected back into the lens is typically measured to be ∿ 10 - 20%.

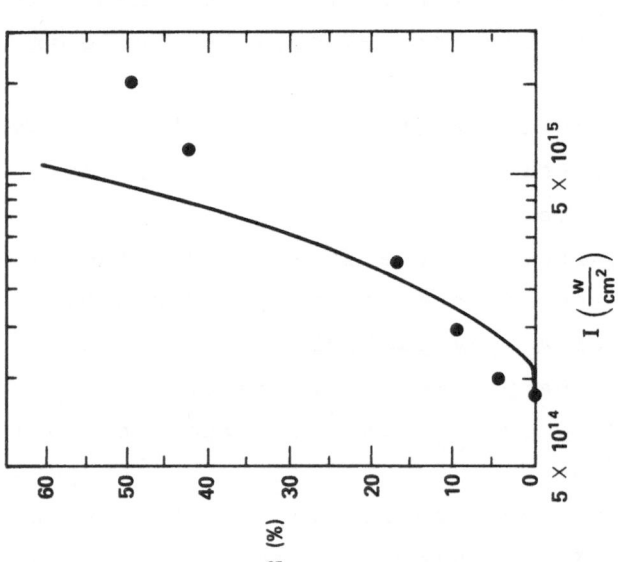

FIG. 6 The laser light reflection due to the Brillouin instability versus intensity. The computer simulation results are averaged over a time of ~ 300 light periods.

However, the experiments with very intense light ($\sim 10^{16}$ W/cm^2) are generally characterized by rather short density gradients (10 - 30 μ) and so are not very severe tests. This is particularly true when one takes into account that there are a number of long-term effects which can limit the time over which the strong reflection occurs.

Long-term ion heating is especially effective in limiting the reflection in plasmas with short gradients. In the reflection process, a small fraction of the light energy is deposited in the ion waves and then damped into the ions. As shown by the Manley-Rowe relations, this fraction is $\Delta\omega/\omega_o$, where $\Delta\omega$ is the ion wave frequency. The massive ions transport energy very slowly, and so even a small deposition of energy drives them to a very large effective temperature, which can turn off the instability. Indeed, recent time-resolved measurements[5] of back reflected light at the University of Rochester have shown a large reflection for the first 50 - 100 ps of a 500 ps pulse. A simple calculation shows that 50 ps is roughly the time needed to heat the ions in the underdense plasma to a temperature of several keV, which makes the ion waves heavily damped. Clearly more work is needed to better understand when a large reflection can occur and which long-term effects may possibly limit it, particularly in plasmas with long density gradients.

VIII. SUMMARY

In conclusion, we have given a brief overview of how intense laser light is absorbed. Intense light is not efficiently absorbed classically but can be absorbed by its conversion to electron plasma waves near the

critical density.   The physical mechanisms for this
conversion have been discussed, and some simple estimates of heating by collective processes have been
applied to some recent experiments.   Several effects
which strongly influence the absorption of high intensity light have been emphasized, including a nonlinear steepening of the plasma density profile which
has been demonstrated in computer simulations.   Finally,
the possibility of a strong induced reflection of
light due to instabilities in the underdense plasma has
been pointed out.   Such stimulated reflection can be
particularly important in plasmas with very long density
gradients.

ACKNOWLEDGEMENTS

    I am very grateful for many recent discussions of
laser light absorption with K. Estabrook, B. Langdon,
B. Lasinski, C. Max, J. Nuckolls, J. Thomson and E.
Valeo.

REFERENCES

1. C. Yamanaka, T. Yamanaka, T. Saski, J. Mizue and
    U. B. Kang, Phys. Rev. Letters $\underline{32}$, 1038 (1974).

2. B. Ripin, J. McMahon, E. McLean, W. Manheimer and
    J. Stamper, Phys. Rev. Letters $\underline{33}$, (1974).

3. J. F. Kephart, R. P. Godwin and G. H. McCall,
    Applied Physics Letters $\underline{25}$, 108 (1974).

4. K. Eidmann, C. van Kessel, M. H. Key, P. Mulser
    and R. Sigel, Proceedings of the 5th Conference on
    Plasma Physics and Controlled Nuclear Research
    (Tokyo, 1974), paper F3-1.

5. M. Lubin, E. Goldman, J. Soures, L. Goldman, W.
    Friedman, S. Letzring, J. Albritton, P. Koch and
    B. Yaakobi, ibid, paper F4-2.

6. G. Charatis, J. Downward, R. Goford, T. Henderson,
    J. Hildum, R. Johnson, T. Leonard, F. Mayer, S.
    Segall, D. Slolmon, ibid, paper Fl.

7. J. Dawson and C. Oberman, Phys. Fluids $\underline{6}$, 394 (1963).

8. W. L. Kruer and J. M. Dawson, Phys. Fluids $\underline{15}$, 446
    (1972); J. DeGroot and J. Katz, Phys. Fluids $\underline{16}$,
    401 (1973).

9. V. L. Ginzburg, <u>The Propagation of Electromagnetic
    Waves in Plasmas</u>, (Pergamon Press, New York, 1964),
    p. 260.

10. J. Freidberg, R. Mitchell, R. Morse and L. Rudsinski,
    Phys. Rev. Letters $\underline{28}$, 795 (1972); P. Koch and J.
    Albritton, Phys. Rev. Letters $\underline{32}$, 1420 (1974).

11. A. Ehler, D. Giovanielli, R. Godwin, G. McCall,
    R. Morse and S. Rockwood, Los Alamos Scientific
    Laboratory preprint LA-5611 (1974).

12. W. L. Kruer, Phys. Fluids $\underline{15}$, 2423 (1972).

13. E. Valeo, C. Oberman and F. Perkins, Phys. Rev.
    Letters $\underline{28}$, 340 (1972); D. Dubois and M. Goldman,

ibid $\underline{28}$, 218 (1972).

14. H. Dreicer, R. Ellis and J. Ingraham, Phys. Rev. Letters $\underline{26}$, 1616 (1971); T. K. Chu and H. Hendel, Phys. Rev. Letters $\underline{29}$, 634 (1972).

15. E. J. Valeo and W. L. Kruer, Phys. Rev. Letters $\underline{33}$, 750 (1974); K. G. Estabrook. E. Valeo and W. Kruer, Phys. Letters $\underline{49A}$, 109 (1974); D. Forslund (this meeting).

16. A. B. Langdon and B. F. Lasinski, Lawrence Livermore Laboratory preprint UCRL-75029 (1973).

17. R. E. Kidder, Lawrence Livermore Laboratory preprint UCRL-74040 (1972).

18. R. L. Stenzel, A. Y. Wong and H. C. Kim, Phys. Rev. Letters $\underline{32}$, 654 (1973).

19. E. J. Valeo and E. G. Estabrook, Lawrence Livermore Laboratory preprint UCRL-75936 (1974).

20. D. Forslund, J. Kindel and E. Lindman, Phys. Rev. Letters $\underline{30}$, 739 (1973).

21. W. L. Kruer, K. Estabrook and K. Sinz, Nuclear Fusion $\underline{13}$, 952 (1973).

22. C. S. Liu, M. N. Rosenbluth, R. B. White, Phys. Fluids $\underline{17}$, 1211 (1974).

THE EFFECTS OF FLUID INSTABILITIES ON LASER FUSION PELLETS

W. C. Mead and J. D. Lindl

Lawrence Livermore Laboratory

University of California

I. INTRODUCTION

Recently a number of publications have dealt with the phenomena of Taylor instability in attempts to determine how the classical Taylor picture is modified in the presence of ablation and energy transport, and to estimate the importance of the effect to target implosion design. Stephen Bodner,[1] now at the Naval Research Laboratory, performed model calculations which indicated that ablative stabilization can partially eliminate Taylor growth.

Two different groups have performed instability analyses using one-dimensional linear perturbation codes to calculate growth rates. In these codes, the analytic equations governing fluid dynamics are transformed into a set of linearized equations governing the time evolution of perturbations, decoupled by an expansion in terms of spherical harmonics. These equations are then solved numerically with a combined zero-order/first order code to study the behavior of the instability. Shiau, Goldman, and Weng[2] at the University of Rochester,

published results showing very high growth rates at
long wavelengths and concluded that instabilities do
grow in situations involving ablation. Henderson,
McCrory, and Morse[3] at Los Alamos Scientific Laboratory,
published results showing small or zero growth rates at
long wavelengths and concluded that an ablation surface is positively stable.

In the present work, we have used an entirely different technique to study fluid instabilities, based
upon direct two-dimensional simulation of the fluid
flow and plasma physics. The computer code LASNEX,
written by G. Zimmerman[4] of Lawrence Livermore Laboratory, is a direct numerical solution of the two-
dimensional fluid dynamics problem. The code models the
plasma phenomena of laser light absorption by inverse
bremsstrahlung and plasma instabilities; energy transport and partition, using flux-limited diffusion and
separate ion, electron, and radiation temperatures; and,
optionally, effects of multigroup photon and particle
transport and magnetic field physics. The fluid dynamics itself is Lagrangian, with an equation of state
used to determine pressure, energy, and opacity as a
function of density and temperature. Thermonuclear
burn of compressed matter is included to permit evalua-
tion of output to input energy ratios.

The fluid instabilities are studied by applying a
perturbation to a spherical shell or solid drop, then
observing the amplitude of the disturbance as a function
of time and position as the acceleration or implosion
takes place. In the discussion to follow, unless otherwise specified, all LASNEX perturbation amplitudes will
be RMS deviations of the Lagrangian quantities from
average values along a symmetry direction, e.g.,

$$ZRMS_L = \left[\frac{\sum\limits_{K=1}^{KMAX} (Z_{K,L} - \overline{Z}_L)^2}{KMAX} \right]^{1/2}$$

is the perturbed z-coordinate. This quantity may be plotted at a given time as a function of space or as a function of time for either a constant material point or for matter at a fixed position relative to the ablation front.

In the following, we shall discuss code tests and anomalies, comparison with previous work, our current understanding of fluid instability in the presence of ablation, and the implications of these results for laser fusion target design.

II. CODE TESTS

Consider now test cases consisting of slabs accelerated by a pressure source. Such problems can readily be calculated by LASNEX using an ideal gas equation of state. By using a large Γ, the material becomes essentially incompressible. If the material's temperature is large enough, the sound speed allows communication over distances of order of a wavelength in times short compared to the growth time for perturbations. That is, for incompressibility to hold, these problems satisfy $kc_s/\gamma \gg 1$, and are free of shock effects. The results for twenty problems with Atwood number 1.0, having a wide range of parameters, show that the ratio of growth rate computed by LASNEX to that computed from classical Taylor instability theory is .9 ± .2, as indicated in Figure 1.

The next two illustrations show the result of acceleration of compound slabs having an interface with

| Slab geometry
 | Classical Taylor

 $\eta = \eta_o e^{\gamma t}$ Amplitude
 $\gamma = \sqrt{\alpha k a}$ Growth rate
 $\alpha = (\rho' - \rho)/(\rho' + \rho)$ Atwood number | LASNEX results, $\alpha = 1$

 20 cases, $\dfrac{\gamma_{LASNEX}}{\gamma_{Taylor}} = .9 \pm .2$ $\dfrac{k c_s}{\gamma} \gg 1$

 $k\eta < 1,\ kd \gg 1$ |
| Spherical geometry
 | Plessett solution

 J. Appl. Phys. <u>25</u>, <u>96</u> (1954) | LASNEX results

 12 cases, $\dfrac{\gamma_{LASNEX}}{\gamma_{Plessett}} = .8 \pm .2$ |

Figure 1: Summary of LASNEX code tests in plane and spherical geometry.

Atwood number equal to ± .5. Figure 2 shows a comparison
of the Atwood number + .5 interface LASNEX amplitude vs
time with that of Taylor incompressible theory. The
LASNEX growth rate is 23% low, and clearly shows reduced growth from the unstable Atwood number 1.0 interface. Figure 3 shows the amplitudes vs time for the
same slab with the acceleration reversed. In this
example, the Atwood number - .5 interface is correctly
shown to be stable by LASNEX, exhibiting only a weak
oscillation in perturbation phase, with no growth in
magnitude.

     Another test class of problems is the spherical
shell driven by an external pressure source. An analytic solution for this geometry was presented by M. S.
Plesset[5] for an Atwood number 1.0 spherical interface.
The results of twelve test cases show the average ratio
of LASNEX to analytic growth rates to be .8 ± .2.

     The various non-ablating test problems have been
examined to determine the severity of some sources of
systematic error in the growth rates. The effect of
primary concern is the growth rate reduction of about
20% caused by zoning the waveform as a ramp with 4 zones
per wavelength. This perturbation form has been used
frequently in problems with LASNEX, since the waveform is
quite stable, allowing single wavelength studies to be
made without growth of shorter wavelengths which are
damped by the code's artificial viscosity. Calculations
begun with more than 4 zones per wavelength initially
show improved agreement with classical theory, but eventually shift to shorter wavelengths which emerge as a
result of noise in the impressed waveform and the higher
growth rates for short wavelengths. The remaining zoning modifications tested were found to be insignificant
for the intended uses of the code.

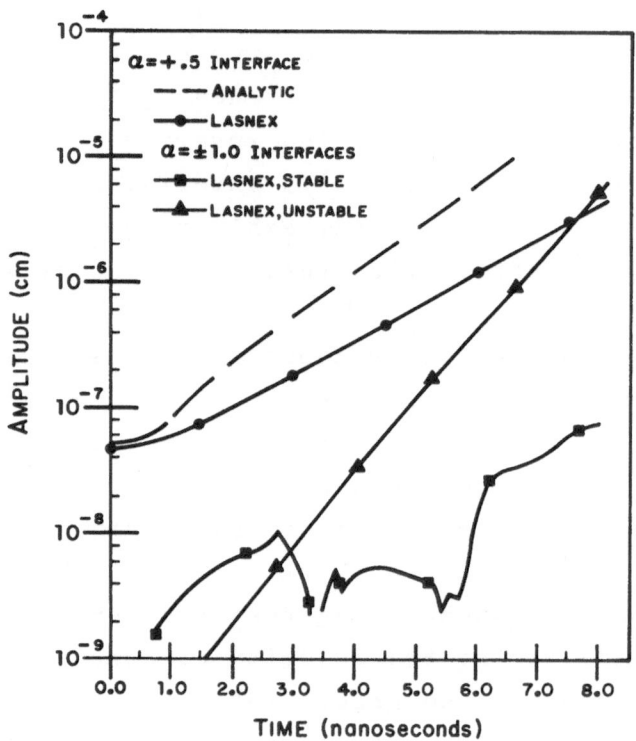

Fig. 2:   Amplitude (ZRMS) <u>vs</u> time for slab accelerations
          (unstable). The compound slab consisted of two
          .02 cm thick DT regions of density .105 g/cm^3
          and .21 g/cm^3, yielding an Atwood number of .5
          at the common interface. A pressure source of
          50 Mb was applied to the free boundary of the
          low density regions. The amplitude <u>vs</u> time is
          shown for all three interfaces in the test pro-
          blem together with the amplitude predicted by
          incompressible Rayleigh-Taylor theory.   The
          perturbation wavelength was .036 cm.

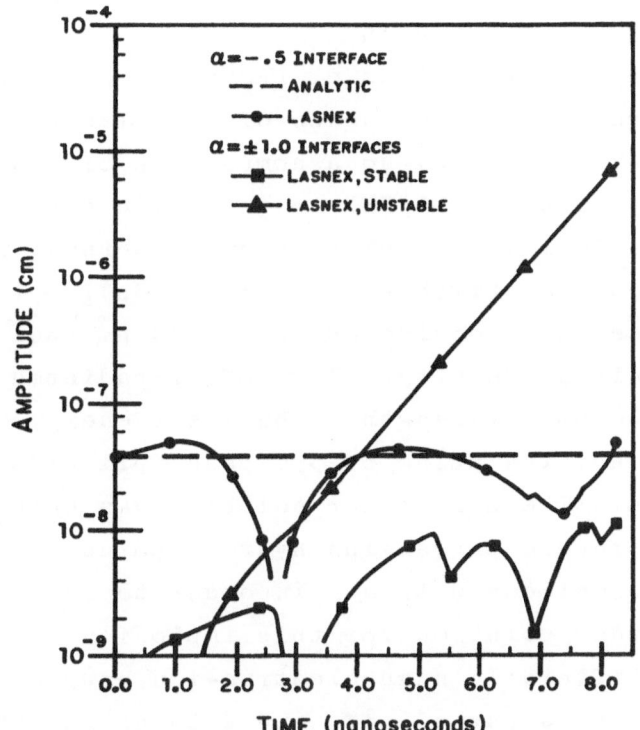

Fig. 3:   Amplitude (ZRMS) vs time for slab acceleration
          (stable). This problem had the same parameters
          as that of Figure 1, but the pressure source
          was applied to the high density free boundary.
          The acceleration was thus in the stable direction
          for the Atwood number -.5 interface.

III. FLUID INSTABILITY AT AN
ABLATING INTERFACE

We now turn to consideration of cases with ablation. Here theory becomes extremely difficult and the
theoretical results available have limited application
to the problems of interest.

Our early applications of LASNEX to laser pellet
implosions gave little evidence of ablative stabilization of Taylor growth, so a comparison of considerable
interest was to apply LASNEX to the solid drop, appar-
ently stable implosion published by Henderson, et al.,[3]
of Los Alamos Scientific Laboratory (LASL).

The case they considered was a 500 μm radius DT
sphere of initial density .21 g/cm^3, irradiated by a
laser of 1.06 μm wavelength. The laser energy was
50 KJ in a Gaussian pulse of 550 psec full width at
half maximum. The zero order solution was calculated
in a 1 temperature Lagrangian hydrodynamics code with
electron thermal conduction. In order to reproduce the
LASL zero order solution for this implosion, LASNEX was
run with artificially high electron-ion coupling, with
all flux limiters turned off, and with no radiation
physics. The resulting LASNEX calculation, shown in
Figure 4, agrees fairly well with the published results
of LASL, shown for comparison purposes. Plotted here
are the density vs radius profiles at t = + .24 ns and
t = + .53 ns with respect to the peak of the Gaussian
pulse and the temperature profile at t = + .24 ns.

The first order perturbed quantities are presented
by LASL only for the relatively long wavelength modes
ℓ = 2, 3 and 5. The modes ℓ = 2 and 5 were run with
LASNEX. The pellet surface was initially perturbed with
a spherical harmonic of amplitude .5μm RMS. In

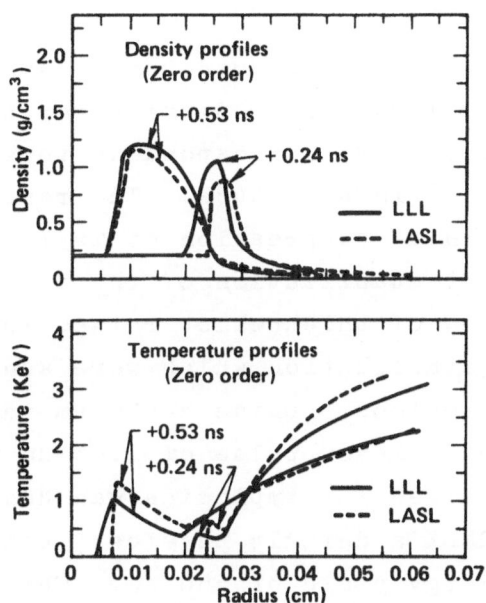

Fig. 4:   Zero order density and temperature profiles.
For a DT sphere having initial density
.21 g/cm^3 and radius of .05 cm, irradiated by
a 1.06 μm laser pulse with a Gaussian time profile (FWHM = 550 psec), this figure shows
sample snapshots of density and temperature.
The solid curves show LASNEX results with parameters adjusted to simulate the results of
Henderson, McCrory and Morse.   The two comparison times are +.24 ns and +.53 ns with respect
to the time of peak laser power.

Figure 5 are shown the ℓ = 5 perturbed density <u>vs</u> position and the perturbed temperature <u>vs</u> position at the
same implosion times as the zeroth order profiles.
Agreement between the two codes is qualitatively good,
and quantitative within an overall factor of three in
amplitude.

We also ran the ℓ = 100 mode with LASNEX, for
which the classical Taylor prediction for the number of
e-foldings is about 8.9, corresponding to a growth
factor of about 10^4 in amplitude. The results of this
run indicate definite suppression of Taylor growth and
thus a significant stablization of the interface.

The implosion of this pellet raises interesting
questions about stablization which were addressed by a
series of test problems. Using a 450 μm radius bare
DT drop and a 60 kJ laser pulse of 1.06 μm wavelength,
a partially optimized 1-D implosion was designed which,
according to LASNEX's default physics, achieved a
thermonuclear energy yield of 550 kJ. The pulse for
this implosion had the time profile shown in Figure 6a.
A series of 2-D LASNEX calculations was then made, pro-
gressing from a highly chopped pulse, which included
only the high energy tail, towards the fully optimized
pulse. The most severely chopped pulse (A) produces an
implosion which has a basic similarity to the previous
case: the density of compressed matter is essentially
constant during the implosion, while its temperature
continually rises. This is evident in Figure 6b,
which shows the material adiabats for the various pulses.
As the pulse is extended to lower initial power levels,
the material achieves increasingly isentropic compressions, moving towards the optimized case which achieves
a peak density of 300 and a ρr of 1.2 g/cm^2.

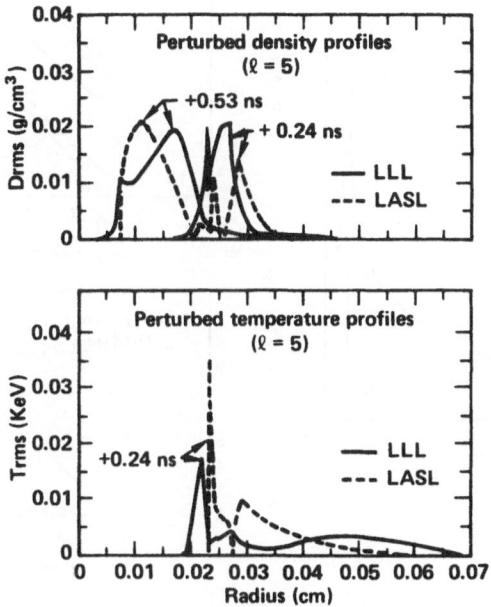

Fig. 5:   First order perturbed density and temperature
          profiles. For the same case as Figure 3, this
          figure presents snapshots of perturbed densities
          and temperature at corresponding times. The
          LASNEX results, shown as the solid curve, are
          RMS amplitudes for an ℓ = 5 perturbation of
          initial RMS amplitude .5 μm. The profiles pub-
          lished by Henderson, McCrory and Morse are
          shown for comparison. Note that the time evo-
          lution and density profiles are very similar,
          but a factor of 2 discrepancy in initial ampli-
          tude does not show in the figure, since the
          calculation of Henderson, et al., was done with
          1μm initial amplitude.

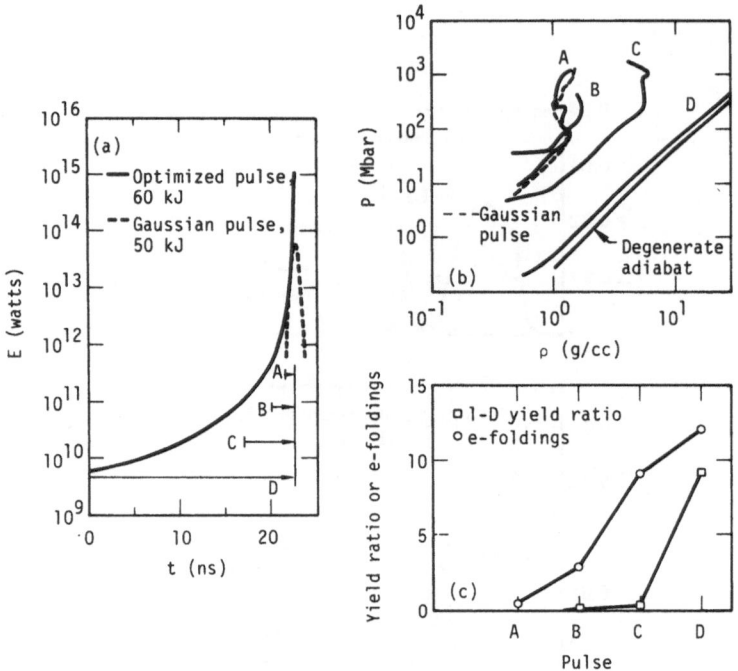

Fig. 6a: Pulse shapes for stability example. The
           solid curve shows the semi-optimized time
           profile of a 60 kJ, 1.06μm laser pulse for a
           .045 cm radium DT sphere. The pulse (D) yields
           peak density of 300 g/cm^3 and about 550 kJ of
           thermonuclear energy in a 3-temperature, 1-D
           model. The time intervals indicated by A, B
           and C show truncations used in studying stabi-
           lization of the implosion. Pulse shape A
           gives stability comparable to the 50 kJ, 550
           ps FWHM gaussian, shown as a dashed line.

      6b: Implosion adiabats.

      6c: 1-D energy yield ratios and 2-D calculated
           e-foldings.

The plot of Figure 6c shows the 1-D yield ratio
and the number of e-foldings observed in the LASNEX calculated perturbed spherical radius as a function of
pulse start-time for the $\ell = 100$ mode. The most extreme shortened pulse shows characteristics very similar to the LASL case. As the pulse is extended, the
instability begins to grow, until, for the optimized
case, the amplitudes exhibit enough growth to make the
implosion calculation fail before completion. It is
evident that the thermonuclear yield is reduced significantly before any significant instability growth
rate reduction is achieved.

This example and others point out the general
characteristics of the only known class of strongly
stabilized ablative accelerations. The stable situation
occurs under these conditions:

1. accelerated matter forms a more or less
   constant density shell
2. the shell is propagating as a shock of
   increasing strength through cold matter,
   hence the material in the shocked-up shell
   has a continually worsening adiabat
3. ablation rate is nearly equal to the rate
   of mass intake at the front of the shock
4. thermal conduction times across the
   shocked shell are short enough to readily
   transmit thermal energy to the shock front.

In light of condition 2, it appears that the stabilization evidenced may be difficult to apply to a target
implosion having significant thermonuclear yield.

IV. EFFECTS OF FLUID INSTABILITY
ON LASER FUSION TARGETS

Next, consider the implications of these findings
for laser-fusion targets. We consider here shells of
pure DT with aspect ratios ($r/\Delta r$) varying from 60:1
to 1:1 where r is the radius and Δr is the thickness of
the shell. We limit present discussion to laser energy
in the 100 kJ range and to targets which give a yield
ratio (fusion energy out/laser energy in) in the range
of 20 to 60.

Interest in the use of hollow shells stems from
the fact that they can be imploded using a lower laser
power and less severe pulse shaping. This arises from
the fact that one must do a certain minimum of work on
the DT fuel to compress and ignite it. This work is
$W = \int PdV$ where P is the applied pressure and V is the
volume. By increasing the volume, you decrease the required pressure and laser power. Lower power is important
for two reasons:

   a) Lower power means lower cost for the laser.

   b) The existence of parametric plasma instabilities
       and resonance absorption processes lead to the
       production of very energetic electrons when a
       threshold laser intensity is reached. These
       energetic electrons result in preheat of the
       fuel and a drop in the driving pressure be-
       cause of decoupling.

In most cases of interest to laser fusion, the
Atwood number is above one so that for classical
Rayleigh-Taylor instability we have $\gamma \sim \sqrt{ka}$, k is the
wavenumber of the perturbation and a is the acceleration.
Three ranges of wavelengths and physical effects are
important.

a) $\lambda >> \Delta x_{min}$ where Δx_{min} is the minimum shell
thickness: The effect of perturbations at
these wavelengths is to reduce the overall symmetry of the implosion. With convergence
ratios, given by the ratio of the initial radius
to the final radius, on the order of 100, the
symmetry and uniformity of implosion velocities
must be maintained within a percent or so in
order to get good spherical convergence and
conversion of kinetic energy to thermal energy
at the end of the implosion. Since long wavelengths have small growth rates, they generally do not cause a problem if the effects of
shorter wavelengths can be tolerated.

b) $\lambda \sim \Delta x_{min}$: Wavelengths of this size result in
a breakup of the shell and a gross mixing of
high and low density matter. Perturbations
of this size have high growth rates compared
to the wavelengths which affect the overall
symmetry and require much smaller surface perturbations. They are consequently much more
difficult to live with.

c) $\lambda << \Delta x_{min}$: Short enough wavelengths are stabilized by viscosity and density gradient effects.
But there is a range of wavelengths which have
even higher growth rates than those for
$\lambda \sim \Delta x_{min}$. These wavelengths reach the nonlinear bubble and spike phase with an amplitude
about equal to a wavelength and only grow
linearly in time beyond this point. Perturbations at these wavelengths do not become as
large as the shell thickness before being

overtaken by perturbations at longer wavelengths which are still growing exponentially.
The primary effect of these wavelengths is
expected to be a modification of matter and
energy transport at the ablation surface.
We are not able to study this effect directly
with LASNEX because a Lagrangian code cannot
handle the non-linear turbulent stage of evolution.
We deal at present with the wavelength range (b), and
with the picture of fluid instability for which LASNEX
predicted growth rates are in the range of 50 - 100%
of classical Taylor values.

In general, there are two sources for perturbations
that can be amplified by Rayleigh-Taylor instability:
1) surface perturbations due to imperfections during
the manufacturing process; and 2) laser irradiation
non-uniformity. Non-uniform illumination is essentially
equivalent to a surface perturbation because the difference in intensity across the surface results in the
imprinting of a surface perturbation. After the initial
imprinting has occured, the exponential growth due to
the Rayleigh-Taylor instability quickly dominates the
effect of a non-uniform intensity. Use of a preheated,
low density atmosphere can greatly reduce the effects
of such a laser perturbation, because lateral heat conduction will smooth out the variation.

Consider a 30-1 shell whose initial radius was
1.5 mm, which had an atmosphere density of 3×10^{-4} gm/cc
extending to 4.2 mm, preheated to 1 keV by a 7.5 kilo-
joule (kJ) prepulse of 4 μ light, as shown in Figure 7.
A 4.5° per wavelength variation in intensity ($\ell = 80$)
was equivalent to .0043 Å surface perturbation per

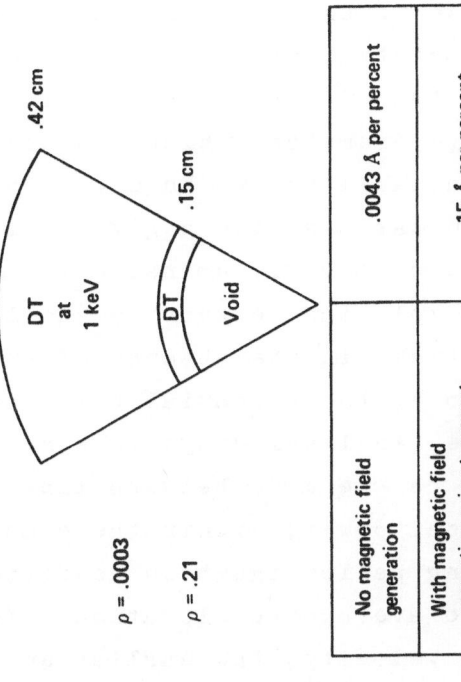

Fig. 7: Effects of Non-Uniform Illumination. The 30:1 shell shown here has a pre-pulse heated atmosphere. Laser illumination non-uniformity imprints an initial effective surface perturbation which subsequently grows with the same rate as a surface perturbation with uniform illumination.

percent variation when magnetic fields are not induced.
When the production of magnetic fields by the non-uniform
laser is included, and the transport coefficients of
Braginskii are used, the laser intensity variation is
equivalent to .15Å per percent at the same wavelength.
For longer wavelengths, lateral heat conduction is not
as effective at smoothing the intensity variation. However the growth rate is smaller so there is a tradeoff
which varies from case to case and depends on the
temperature, density distribution and radius of the
atmosphere. Because the effects of non-uniform illumination depend on many parameters which are not relevant
to Rayleigh-Taylor instability, we have concentrated
most of our effort on targets with a given initial surface perturbation and uniform illumination.

To optimize an implosion to survive shell break-up
from fluid instabilities (in the absence of strong
ablative stabilization), the essential criteria are:

1) minimize the final velocity: to some extent
   a trade-off can be made between final velocity
   and peak laser power, within the constraint
   of satisfying pellet ignition conditions;

2) maximize the average acceleration: for a
   given final velocity, the earlier and larger
   the acceleration, the fewer instability
   e-foldings will result, but the maximum
   acceleration obtainable is limited by requirements for an isentropic compression;

3) maximize the minimum shell thickness: since
   longer wavelengths grow more slowly, considerable reduction in growth can be achieved by
   keeping the shell thickness as large as possible, particularly during the late parts of an
   implosion.

Figure 8 shows the characteristics of the implosion of a 30:1 shell optimized according to these
criteria. The pulse shape is shown in Figure 8a and
the velocity history in Figure 8b. Amplitudes for
various perturbation modes are shown as solid curves
in Figure 8c. The dashed curve shows the shell thickness history. With its outer surface perturbed by an
initial amplitude of 7 Å RMS, the shell could not be
successfully imploded at modes higher than $\ell = 160$. The
growth rate increases more slowly than the $\sqrt{\ell}$ scaling
to be expected from the classical incompressible
solution. For $\ell = 80$, the growth rate shown is 80% of
the Plessett prediction, while for $\ell = 640$, the rate is
50% of the Plessett value. The plot of Figure 8d shows
isodensity contours at the time of break-up of this
30:1 shell, for an $\ell = 320$ mode. When the shell has
disintegrated to this extent, no significant thermonuclear yield can be obtained.

Figure 9 indicates the ratio of the number of e-
foldings expected at the worst wavelength to the maximum tolerable number of e-foldings as a function of
$r/\Delta r$. This ratio is estimated by taking the square
root of the ratio of the worst wavenumber to the
largest wavenumber at which success was achieved. The
maximum tolerable number of e-foldings is calculated on
the basis of a 10 Å initial perturbation. Each of the
numbers indicated is the best case at that $r/\Delta r$, which
in each case was achieved with the most rapid acceleration possible, consistent with high gain. The last
entry, for a 2 1/3 - 1 shell is for a continuously accelerated shell that successfully imploded with a 1 Å
surface finish at the worst wavelength.

Instead of applying a continuous power source, one

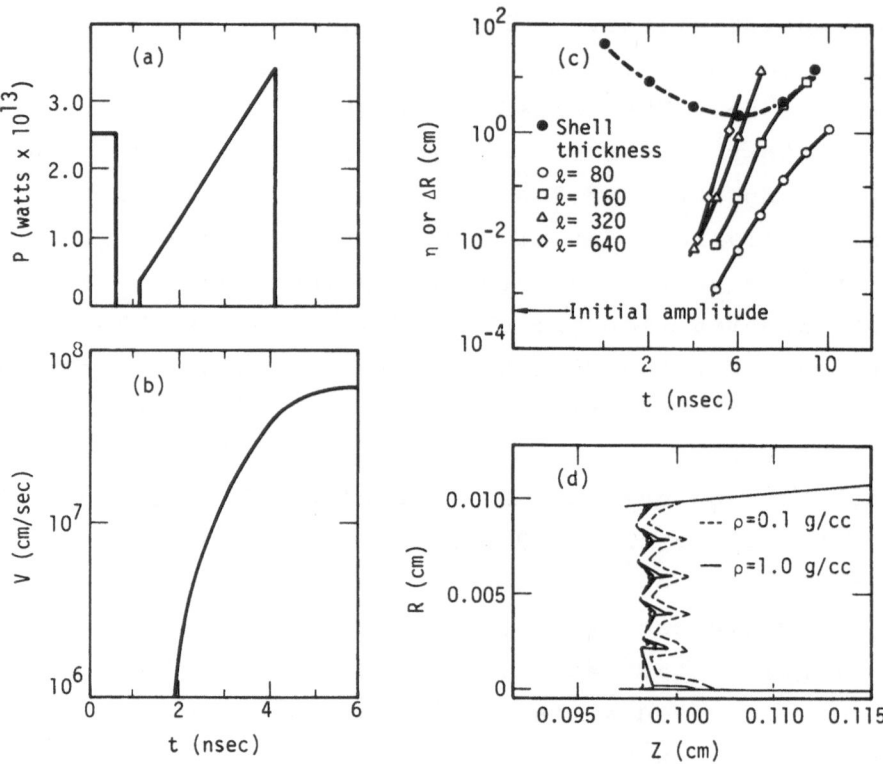

Fig. 8:   30:1 shell a) Laser pulse shape. b) Velocity
          history. The shell for this calculation had
          no atmosphere and was uniformly illuminated.
          The initial radius was .15 cm. c) Amplitude
          and shell thickness history. The "shell thick-
          ness" is the distance separating the two points
          at which density is 30% of the maximum density.
          d) Iso-density plot. The mode ℓ = 320 fails
          at time 4.3 nsec, well before the shell reaches
          the origin.

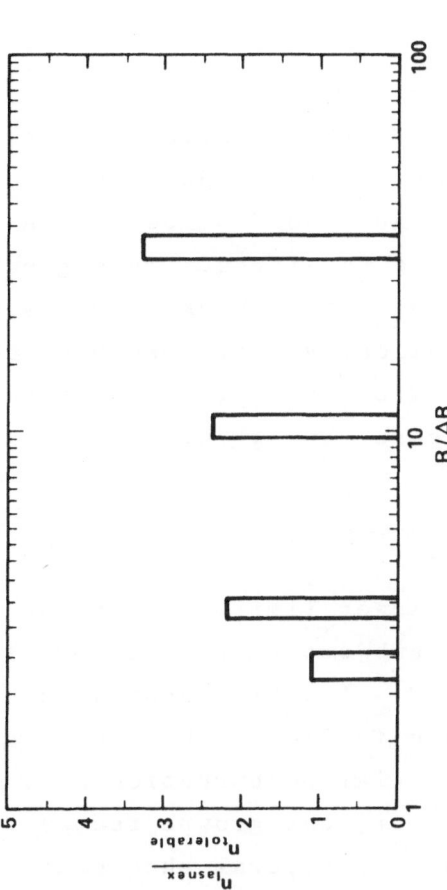

Fig. 9: Estimated $n_{LAXNEX}/n_{TOLERABLE}$ $\underline{vs}$ aspect ratio. The ratio of LASNEX calculated e-foldings (n_{LAXNEX}) to the tolerable number ($n_{TOLERABLE}$),
assuming an initial 10 Å surface perturbation improves as the aspect
ratio($r/\Delta r$) decreases. These results are scaled from 2-D calculations
of optimized implosions at the indicated aspect ratio.

can impulsively accelerate the target by turning the
laser on and off. In this way, the target is subjected
to bursts of very rapid acceleration followed by near
coasting. After the passage of each impulse, the perturbations do not grow exponentially but they do grow
linearly. As given by Richtmeyer,[6]

$$\dot{a} = k \Delta v \alpha a_o$$

where Δv is the velocity of the material behind the
shock relative to that in front of the shock, a_o is the
initial amplitude, α is the Atwood number and $\dot{a}$ is the
time rate of change of the amplitude. This growth arises
because of shock focusing as the shock passes a perturbed surface. The smallest growth possible occurs
when the shell receives its entire velocity from a
single shock. In this case, the growth factor is given
approximately by

$$\frac{a}{a_o} = kR.$$

This growth factor is a lower limit to what one can
achieve with implosions subject to Rayleigh-Taylor instability. For the 2 1/3 - 1 shell considered below,
this factor is 160 or 5 e-foldings. With such growth,
one could tolerate an initial perturbation of a couple of
hundred angstroms. However, the growth factor increases
as more shocks are used, and several shocks are necessary to maintain near adiabatic compression. In the
limit of a large number of weak shocks, the growth
factor goes over to the Rayleigh-Taylor value.

By suitably timing the several pulses and keeping
the ratio of magnitudes of succeeding shocks within a
factor of 2-3, one can decrease the number of genera-
tions and maintain isentropic compression and high gain.

Using this technique, we are able to lower the power to 10^{14} watts, an order of magnitude lower than for a typical solid sphere and survive with a 15 Å surface finish. The pulse shape and velocity history for this implosion are shown in Figures 10a and 10b. Perturbation amplitude and shell thickness versus time are shown in Figure 10c. The peak laser intensity is about 2×10^{15} w/cm^2 at a peak temperature of 5 keV. This intensity is about an order of magnitude above the calculated threshold for the parametric decay instability at $1/4\mu$, although about 85% of the light is absorbed by inverse Bremsstrahlung.

We expect to be able to live with this intensity by seeding with a higher Z material. Improvements in the impulsive acceleration technique may allow us to further lower the intensity.

V. CONCLUSIONS

Our tests of LASNEX indicate that the code can be successfully used to study fluid instability. We expect final amplitudes to be accurate to within a factor of 3 or 20% e-foldings, whichever is larger, on the basis of known systematic errors.

We find that suppression of fluid instability growth does occur for a class of highly non-adiabatic implosions, but that the effect is not significant for isentropic implosions of primary interest for laser fusion.

For isentropic compressions, LASNEX predicts growth rates for wavelengths on the order of the shell thickness in the range of 50 - 100% of classical Taylor/Plessett values, with systematic reduction in the ratio of LASNEX to classical growth rates at shorter wavelengths.

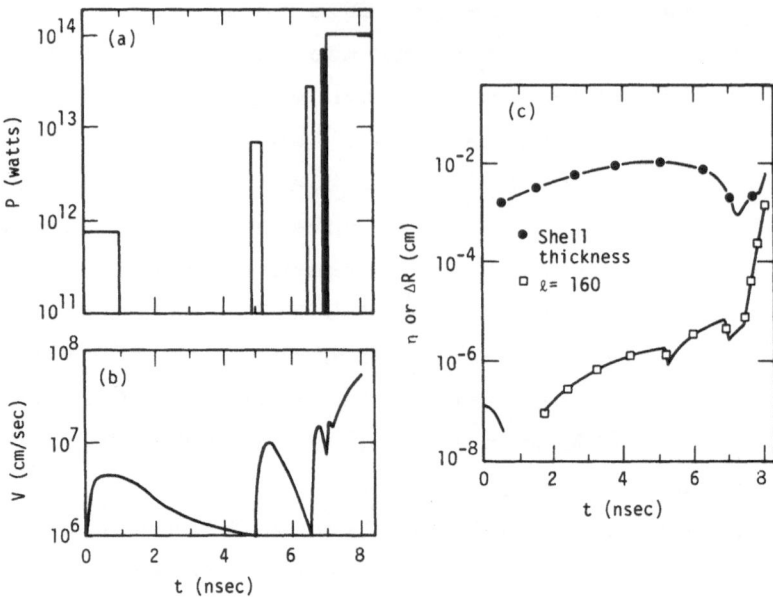

Fig. 10:   2:1 shell a) Laser pulse shape. b) Velocity
            history. The shell for this calculation in-
            itially extended from .04 cm to .07 cm at a
            density of .21 g/cc. The impulsive accelera-
            tion helps reduce fluid instability growth.
            c) Amplitude and shell thickness histories.
            Calculations show that this shell could be
            successfully imploded at any wavelength with
            an initial amplitude of 15 Å or greater.

With this instability behavior, the parameter space
for successful laser fusion target design is restricted
to some extent.

Our calculations show, however, that even in this
quite pessimistic view of instability growth, shells of
low aspect ratio can be imploded isentropically. Particularly, if impulsive acceleration is used during the
early part of the implosion, an initial surface perturbation of a few tens of angstroms can be tolerated
at the worst shell break-up wavelengths. It looks possible to achieve these surface finishes. Further, it
should be possible to produce an atmosphere around the
pellet which will reduce the imprint from non-uniform
illumination to tolerable levels.

ACKNOWLEDGEMENT

The authors would like to express their appreciation
for many helpful discussions with John Nuckolls. Many
thanks are due George Zimmerman for cooperating with
our use of LASNEX.

REFERENCES

1. S. Bodner, Phys. Rev. Lett. $\underline{33}$, 761 (1974).

2. J. N. Shiau, E. B. Goldman, and C. I. Weng, Phys. Rev. Lett. $\underline{32}$, 352 (1974).

3. D. B. Henderson, R. L. McCrory, and R. L. Morse, Phys. Rev. Lett. $\underline{33}$, 205 (1974).

4. G. B. Zimmerman, University of California Report UCRL-74811 (1973).

5. M. S. Plesset, J. of Appl. Phys. $\underline{25}$, 96 (1954).

6. R. D. Richtmeyer, Comm. on Pure and Appl. Math, $\underline{XIII}$, 297-319 (1960).

HF CHEMICAL LASERS*

Reed J. Jensen

Los Alamos Scientific Laboratory

University of California

OVERVIEW OF PROPOSED SYSTEM

The very rapid reaction rate of the H_2 + F_2 chain reaction coupled with the efficient formation of chain carriers by fast electron beam radiolysis of F_2 and SF_6 provide laser power of about 10^{12} watts/liter and qualify the HF chemical laser as a possible laser for driving compression fusion experiments. This laser differs radically from all other possible fusion lasers in that optical energy storage is not necessary. It derives its high power from the real time burn rate of the H_2 + F_2 fuel. The laser gain is very high, but of short duration making the amplifier system much simpler than those envisioned for low gain high storage lasers. Figure 1 shows the scheme. It has an oscillator, amplifier 1, and amplifier 2. The beam is split into the number of beams to be used in the experiment after amplifier 2. In the figure the beam is split into four. Each of the beams, C, D, E, and F, are

* Work performed under the auspices of the United States Atomic Energy Commission.

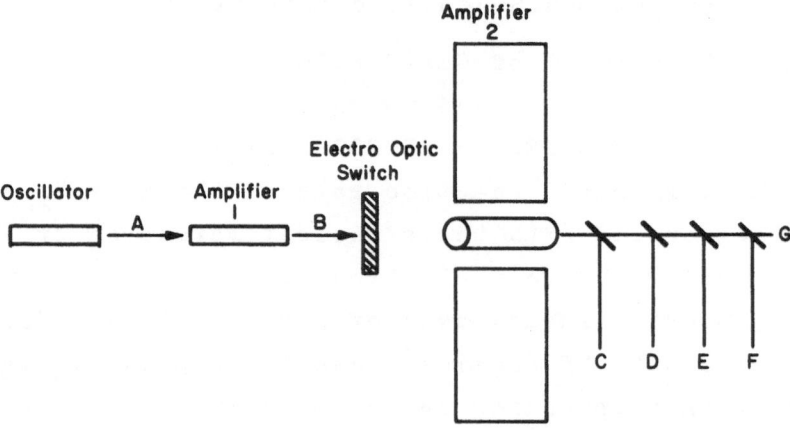

Figure 1   Schematic of HF laser system. Beams C, D, E,
           and F are routed to final amplifiers.   Beam G
           is for triggering spark gaps.

amplified by a final amplifier and beam G is used for triggering spark gaps for the e-beam transmission lines for the final stage amplifiers.

The oscillator and amplifier 1 will operate on the nonexplosive fuel SF_6 - HI. The spectrum of the HF laser operating on this fuel will be discussed below. The amplifier 2 and the final stage amplifiers will operate on H_2 + F_2 fuel and will derive their high power from the chemical energy boost provided to the laser. The output from the oscillator will be approximately 10 mJ in 100 ns. After amplifier 1, there will be about 2 J per 100 ns pulse. A fast electro-optic switch will gate out a 150 ps rise time pulse about 1 ns long, containing 40 mJ. This short 40 mJ pulse will be amplified to about 4 kJ by amplifier 2. This assumes a gain of 0.2 cm^{-1} for 60 cm. The diameter of the amplifier 2 should be about 30 cm in order to transmit the 4 kJ through the sapphire output window. Each of the beams, C, D, E, and F, should contain at least 50 J as input to the final amplifiers. The remainder of the beam can be used for triggering spark gaps.

FRONT END (OSCILLATOR, FIRST STAGE AMPLIFIER
AND ELECTRO-OPTIC GATE)

Beam quality is an important consideration in any laser fusion system. Use of a stable oscillator (500 cm radius total mirror with a sapphire flat output coupler) with spatial filters and one stage amplification by an extended electrode discharge device currently provides about 1 J of diffraction limited lowest order mode laser output. A schematic diagram of the system is shown in Figure 2. Figure 3 shows a typical burn spot from the laser system, and Figure 4 shows the

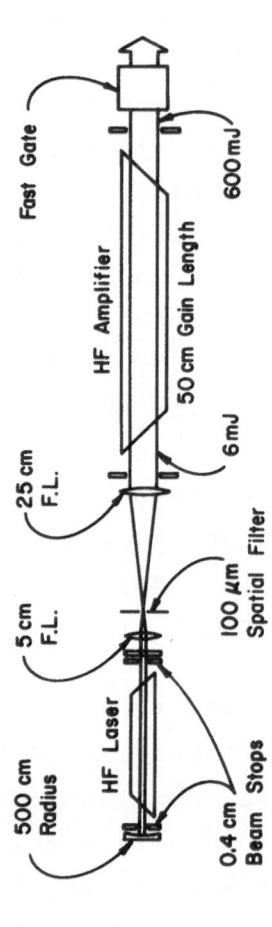

Figure 2 Schematic of oscillator, first stage amplifier, and gate.

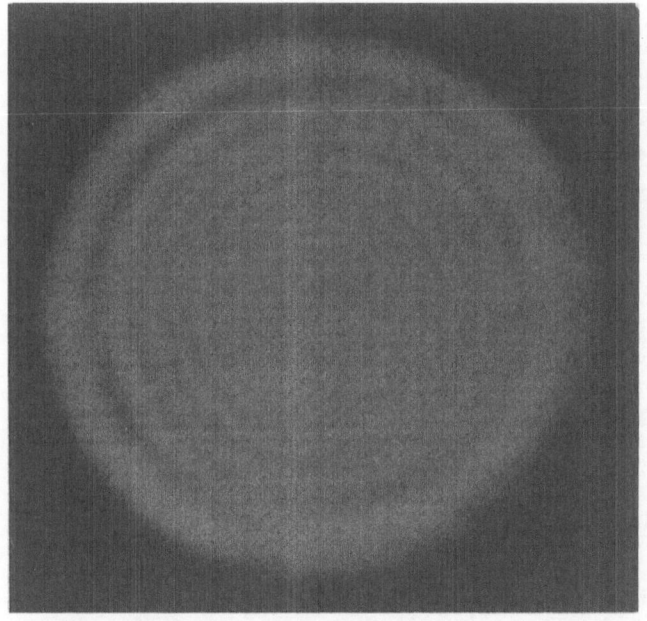

Figure 3   HF laser spot.

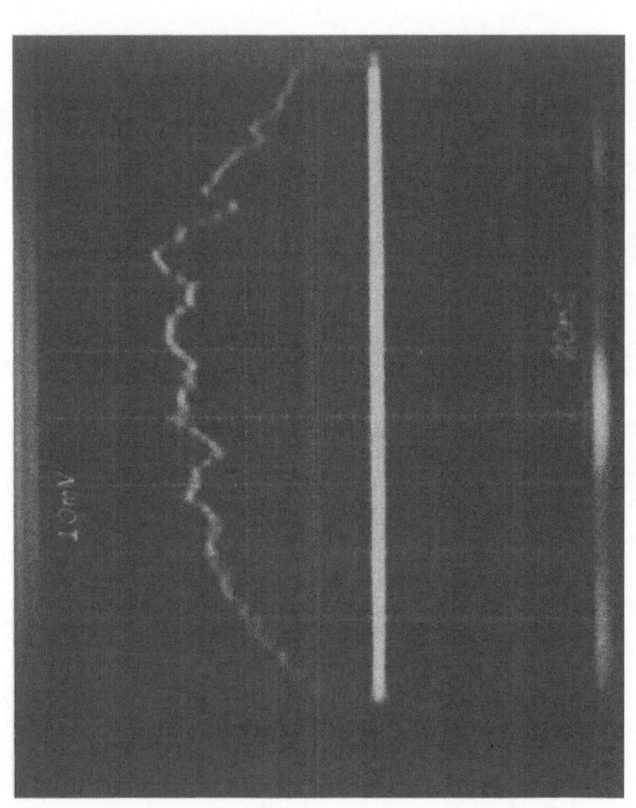

Figure 4   Intensity vs time for HF laser front end.   Figure shows absence of
modulation.

temporal dependence of the laser intensity. It will be
noticed that there is very little temporal modulation
in the pulse. This is indicative of a high degree of
mode purity. The spectrum of the H_2 + F_2 laser contains lines P(3) to P(10) of the transitions v=2 ← v=3,
v=1 ← v=2, v=0 ← v=1 and lines near P(5) for the
transitions v=3 ← v=4, v=4 ← v=5, and v=5 ← v=6. This
is the spectrum that must be matched for the oscillator
and amplifier. However, it is desirable to avoid the
use of F_2 + H_2 fuel in the front end because of the
necessity of having that part of the system in a closed
room near the central controls. It has been found that
the fuel mixture SF_6 + HI gives the spectrum shown in
Figure 5. This spectrum essentially overlaps the
spectrum of the H_2 + F_2 explosion laser,[1] and the SF_6
+ HI is a relatively benign set of chemicals with
which to work. There is no explosion hazard, and gas
toxicity is reduced to a minor level.

AMPLIFIER CHARACTERISTICS

The scaling demonstration[2] for the H_2 + F_2 explosion chemical laser amplifiers was performed in a joint
LASL-Sandia experiment using the REBA electron accelerator to ignite the laser. A schematic of the experimental setup is shown in Figure 6. In this demonstration
over 2.3 kJ of laser energy was obtained in a 35 ns
pulse that was ignited by an electron beam deposition
energy of less than 1500 J. This demonstration was
run at initial gas pressures of about one atmosphere
in a cell that was 167 cm long by 15 cm in diameter.
A summary of the experimental results from the experiment is given in Table 1.

We have developed an analytical model for this

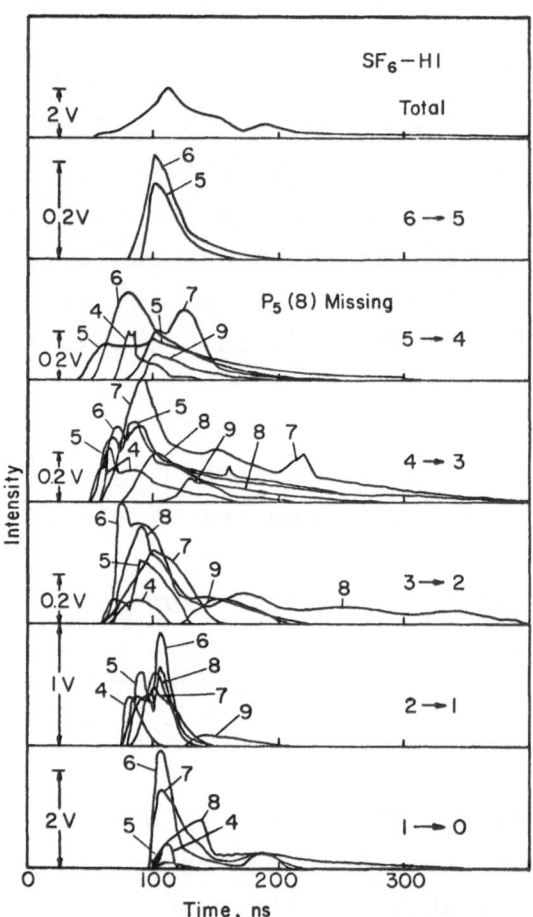

Figure 5   Spectra of SF_6 + HI laser.

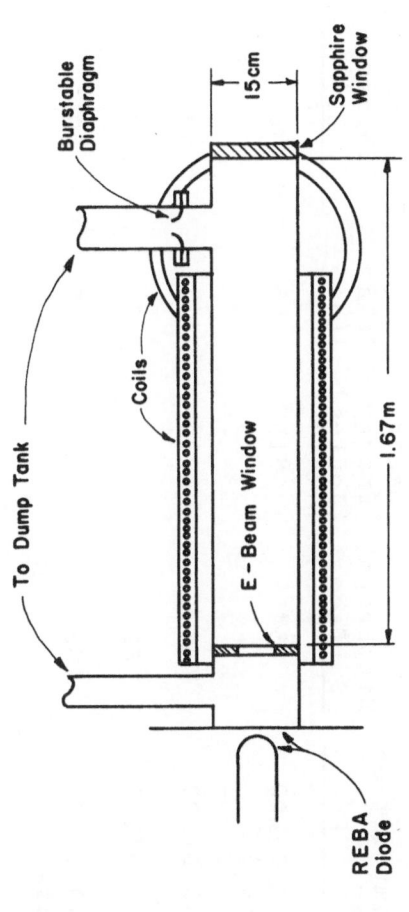

Figure 6   Diagram of experimental layout for H_2 + F_2 laser scaling demonstration.

Table 1

Laser Characteristics for Various Mixtures

of F_2, O_2, SF_6, and H_2 Initiated by REBA Electron Beam

η_e and η_c are the Electrical and Chemical Efficiencies, Respectively

Mixture Composition $F_2/O_2/SF_6/H_2$	Diaphragm Surface	Laser Energy (Joules)	Delay (ns)	Pulse Width, 10% max (ns)	η_e %	η_c %
1. 360/140/0/100 (7 shots)	SS*	815	25	46	133	1.1
2. 360/140/0/100	Au	1228	17	43	200	1.8
3. 360/140/0/100	SS	370	35	~110	60	5.2
4. 360/140/100/100 (4 shots)	SS	1890	32	69	145	2.6
5. 360/140/100/100 (2 shots)	Au	2340	30	57	178	3.3

* stainless steel

laser amplifier,[3] and have calculated the expected performance of this laser at high initial gas pressures.
Figure 7 displays some of the results of these calculations. In Figure 7 the rate of formation of HF is
plotted vs time. In this calculation the reaction was
initiated by an electron accelerator that falls easily
within current state-of-the-art. It delivers the
energy in a 40 ns pulse (the dotted line). The rate of
power development of the HF laser is seen to be greater
than 10^{12} watts/liter. The initial pressures in this
calculation were 13 atm F_2, 3 atm H_2, and it was assumed
that 1.5 laser photons would be obtained per HF molecule as observed in past work.[3,4] It is to be stressed
that the curve is for a rate, not for energy storage.
Even if superradiance depletes or tops the developed
energy until the oscillator pulse arrives, laser energy
is still supplied at the rate given by the curve.

Present design philosophy for the HF laser calls
for very large energy per channel, so that very high
pulse energy can be obtained with relatively few channels. Present considerations of amplifiers indicate
a 70 cm diameter cell 70 cm long having a volume of
270 liters. This cell size utilizes the full range of
2 MeV electrons and should provide laser pulse energies
near 50 kJ for the first ns of the pulse duration.

A schematic of the final amplifier is shown in
Figure 8. Very high energy systems can be constructed
by running a number of these amplifiers in parallel.
Experiments and calculations show that the laser system
should run with electrical efficiencies near 100%.
The very high electrical energy deposition can be pro-
vided by a system similar to the one depicted in
Figure 9, where four parallel electron accelerators

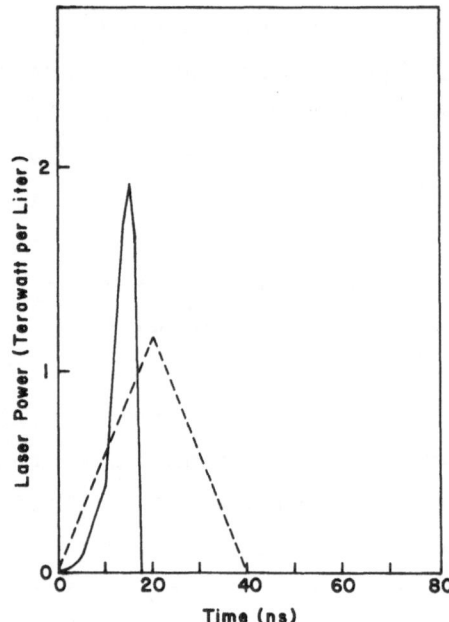

Figure 7   Calculation of power vs time for H_2 + F_2
             laser amplifier.

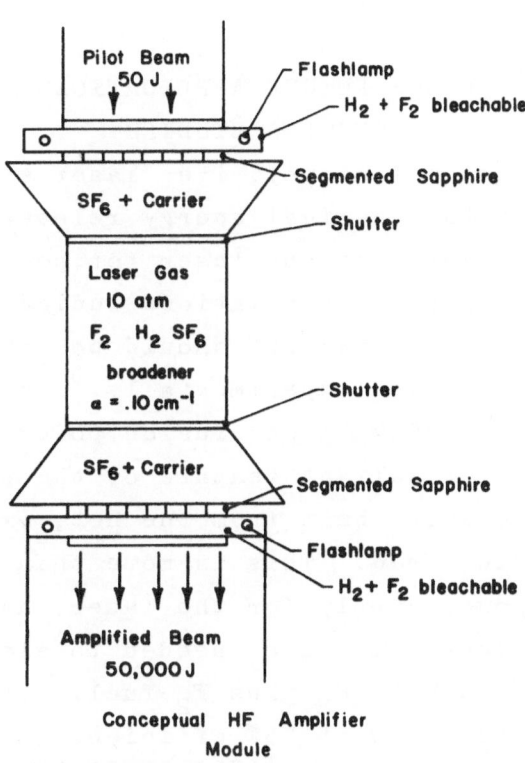

Conceptual HF Amplifier
Module

Figure 8   Schematic of 50 kJ amplifier.

simultaneously irradiate the laser gas providing a
strong, rapid ignition of the lasing explosion. Each
of the four beams will have to deposit about 12 kJ in
40 nsor less to provide the necessary ignition energy
for the amplifier module. The simplicity of the laser,
along with its very high electrical efficiency make it
attractive for near term experiments.

LONG RANGE FUTURE APPLICATIONS
AND SYSTEMS STUDIES

The chemical efficiency, i.e. laser energy extracted divided by total chemical energy released, is the
most important aspect of the laser for power plant
applications. Present calculations indicate chemical
efficiencies greater than 10% should be achievable, and
6.3% has been observed experimentally.[5] Figure 10 shows
a schematic of a 730 MW laser fusion power plant driven
by an HF laser. A salient feature of the plant is
that 25% of the waste heat from the hot, exploded
laser gas is recovered. This is more than sufficient
to drive the power supply for the laser, and it contri-
butes to the electrical power needed to electrolyze
the HF product back to H_2 plus F_2 fuel. If the electro-
lysis can be performed at 75% efficiency, the net
laser efficiency ($\frac{\text{laser power out}}{\text{electrical power to laser system}}$) will
be 8.5%, including all fuel reprocessing and the re-
circulating power in the plant will be a modest fraction
of the total plant capacity.

ACKNOWLEDGMENT

Credit is due the entire HF laser group at Los
Alamos Scientific Laboratory for the technical work

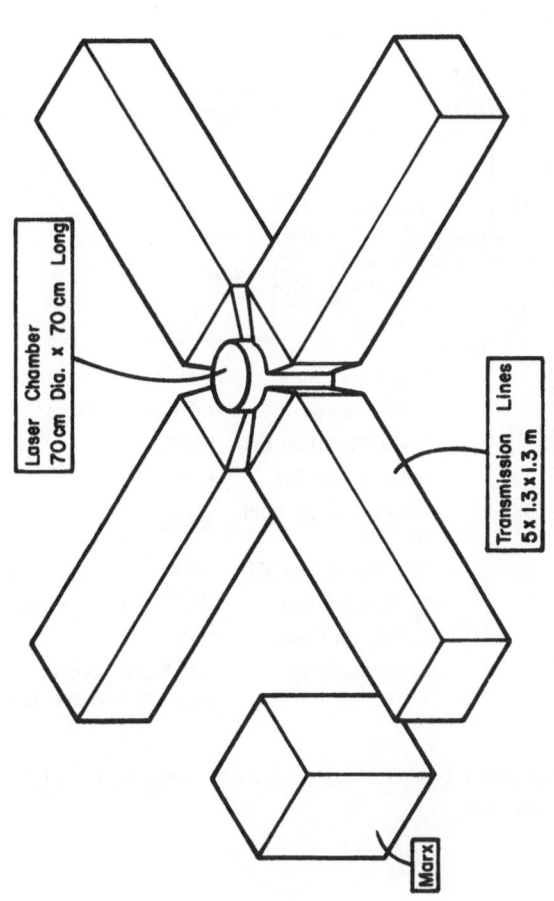

Figure 9   Schematic of e-beam configuration for 50 JK amplifier.

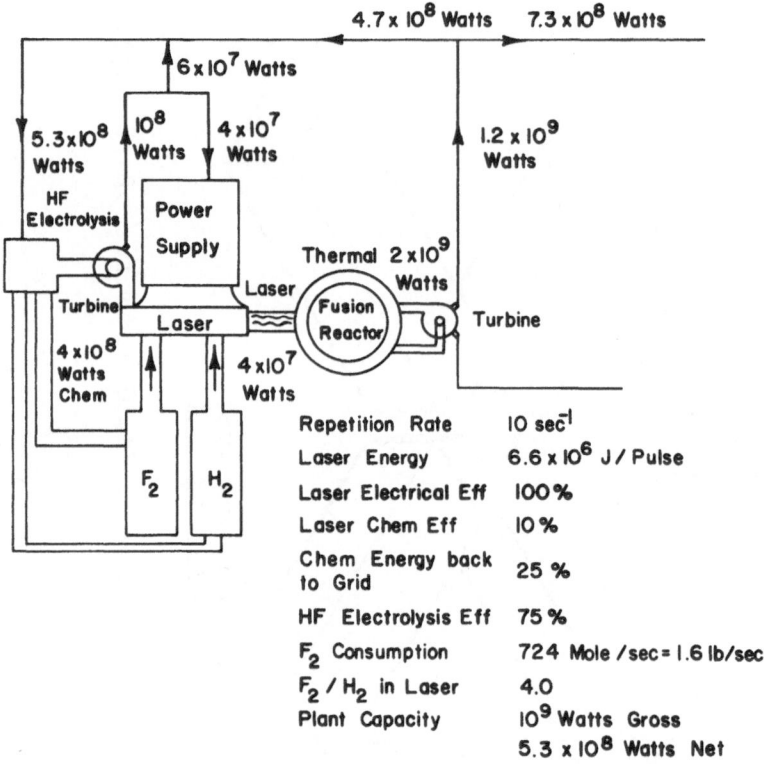

Figure 10   Schematic of HF laser fusion power plant
             system.

upon which this proposed laser system is based.   A
great deal of work on chemical stability, compatibility,
and handling was not mentioned, but along with laser
development was necessary as a basis for this laser
system.

REFERENCES

1.   N. R. Greiner, IEEE J. Quantum Electron QE-8(12),
      872 (1972).

2.   N. R. Greiner, L. S. Blair, E. L. Patterson, and
      R. A. Gerber, "A 100 Gigawatt H_2 - F_2 Laser
      Driven with an Electron Beam", VIII International
      Quantum Electronics Conference, San Francisco,
      CA., June 1974, postdeadline paper; R. A. Gerber,
      E. L. Patterson, N. R. Greiner, and L. S. Blair,
      Appl. Phys. Letts. 25, 281 (1974).

3.   W. D. Breshears, Internal memorandum, L-3-74;41,
      Los Alamos Scientific Laboratory.

4.   W. H. Beattie, G. P. Arnold and R. G. Wenzel,
      Chem. Phys. Letts. 16, 164 (1972).

5.   R. Hofland, The Aerospace Corporation, private
      communication.

EXPERIMENTS IN LASER FUSION

Robert Hofstadter

Department of Physics, Stanford University

and

KMS Fusion, Ann Arbor, Michigan

Based on the work of: G. Charatis, J.
Downward, R. Goforth, B. Guscott, T.
Henderson, S. Hildum, R. Johnson, K.
Moncur, T. Leonard, F. Mayer, S. Segall,
L. Siebert, D. Solomon, and C. Thomas

The subject of laser induced fusion has advanced by some dramatic steps in the last few years. The following article by the staff of KMS Fusion outlines some of these events. I have had the pleasure of being associated with the group while many of the experiments were being carried out. I feel that the high values of volumetric compression achieved speak for the eventual success of this method of generating large numbers of neutrons. The neutron yield has increased from a few thousand to 7×10^6 in a comparatively short time and the various tests that have been conducted lead me

to believe that the neutrons are of thermonuclear
origin.

The data presented in the KMS Fusion article have
been exceeded even very recently. In the case of a
gas filled microballoon containing equal mixtures of
dueterium and tritium at pressures of approximately 15
atmospheres apiece, a compression of 650 was reached.
In the case of an empty microballoon a compression of
approximately 2000 was achieved. In order to have
successful fusion yields the temperature of the ions
must be high in addition to achieving high compression.
I anticipate that higher core temperatures will be
attained as soon as higher laser fluxes are available.
In the meantime, it is very promising that significant
high compression values have been achieved without the
appearance of any evidence of instabilities.

The following experiments speak for themselves.

1. INTRODUCTION

The experiments at KMS Fusion are designed to test
the theoretical concepts of laser driven fusion [1] [2].
These theoretical concepts indicate that spherically
symmetric target heating by intense laser irradiation
can produce ablation driven compression and heating of
the target core to thermonuclear conditions. A number
of investigators have reported neutron generation from
CD_2 and LiD spheres [3] [4] and solid deuterium targets
[5] [6]. The KMS Fusion experiments have provided the
first direct observation of target compression and core
heating resulting in neutron generation. The neutron
generation from deuterium-tritium pressurized spherical

glass shells is experimentally correlated with target
compression inferred from X-ray pinhole photographs.
The coupling of laser energy into target compression is
studied by X-ray and particle kinetic energy measurements, and the partition of on-target laser energy into
radiation and plasma kinetic energy is experimentally
determined.

2. THE KMSF LASER SYSTEM

The KMSF laser system consists of a mode locked
YAG oscillator, a CILAS VK640 laser, an additional 80 mm
rod amplifier, and seven 10 cm-clear-aperture GE disk
amplifier units. A single 30 psec pulse is selected
from the oscillator by a laser-triggered spark gap and
double Pockels cell combination. It is then divided
into a predetermined number of temporally delayed pulses,
which are suitably attenuated and spatially recombined
into a tailored pulse shape. The full width half maximum (FWHM) of the stacked pulse used in the laser target
experiments has ranged from 0.03 to 1.0 nsec with on-
target laser energies up to 230 joules.

The tailored pulse is amplified by the VK640 laser
and the 80 mm amplifier and then travels 30 meters before entering the GE disk amplifiers (Fig. 1). The pulse
emerges from the amplifier train horizontally polarized.
Successful operation of the KMSF laser system requires

a) Apodized apertures to minimize the formation
   of diffraction rings in the beam.

b) Isolating Pockels cells to prevent self-oscil-
   lation and aid in target isolation (contrast
   ratio $> 10^4$).

c) Faraday rotators to protect the laser from
   target-reflected energy.

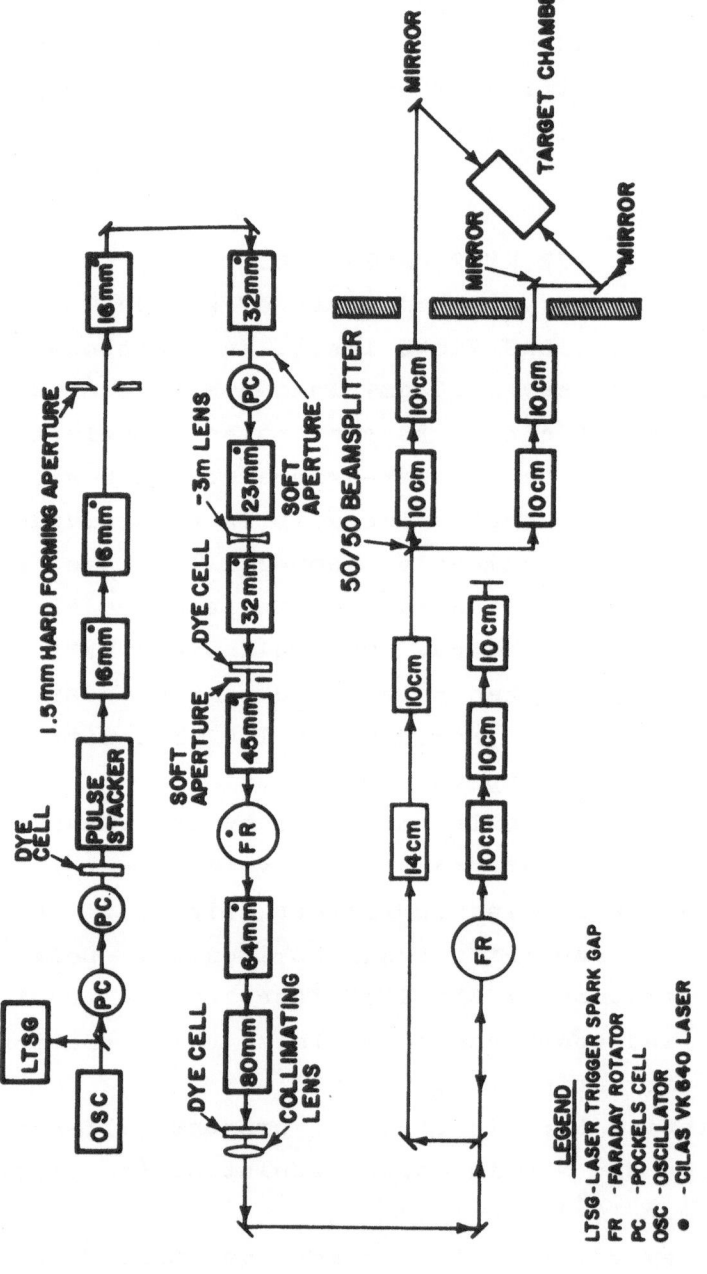

Fig. 1   The KMSF laser system.

   d)   Saturable dye cells for target isolation from
        laser fluorescence (amplified spontaneous emis-
        sion).

Fluorescence energy incident on the target is moni-
tored with an integrating photodiode. Normally, targets
are exposed to 0.2 - 5 millijoules of fluoresence pre-
ceding the main pulse. Fluorescence levels above 5
millijoules can influence the target response.

3. TARGET ILLUMINATION AND ALIGNMENT

The target illumination system delivers nearly uniform laser energy to the targets, at essentially orthogonal incidence over the spherical surface, using two
aspheric f/0.6 lenses and two ellipsoidal mirrors [7].
The target is placed at the common focus of the two
mirrors (Fig. 2). The mirrors effectively increase the
80° cone angle of each f/0.6 lens to a 144° cone angle
at the target. Small longitudinal displacements of the
lenses and mirrors away from their coincident-focus
position are used to introduce intentional aberrations
which make the target illumination nearly uniform over
the complete target surface.

The target is aligned with a continuous-wave YAG
laser which is collinear with and is divergence-method
to the main laser. A television camera views the YAG
light reflected from the target via a 1% beamsplitter in
each channel. A second television camera monitors the
target focal position along the optic axis.

4. THE TARGETS

The targets used in this series of experiments are
spherical glass shells filled with deuterium or deuterium-
tritium (DT) gas mixtures and mounted on 3µm alumina

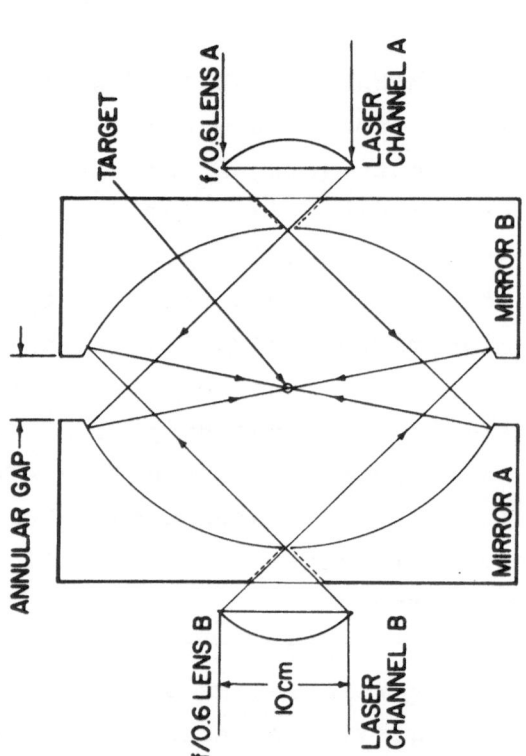

Fig. 2   The KMSF two-beam ellipsoidal mirror target illumination system.

fibers.

Optimization of laser-target experimental parameters requires that there be available targets posses-
sing an extensive range of diameters, wall thicknessess,
and partial pressures of the fuel gases. The outside
diameters vary from 30 to 700μm; the wall thicknesses
vary from 0.5 to 12μm with an observed minimum variation
of the wall thickness of any single target of 5%. The
total fuel gas pressures vary from 1 to 100 atmospheres.

Shells to be used as targets are characterized
individually. The characterization consists of:

a)   A determination of the mean wall thickness and
     wall nonuniformity by a computer analysis of
     the optical density measured on a contact micro-
     radiograph of the shell (Fig. 3a).

b)   A scanning electron micrograph of the target
     surface (Fig. 3b).

c)   Measurement of the shell's outer diameter and
     circularity using an image-splitting eyepiece
     and gauging unit (Fig. 3c).

The shells are filled to the desired pressures by
permeation of the gases through the shell walls. To
accomplish this, the shells are inserted into an autoclave which is then pressurized at room temperature with
gas or a mixture of gases to pressures equal to the desired fill pressures. Filling is achieved by maintaining the autoclave at an elevated temperature long enough
to assure that the internal pressures of all the shells
are within a predesignated fraction of the autoclave
pressure. Upon completion of the filling process, the
shells are removed from the autoclave and stored at room
temperature.

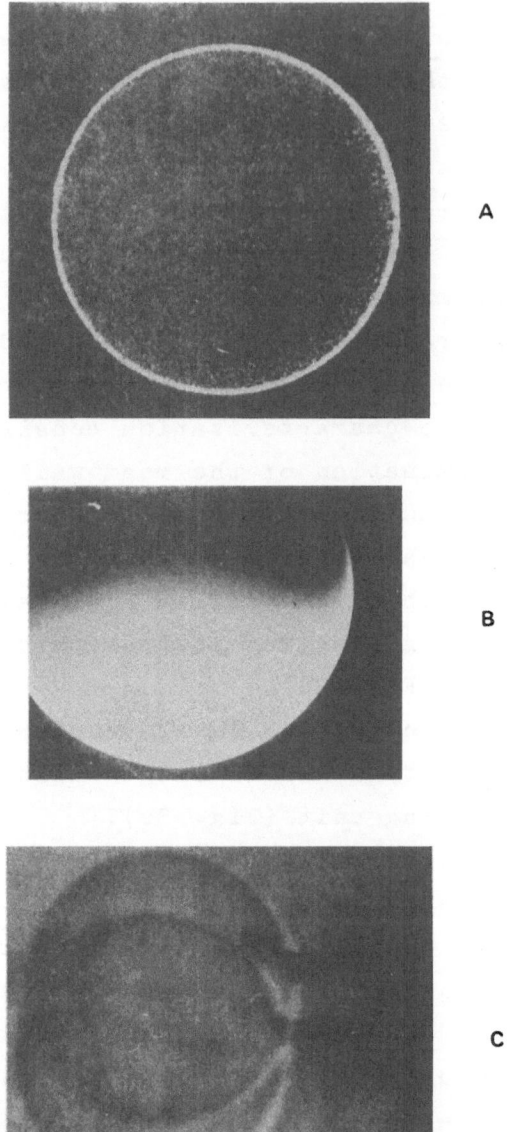

Fig. 3   Stages of target characterization.

a) radiographic negative
b) scanning electron micrograph
c) image splitting eyepiece

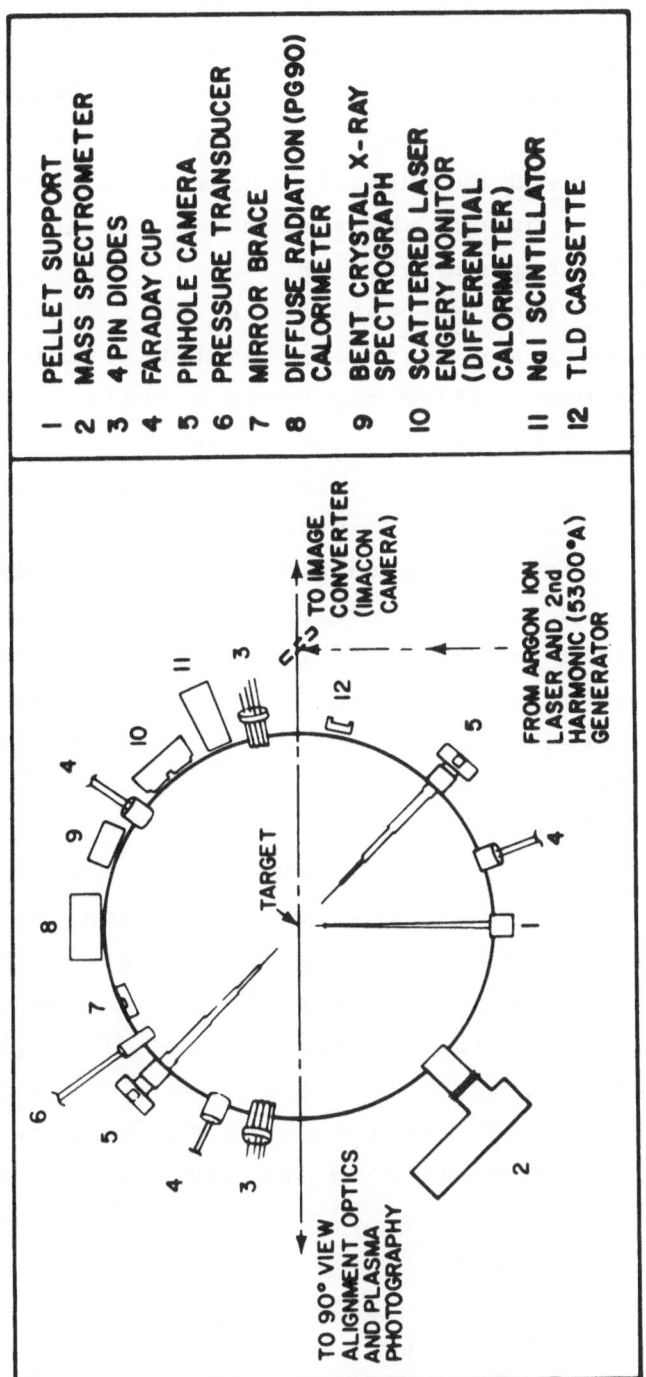

Fig. 4   Diagnostics located in the DVDR.

The pressure and composition of the gas in a batch
of shells are obtained by rupturing a statistical sample
of shells, and measuring or analyzing individually
the gas released by each ruptured shell. Total gas
pressure is measured by rupturing a shell in glycerol,
thereby forming a bubble whose diameter can be related
to the initial gas pressure. The total quantity of
tritium is determined by rupturing a shell inside an
ionization chamber and measuring the total radioactivity.
Gas ratios are measured by rupturing a shell in the
ionizer of a quadrupole mass spectrometer.

The quantity of gas in an individual shell is measured nondestructively by freezing the fill gas with a
cryogenic apparatus, and then optically measuring the
dimensions of the frozen DT within the shell. This
optical-cryogenic technique is useful for determining
whether fuel gas remains inside a shell that has been
exposed to fluorescence (or other prepulse energy) but
not to the main laser pulse.

5. LASER TARGET DIAGNOSTICS

The 50 mm annular gap between the ellipsoidal mirrors provides a direct-view diagnostic region (DVDR).
Plasma, optical, and x-ray diagnostics are arranged
circumferentially in this region as shown in Figure 4.
The annular gap at the periphery of the mirror subtends
a solid angle of 0.53π steradians at the target. Target diagnostics not located in the DVDR are the neutron-
detection diagnostics (Fig. 5) and the target-reflec-
tivity diagnostics (Fig. 6).

5.1 Optical Energy Measurements

The laser radiation is measured by fast photodiodes

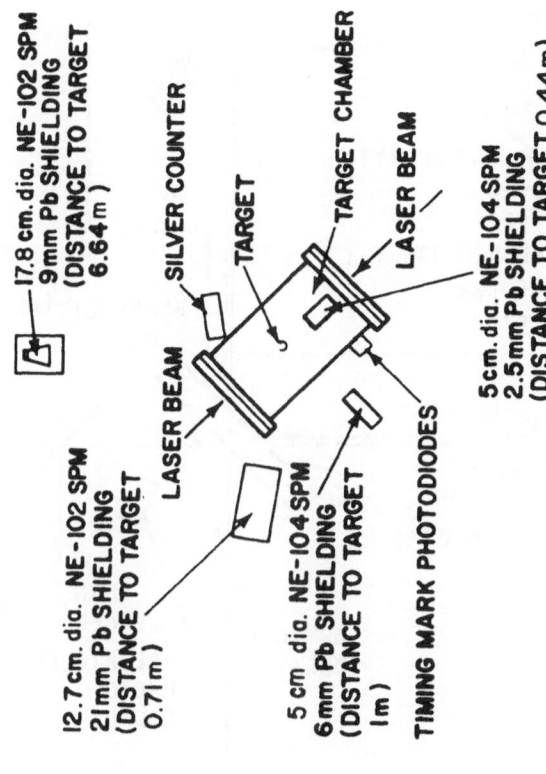

Fig. 5 Location of neutron detectors.

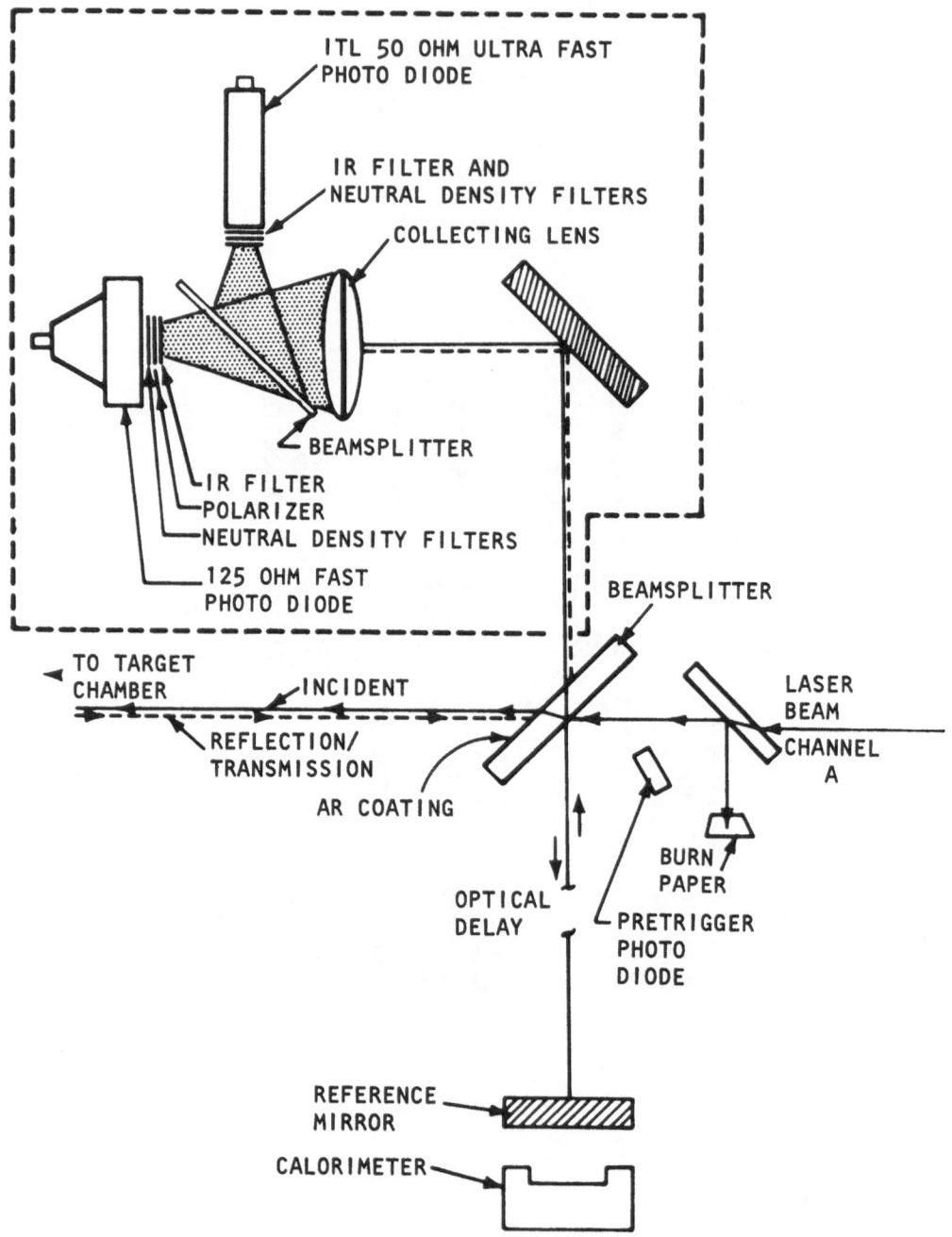

Fig. 6 Pulse shape, laser energy, and target reflecti-
        vity diagnostics.

and calorimeters in both laser channels to determine the
laser pulse shape, the energy on the target, the energy
reflected and transmitted back through the mirror illumination system, and the energy refracted and scattered at 90° to the optic axis that escapes through the
DVDR (Fig. 4). One percent of the incoming horizontally
polarized radiation is split off with a beamsplitter
(Fig.6) and transmitted to a reference mirror. Part of
the radiation passes through this mirror into a calorimeter. The energy measured by the calorimeter, corrected for the transmission losses through the optical system, provides a direct measure of the on-target energy.
The light reflected from the reference mirror is returned to the fast photodiodes 6 nsec before the reflected light from the target. One fast photodiode
covered with an infra-red polarizer detects horizontally
polarized light. The other ultrafast photodiode (without a polarizer) detects the sum of the horizontally
and vertically polarized back-reflected radiation.
These photodiodes measure the total fraction of the
incident energy returned through the mirror system. The
laser pulse width (FWHM) is obtained using the ultrafast (≤ 0.1 nsec rise-time), 50-ohm ITL photodiode[1]
coupled directly to the deflection plates of a Tektronix
7904 oscilloscope (≤ 0.2 nsec rise-time). The photodiodes and oscilloscope have a combined measured rise-
time of 0.22 nsec with less than 10% aberration. Improved time resolution (~ 10 psec) is obtained using an
Imacon 600 image converter-intensifier streak camera.

    5.2 X-ray Energy and Spectral Measurement
    Total emitted X-ray energy, both line and continuum,
is measured by $CaF_2(Dy)$ thermoluminescent dosimeters

(TLD)[9].   The spectral distribution of the continuum
radiation above 3 kev is determined by the X-ray fil-
tration technique [10] using thin foil attenuators
covering the TLDs and silicon PIN X-ray diodes.   Detector readings are then compared with temperature-
parameterized theoretical curves.

TLD detectors covered by thin K-edge foils of 25μm
or less observe the lower energy portion of the spectrum,
and silicon X-ray diodes covered by thicker foils or
foils of higher atomic number sample the harder portion
of the X-ray spectrum.   Figures 7a and 7b show typical
TLD and diode data.   The TLD data in Figure 7a are best
fitted by a 1.2 keV temperature; the diode readings are
best fitted by a 10 keV temperature.   Total X-ray energy
as measured by unshielded TLD detectors increases approximately as the 1.5 power of the laser energy on
target for typical targets such as those shown in
Table I.

Preliminary results from measurements of the X-ray
line radiation using a bent crystal spectrograph (Fig. 4)
indicate strong line emission from one and two electron
silicon ions.   Estimates of electron temperature and
density are made using line ratios of allowed and forbidden lines [11].

5.3 Plasma Measurements

Measurements of the target mass ablation and total
ion kinetic energy are made with several charge collectors located at different azimuthal positions in the
DVDR (Fig. 4).   The charge collectors are biased to col-
lect ion current.   Charge-collector oscilloscope traces
(Fig. 8) are compared to determine the degree of symmetry
of the target disassembly.   The initial sharp spike is

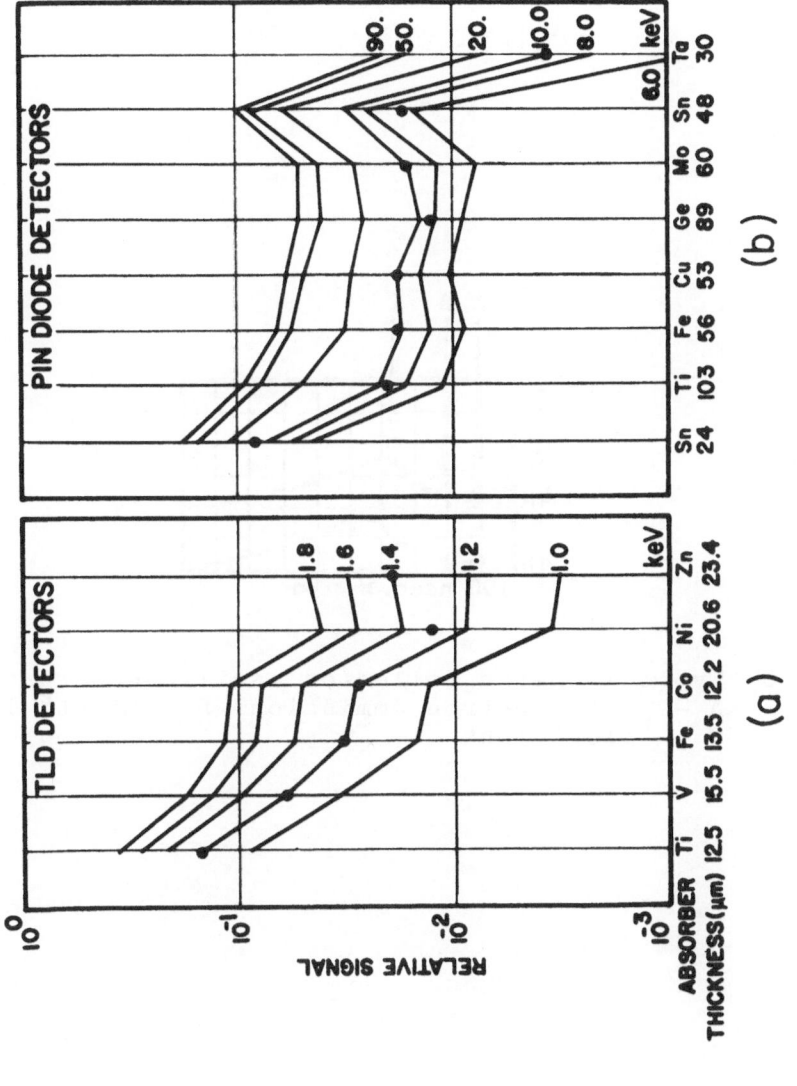

Fig. 7 a) Typical fit of theoretical temperature-parameter curves to TLD data

b) Fit of theoretical curves to silicon X-ray diode data

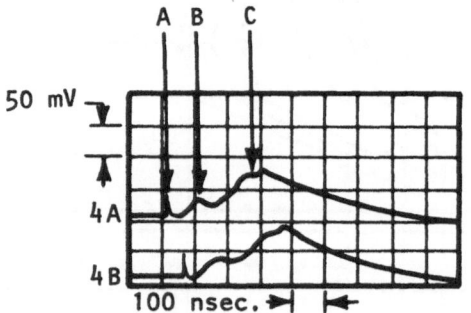

Fig. 8   Typical charge collector oscilloscope traces:
         A - photoelectron emmission, B - "fast" ions,
         C - "slow" ions.

produced by photoelectron emission from the collector
surface and is coincident with the laser-energy deposition. The first small ion peak following the photoelectron spike is produced by "fast" ions whereas the
large ion peak is produced by "slow" ions. The 4π rate
of mass collection is determined from

$$\dot{m} = \frac{\Omega}{\Delta\Omega} \frac{V(t)Am_p}{R_L e Z_{eff}}$$

where $\frac{\Omega}{\Delta\Omega}$ is the inverse of the collector fractional solid
angle, $V(t)$ is the oscilloscope voltage, R_L is the load
resistor (50 ohm), A is the atomic weight, m_p is the proton mass, e is the electron charge, and Z_{eff} is the
effective charge number of the collected ions. The
effective charge includes a correction for secondary
electron emission from the collector surface,[2] and is
given by $Z_{eff} = \overline{Z} + 4.65 \times 10^{-8}$ v, where $\overline{Z}$ is the
average ion charge number and v is the ion velocity
(cm/sec). An estimate of $\overline{Z}$ is obtained from the mass
spectrometer measurements. If the target glass shells
are completely ionized, the collected ions will have an
average charge state of Z = 10.

In a similar manner, the 4π rate of kinetic energy
collection is calculated from $\dot{E}_{ke} = \frac{1}{2}\dot{m}v^2$. Here, v = R/t,
where R is the target-to-detector distance and t is time
measured from laser deposition. Figure 9 shows $\dot{m}(t)$ and
$\dot{E}_{ke}(t)$ curves determined from the lower trace of Figure
8. Notice that the "fast" ions, although they amount
to only $\sim 8\%$ of the ablated mass, carry $\sim 45\%$ of the
target kinetic energy. The total mass and kinetic energy
are determined from numerical integration of the $\dot{m}$ and
$\dot{E}_{ke}$ curves.

The mass spectrometer, shown in Figure 10, measures

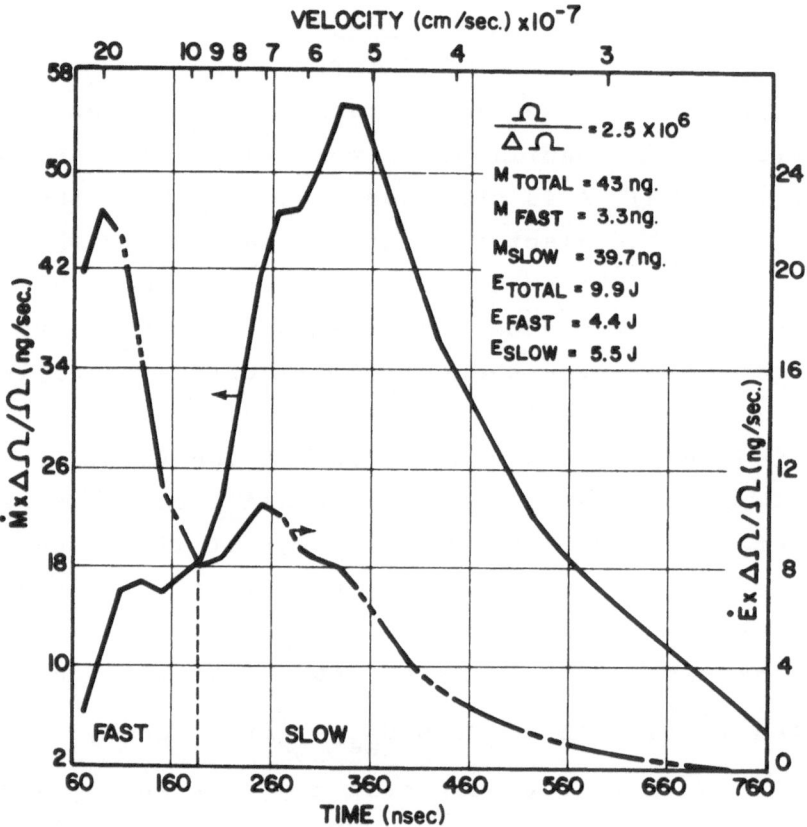

Figure 9 - Mass and energy collection rate.

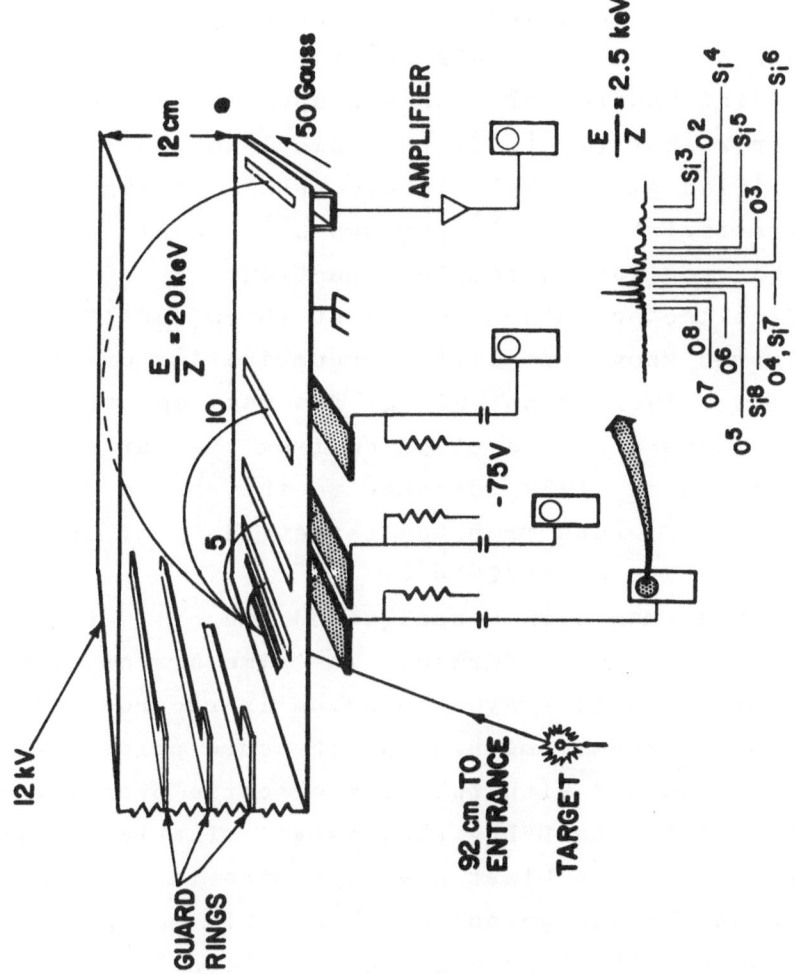

Fig. 10 The mass spectrometer and a typical spectrum.

ion species and energy in the blowoff plasma. Ions
entering at 45° to a static 1 kV/cm electric field
follow parabolic trajectories characterized by the ion
energy per unit charge (E/Z) and impinge on several
Faraday ion detectors. The output of each ion detector
is a time-of-flight sequence of A/Z pulses, where Z is
the ion charge state. By this method, several points
on the E/Z distribution of each A/Z species can be
determined per shot [12][13]. A typical ion spectrum
from a DT-filled glass shell is also shown in Figure 10.
Pulses characteristic of O^{8+} through O^{2+} and Si^{8+} through
Si^{3+} are observed in the complex spectrum.

Charge-collector data obtained at 18 cm and 120 cm
from the target show that little recombination or charge
exchange occurs over 18 to 120 cm. The mass spectrometer
with its entrance aperture at 92 cm from the target can
therefore obtain an <A/Z> relevant to the charge collectors at 18 cm. Data from the target shots of Table I
provide an estimate of <A/Z> $\cong$ 4.

Target experiments in a background gas of 20 torr
helium have also been performed. The time history of
the self-luminous blast-wave expansion is recorded with
the Imacon 600 streak camera. The pressure pulse from
the shock is recorded with two piezoelectric pressure
probes (Fig. 4) in the DVDR. The target kinetic energy
as determined from the blast wave dynamics is in good
agreement with the charge-collector kinetic energy
measurements when the long range of the "fast" ion energy
deposition in the helium gas is taken into account [14].

An ultrafast shadowgraph/schlieren camera has been
developed that utilizes a 30 picosecond pulse from the
main oscillator. The 30-psec pulse is amplified, frequency doubled, delayed in time and used to back-illuminate

the laser targets in a direction perpendicular to the
main-beam optical axis. (Fig. 4) The pictures so
obtained are used to monitor target-disassembly symmetry and blast-wave formation in helium gas.

5.4 Energy Balance

The laser energy refracted and scattered (E_s) at
90° to the optical axis is measured with a calorimeter in
the DVDR. This energy plus the vertically (E_v) and
horizontally (E_h) polarized reflected energy is the
total measured nonabsorbed laser energy. In Figure 11,
this nonabsorbed energy, normalized to the incident
energy (E_t) is plotted vs. the average intensity on the
target. The average intensity is calculated from the
measured energy on target, the illuminated area of the
target, and the width of the laser pulse (FWHM) as
measured with the ultrafast photodiode. The on-target
energy minus the nonabsorbed energy, is an upper bound
on the sum of target kinetic energy and X-ray energy.

The X-ray energy is measured by the TLDs plus the
kinetic energy as measured by the charge collectors
represents the amount of laser energy coupled to the
target. A second independent measurement of the target kinetic energy plus X-ray energy is made by a
matched pair of differential calorimeters[3] located in
the DVDR (Fig. 4). One calorimeter is covered by a
1.06μm bandpass filter and measures the 1.06μm laser
energy scattered by the target at 90° to the optical
axis. The other calorimeter is uncovered and measures
the scattered laser light as well as the X-ray energy
and the target kinetic energy. The difference between
the two calorimeter measurements is the particle energy
plus X-ray energy. The differential-calorimeter

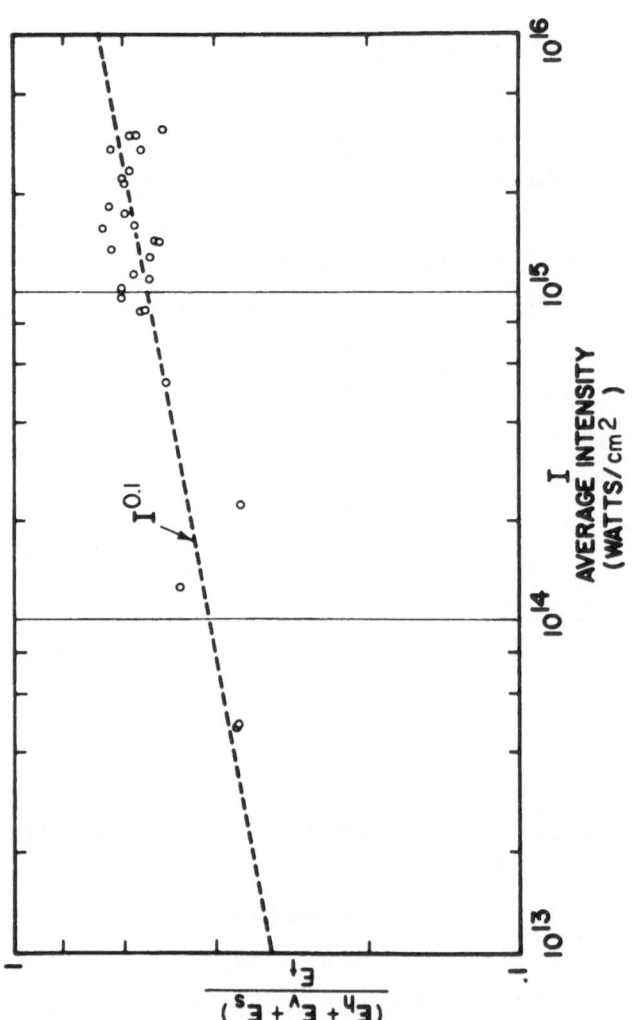

Fig. 11 Fraction of the incident on target energy not absorbed by the target vs. the average intensity

measurement is then compared with the charge-collector-
plus-TLD measurements. The agreement is usually better
than 20%.

Energy balance is determined on each laser shot by
summing the individual energy partitions: 1) horizontally polarized reflected 1.06μm energy, 2) vertically
polarized reflected 1.06μm energy, 3) 1.06μm energy
scattered out between the ellipsoidal mirrors, 4) target
kinetic energy and 5) X-ray energy. A comparison of
this sum with the energy on target gives agreement to
within about 15%. For example, on the average, for the
shots in Table I, 45% of the energy was reflected horizontally polarized, 8.8% was reflected vertically
polarized, 9.6% of the energy was scattered out between
the ellipsoidal mirrors, and 23% of the laser energy
appeared as particle kinetic energy plus X-ray energy.

5.5 Target Compression Measurements

Target compression measurements are made with two
X-ray pinhole cameras located in the DVDR (Fig. 4).
Both camera pinholes are 3 cm from the target. The
upper and lower cameras have magnifications of 4 and
15 respectively. A typical camera resolution at the
target is 14μm for a 10μm diameter pinhole and magni-
fication of 4. The upper camera (5A) records the magnified target X-ray image filtered through 17.5μm of aluminum on Kodak No-Screen X-ray film. Diametric scans of
the pinhole photographic image are made with a micro-
densitometer. The film-density record, pinhole optics,
and the film sensitivity are then used to determine the
time-integrated spatial distribution of the X-ray emission. The lower camera (5B) records the magnified
target X-ray image filtered through 50μm of beryllium

and intensified ($\times 10^6$) by a Chevron electron-multiplier
array[4]. The image is photographed by a conventional
35 mm single-lens reflex camera.

Two basic mechanisms produce strong X-ray emission
from the glass shell targets. First, the hot dense region heated by electron thermal conduction during the
laser pulse radiates at the shell edge. If the shell
is weakly accelerated, the X-ray emitting region remains at approximately the initial shell radius. Experimentally, this occurs when the on-target energy is less
than about 0.7 joules per nanogram of target mass. An
example is shown in Figure 12a.

The second mechanism for strong X-ray emission occurs
when the imploding shell is turned around by the back
pressure of the highly compressed and heated DT gas.
When the inward-directed energy of the shell is converted to thermal energy at the turn-around, the shell
again radiates strongly. Hence, an X-ray pinhole image
consisting of two concentric rings is expected. Such
an image is shown in Figure 12b. The observed volume
compression, determined from the diameter of the inner
ring, is about 125.

The X-ray emission due to each of the two mechanisms
will be radially smeared by movement of the shell wall.
A rapidly accelerated shell will therefore show a
broadened outer ring in the X-ray pinhole image. Also,
greater kinetic energy in the inward motion of the shell
will lead to a smaller turn-around radius and an inner
ring which is smeared toward the center. If the turn-
around radius is less than the resolution limit of the
pinhole camera, the inner ring will fill in to a smooth
central peak, such as that shown in Figure 12c. Pinhole
photographs similar to Figure 12c are characteristic of

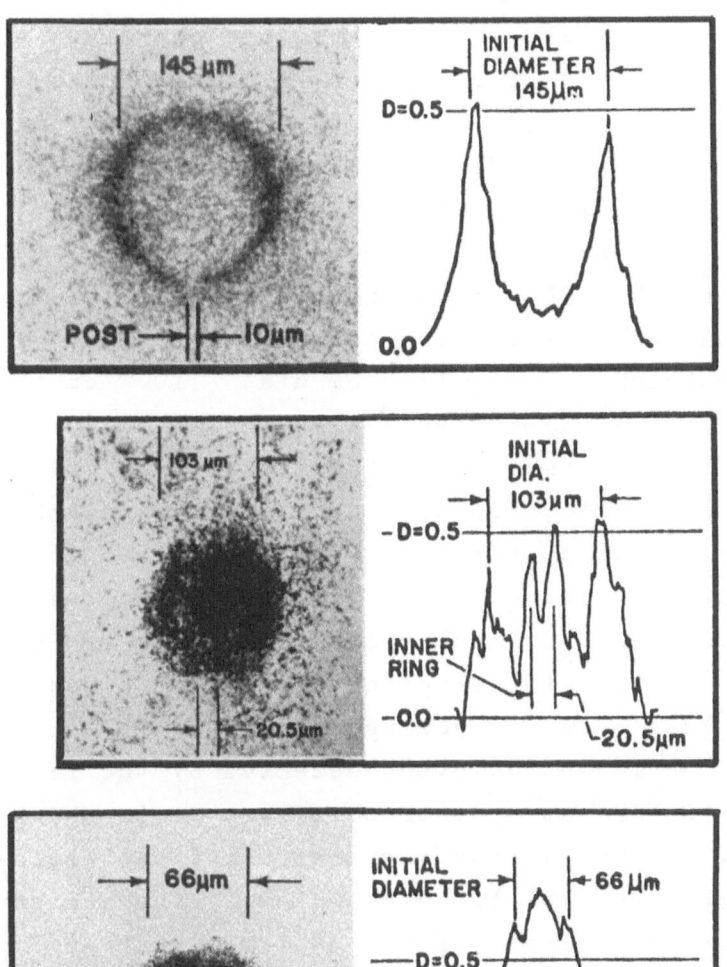

Fig. 12   X-Ray pinhole photographs
          a) Energy on target per unit mass (0.2 J/ng)
          b) Energy on target per unit mass (1.0 J/ng)
          c) Energy on target per unit mass (1.5 J/ng),
             neutron event 1 in Table I

target experiments having the highest neutron yields.
In these high-yield experiments, the on-target energy
exceeds 0.7 joules per nanogram of target mass.

5.6 Neutron Measurements

Neutron-yield and time-of-flight measurements are
provided by four scintillator-photomultiplier detectors
(SMPs); two NE-104 scintillators (5.1 cm dia. × 2.5 cm
thick) and two NE-102 scintillators (one 12.7 cm dia.
× 5.1 cm thick and one 17.8 cm dia. × 10.2 cm thick).

The Ne-104[5] SPMs have a time resolution of 2 nsec
and the NE-102 SPMs have a time resolution of 6 nsec.
These SPMs are located outside the target chamber
(Fig. 5). Neutron time-of-flight to the SPM is determined with reference to a photodiode timing mark (Fig.
13) which indicates the arrival of the laser energy on
target. A standard silver counter [15] is used to
measure total neutron yeilds above 10^6.

A typical neutron event detected by the 17.8 cm
dia. NE-102 SPM at a distance of 6.64 m is shown in
Figure 13. The neutron signal follows the bremsstrahlung signal by 106 nsec which is the correct time-
of-flight difference for the 14.1 Mev neutrons from
the DT reaction. The neutron signal in Figure 13
represents the output from six neutrons arriving
simultaneously.

Some of the target shots with higher neutron yields
are shown in Table I. Within the time resolution limit
of the SPMs, the measured neutron flight times from the
target to the SPMs agree with the calculated time of
flight for 14.1 Mev neutrons. This is strong evidence
that the neutrons were created at the target and not
at some secondary source position in the vacuum chamber
[16].

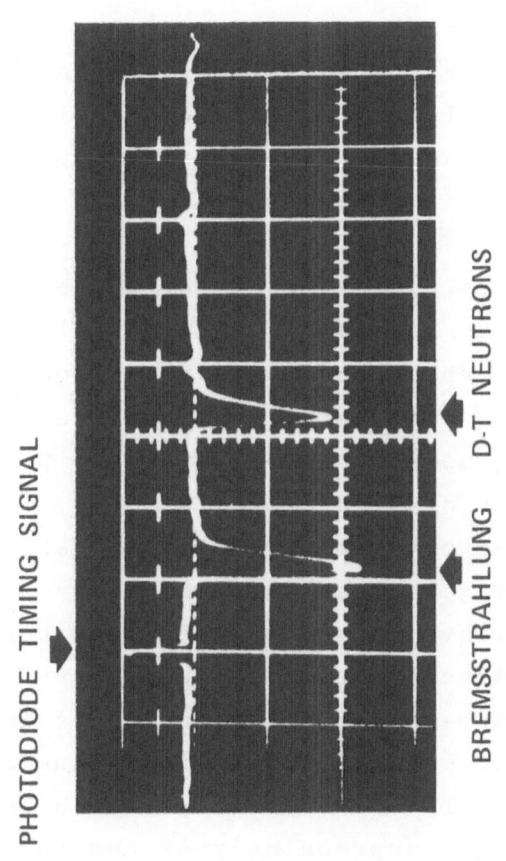

Fig. 13   Neutron detector oscilloscope trace (NE-102-SPM).

6. DISCUSSION OF RESULTS

In these experiments, spherical glass shells are irradiated with intensities from 4×10^{13} to 4×10^{15} watts/cm^2. Over this range, no onset of anomalous absorption or enhanced scattering is observed (Fig. 11). The nonabsorbed laser energy follows a slow ($\sim I^{0.1}$) power law scaling with intensity. Although the "fast" ions represent a substantial fraction of the absorbed laser energy, this fraction shows only a slight tendency to increase at higher intensities, again with no threshold behavior. The low temperature component of the X-ray spectrum remains less than 1.3 keV over this range of intensities.

Table I summarizes the target experimental parameters for selected target shots. The target dimensions are in the range for which the largest neutron yields are obtained with the designated laser pulses. All of these target shots generate an X-ray pinhole photograph similar to that of Figure 12c.

The charge collector symmetry shown in Table I is indicative of the target disassembly symmetry required for neutron yields above 10^5. Degraded compression symmetry and reduced neutron yield are observed whenever asymmetry of the laser-target illumination is intentionally introduced. For example, one-sided illumination of the target results in crescent shaped pinhole photographs with no central compression peak and no observable neutron yield. Also, delaying the laser energy in one channel 40 to 120 psec with respect to the other results in off-center, degraded compressions and substantially reduced neutron yields.

X-ray pinhole photographs from experiments on neon-DT-filled shells[6] (17% neon by pressure) showed a factor-of-4 enhancement in the radiation from the central peak over that obtained in experiments on otherwise identical DT-filled shells. These photographs provide a direct observation of compression and hydrodynamic heating of the neon gas. Assuming that the neon contamination does not affect the compression hydrodynamics, these experiments indicate that the amount of shell material mixing with the DT gas is less than the amount of neon contaminant.

The core temperature obtained in the target compression can be estimated with a simple model of the shell-driven gas-compression-expansion cycle [17]. The neutron yield obtained using this model is given by,

$$N_{DT} = 3 \times 10^{-28} \; n_i^2 \; R_i^4 \; C^{2/3} \; \theta_o^6$$

where n_i is the initial deuterium or tritium particle density, R_i is the initial sphere radius (cm), C is the volume compression and θ_o is the core temperature (keV) at peak compression. Taking R_i = 35μm, n_i = 1.5 × 10^{21} atoms/cm^3, a compression (pinhole resolution limited) of C = 100, and setting N_{DT} = 3 × 10^5, the core temperature is calculated to be θ_o = 700 eV. A higher compression implies a lower temperature; if C = 1000, then θ_o = 540 eV. The target shot corresponding to the pinhole photograph of Figure 12b produced less than the detection-threshold neutron yield of ⌄ 5 × 10^3. For this target R_i = 51.5μm and C = 125, giving a central temperature less than 240 eV.

The present neutron yield level is not sufficient to experimentally observe the ion thermal broadening of the 14.1 MeV neutron line [18]. A DT neutron yield of 10^7 - 10^8 and a peak temperature of ⌄ 4 keV should allow a direct measurement of the compressed core temperature.

7. CONCLUSIONS

The experiments described above have directly observed laser heating, target compression, and hydrodynamic heating. The observed neutron production is
correlated with compressed targets and is consistent
with core temperatures of 0.5 to 0.7 keV. These results
substantially confirm some of the basic concepts of
laser-driven fusion and provide strong encouragement
to extend the experiments to higher laser energies and
a larger class of fusion targets.

ACKNOWLEDGEMENTS

The authors gratefully acknowledge the inspiration
and leadership of Dr. Keith A. Brueckner, whose theoretical insights led to the inception of the KMS Fusion
program and guided its development. We wish also to
acknowledge with gratitude, the direction of and parti-
cipation in the laboratory program by Dr. Robert Hofstadter, which contributed in large measure to the success of the experiments reported in this paper.

The authors also acknowledge the valuable assistance
of W. Lawrence, R. Sanderson, and J. Vidolich in operating the laser system, and of E. Benn, D. Burgeson,
C. Cheng, P. Fairchild, R. Nolen, R. Sigler and D.
Sullivan in the laser-target experiments.

FOOTNOTES

[1] Instrument Technology Ltd., England.

[2] The secondary emission correction is taken from
the work of LARGE, L. N. - see KREBS, K.H., Fort-
schritte der Physik 16 452 (1968). It is assumed
that the ionized SiO_2 impacting on the copper of
the charge collectors has a secondary emission
coefficient close to Ne^+ impacting on tungsten.

[3] Gen-Tec Ins., Model ED-200 Joulemeter, Quebec,
Canada.

[4] Galileo Electro Optics Inc., Sturbridge, Massachusetts.

[5] Nuclear Enterprise Inc., San Carlos, California.

[6] These experiments were first suggested by P.
Hammerling of the KMS Fusion theoretical group.

REFERENCES

[1] BRUECKNER, K.A., Rev. Mod. Phys. 46 325 (1974).

[2] NUCKOLLS, J., WOOD, L., THIESSEN, A., ZIMMERMAN,G.,
     Nature (London) 239 139 (1972).

[3] BASOV, N.G., BOIKO, V.A., ZAKHOROV, S.M., KROKHIN,
     D.N., SKLIZKOV, G.V., JETP Letters 18 184 (1973).

[4] SOURES, J., GOLDMAN, L.M., LUBIN, M., Nuclear
     Fusion 13 829 (1973).

[5] SALERES, A., FLOUX, F., COGNARD, D., BOBIN, J.L.,
     Physics Letters 45A 451 (1973).

[6] YAMANAKA, C., YAMANAKA, T., SASAKI, T., YOSHIDA,
     K., WAKI, M., KANG, H. B., Phys. Rev. A6 2335 (1972).

[7] THOMAS, C. E., KMSF Report No. U-190, submitted to
     Appl. Opt.

[8] DOWNWARD, J. G., Bull. Am. Phys. Soc. 19 886 (1974).

[9] MAYER, F.J., MONTRY, G. R., BENN, E., 23rd Annual
     Conf. for Appl. of X-ray Analysis, August 7-9,
     1974, Denver, Colorado, KMSF Report No. U-187.

[10] JAHODA, F. C., LITTLE, E.M., QUINN, W. E., SAWYER,
     G. A., STRATTON, T. F., Phys. Rev. 119 843 (1960).

[11] KUNZE, H. J., GABRIEL, A. H., GRIEM, H. R., Phys.
     of Fluids 11 662 (1968).

[12] ALLEN, F. J., Rev. Sci. Instr. 10 1423 (1971).

[13] GOFORTH, R. R., Bull. Am. Phys. Soc. 19 909 (1974).

[14] MAYER, F. J., LEONARD, T. A., Bull. Am. Phys. Soc.
     19 914 (1974).

[15] LANTER, R. J., BANNERMAN, D. E., LASL Report No.
     LA-3498-MS, July 16, 1966.

[16] McCALL, G.H., YOUNG, F., EHLER, A.W., KEPHART, J.F.,
     GODWIN, R.P., Phys. Rev. Letters 30 1116 (1973).

[17] OSBORN, R.K., MAYER, F.J., KMSF Report No. U-206.

[18] LEHNER, G., POHL, F., Zeit. Für Physik 207 83 (1969).

TABLE I-A

Neutron Event	1	2	3	4	5
Target Diameter (μm)	42	50	40	43	59
Wall Thickness (μm)	0.5	0.6	0.5	0.4	0.9
Target Deuterium-Tritium Pressure (atm)	18-13	18-13	18-13	18-13	18-13
E_T - Energy on Target (J)	58.8	67.6	62.8	69.6	44.7
M_I - Target Mass (ng)	7	12	6	6	25
E_T/M_I (J/ng)	8.5	5.7	10.0	12.0	1.8
I - Average Intensity (W/cm^2 x 10^{15})	N.M.	3.6	6.0	5.7	2.4
Laser Pulsewidth (nsec)	N.M.	0.30	0.30	0.35	0.25
Neutron Yield (x 10^6)	7.0	5.2	5.2	4.5	4.0
Reflected Laser Energy (J)	N.M.	27.6	25.2	29.7	18.1
Depolarized Laser Energy (J)	N.M.	5.1	4.0	4.4	3.6
Scattered Laser Energy Into DVDR (J)	6.5	7.6	5.9	6.9	4.8
Target Absorbed Energy (J)	15.3	12.0	9.8	9.9	9.3
Target X-Ray Energy (J)	0.4	0.6	0.5	0.5	0.3
Charge Collector Symmetry (Ion Mass Collected by 4A/ Ion Mass Collected by 4B	0.84	0.76	0.83	0.74	0.93
Laser Fluorescence on Target (mJ)	1.3	2.0	1.6	2.3	2.4

TABLE I

NEUTRON EVENT	1	2	3	4	5
Target Diameter (μm)	66	64	78	64	71
Wall Thickness (μm)	0.9	0.8	0.6	0.9	0.8
Target Deuterium-Tritium Pressure (atm)	18-13	18-13	18-13	18-13	18-13
E_T - Energy on Target (J)	40.9	48.3	55.7	52.0	61.2
M_I - Target Mass (ng)	27	23	25	28	28
E_T/M_I (J/ng)	1.51	2.14	2.21	1.86	2.19
I - Average Intensity (W/cm^2 x 10^{15})	1.35	1.81	.1.24	1.78	1.64
Laser Pulsewidth (nsec)	0.32	0.30	0.34	0.30	0.34
Neutron Yield (x 10^5)	4	3	3	2.5	2.5
Reflected Laser Energy (J)	17.7	22.1	25.3	23.6	28.6
Depolarized Laser Energy (J)	4.5	4.1	4.7	4.3	4.8
Scattered Laser Energy Into DVDR (J)	2.9	4.6	4.4	5.4	3.5
Target Absorbed Energy (J)	8.0	11.4	14.4	14.0	8.2
Target X-ray Energy (J)	0.8	0.8	---	0.9	0.6
Charge Collector Symmetry (Ion Mass Collected by 4A/ Ion Mass Collected by 4B)	0.94	0.87	0.93	0.93	1.00
Laser Fluorescence on Target (nJ)	2.2	5.0	2.5	3.5	2.2

OUTLOOK FOR LASER FUSION*

John H. Nuckolls

Lawrence Livermore Laboratory

University of California, Livermore, California

In 1974 an important experimental milestone was achieved in the development of laser fusion power. Laser driven implosions were achieved in which moderately high compressions ($\sim$ 100 fold) and fusion neutrons were produced. This is perhaps one of the most significant technical milestones since the development of the laser fusion concepts in 1960-61 (Figure 1). Only a few hundred ergs of energy were released in the best of these experiments, and the DT was imploded to densities of $\approx$ 1g/cm^3, whereas fusion energies of 10^{14} ergs and densities of 1000 g/cm^3 must be achieved for power production applications. Nonetheless, the results are significant in at least two respects. First a Lawson number of $\sim$ 2 X 10^{12} and a DT temperature of $\sim$ 1 keV were achieved. This is comparable to the performance of the best magnetic confinement approaches (Figure 2). Second,

---

*This work done under the auspices of the U. S. Atomic Energy Commission.

Concepts 1960 - 61
                (Laser, Implosion calculations, Reactor)

Experimental Program   1963

Efficient Pulsed Lasers   1969 - 71

AEC Programs Accelerated $\times$ 10 1970 - 73

Implosion/Fusion   1974
                10 - 100 J lasers

Micro Explosion 1977 ?
                ($G \approx 1$), 3 - 10 kJ lasers

Net Energy   ?
                ($G \approx 10 - 20$), 10 - 100 kJ, 5 - 10% eff. laser

Reactor   ?
            ($G \approx 10 - 100$), 30 - 300 kJ, 5 - 10% eff. laser

Figure 1

Laser Fusion Milestones

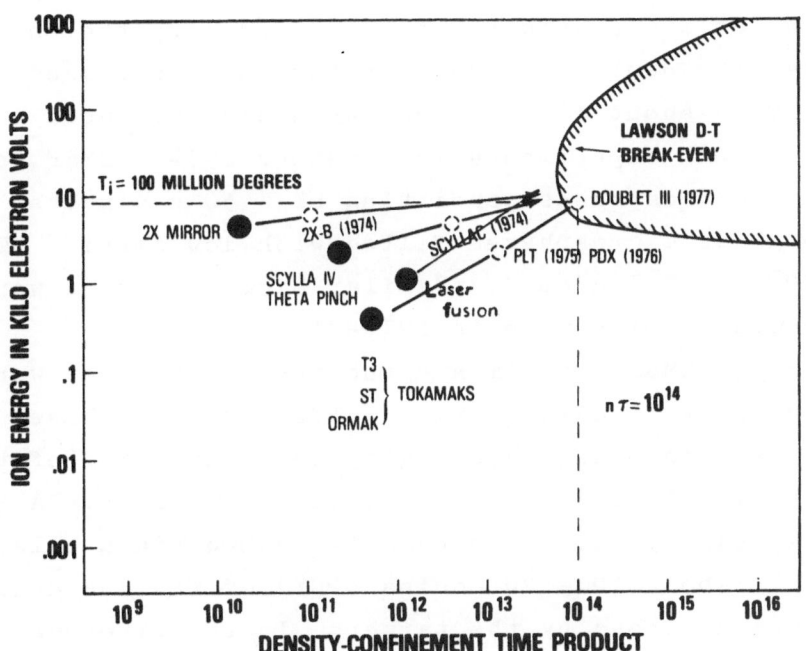

Figure 2

"Break-Even" Plasma Conditions

For Fusion Power

these experiments are substantially in agreement with
our LASNEX laser implosion/fusion microexplosion cal-
culations--which also predict the feasibility of laser
initiated fusion micro explosions with sufficiently
high efficiency for power production.

In the first successful laser implosion/fusion
experiments, which were carried out by KMSF in May
1974, about 10^4 neutrons were produced. By the end
of 1974, the neutron production had been increased to
10^6 - 10^7. About 10^4 neutrons were also produced in
the first LLL experiments in December 1974. KMSF used
fusion micro capsules consisting of ∿ 50 μ diameter
hollow glass microspheres filled with low density DT
gas (10^{-3} - 10^{-2} g/cm^3). Similar micro capsules were
fabricated at Livermore in 1970-71.

In the KMSF experiments the fusion capsules were
symmetrically irradiated with ∿ 100 Joules of laser
light via a complex focussing system consisting of two
laser beams, F0.6 lenses, and elliptic mirrors. A 300
ps FWHM laser pulse was optimally shaped via a pulse
stacker. About 10 - 20 Joules (20%) of the incident
light was absorbed by the target. In the Livermore
experiments the capsules were irradiated assymetrically
with ∿ 10 - 20 Joules of laser energy via a simple
focussing system consisting of one laser beam and an
F1 lens.

A 100 ps Gaussian pulse was used. About 15 Joules
(75%) of the incident laser light was absorbed by the
target. Single beam laser implosions which have so far
failed to produce observable fusion neutrons have been
carried out by the Los Alamos Laboratory.

The Livermore LASNEX computer program predicts about
3 X 10^4 fusion neutrons (after correcting for mix

quenching) for the LLL experiments. This is in re-
markably good agreement with the measured $1 - 2 \times 10^4$
neutrons measured-particularly in view of the complex
physics and target design. This axially symmetric calculation included classical magneto-hydrodynamic multigroup energy transport physics corrected for anomalous
electron conduction, resonance absorption, and non-
Maxwellian superthermal electrons. The calculations
agree out to 10 - 20 KeV with the thermal component of
the measured x-ray spectrum (Figure 3). The measured
non-thermal, decoupled part of the x-ray spectrum indicates that the suprathermal electron spectrum in
the target is substantially softer than has been used
in previous LLL calculations. If this result holds up,
electron preheat and decoupling effects are substantially
weaker than has been assumed.

These successful laser driven implosions are of
the exploding pusher type. The pusher is heated and
exploded by electron conduction from the laser heated
plasma. The inward moving pusher compresses the DT
to nearly the original pusher density (~ 1 g/cm^3) and

Not enough neutrons have been generated in any of
these experiments so that they can be demonstrated to
be truly thermonuclear corresponding to the calculated
1 KeV temperature of the imploded DT fuel. However,
the agreement of the neutron production with the LASNEX
calculation is evidence of their thermonuclear/implosive
origin. This evidence is strengthened by the apparent
agreement of the LASNEX calculations with the KMSF laser
implosion experiments including the KMSF experiments
in which one laser beam was delayed. However, until
KMSF reveals their pulse shape, the quantitative nature
of this agreement will not be known.

These successful laser driven implosions are of
the exploding pusher type. The pusher is heated and
exploded by electron conduction from the laser heated
plasma. The inward moving pusher compresses the DT
to nearly the original pusher density (~ 1 g/cm^3) and

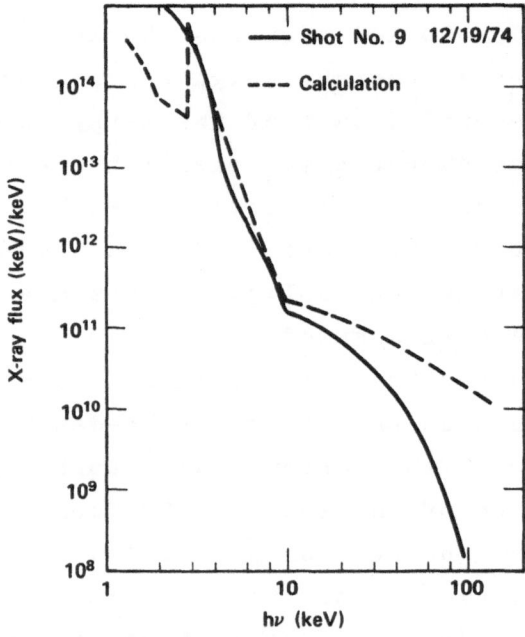

Figure 3

X-ray Spectrum from Fusion Experiment

heats it to temperatures of $\approx$ 1 KeV. Thermonuclear
burn occurs over a period of about 3 ps before decom-
pression and/or quenching by unstable pusher mixing into
the DT. Thermonuclear self-heating is insignificant
because the ρR is so small (10^{-4} - 10^{-3} g/cm^2) compared
to the ranges of the DT alpha particles and neutrons
(0.03 g/cm^2 at 1 KeV and 5 g/cm^2 respectively) and because the burn efficiency is so small ($\lesssim 10^{-6}$);
(Figures 4, 5). Because the ρR is so small--i.e.,
because the hydrodynamic time is so short, in particular
relative to the ion-electron coupling time--significant
ion-electron runaway occurs and the ion temperature is
substantially higher than the electron temperature
(a factor of 2 in these experiments). This greatly
increases the neutron production since $\overline{\sigma v}$ is varying
roughly as the fifth power of the ion temperature in the
one KeV range. The ion temperature is also enhanced by
the relatively high implosion velocity--nearly 5 X 10^7
cm/s. This velocity is approximately given by the
isothermal blowoff of the relatively high density pusher
into the low density DT and is proportional to the sound
speed in the pusher and to the log of the ratio of the
pusher and gas densities. Because the implosion velocity
is so high, and ρR so small, the x-ray Bremsstrahlung
radiation losses are also negligable.

It is important that this exploding pusher implosion
differs in several basic ways from the high density laser
implosions which must be achieved in order to generate
the high efficiency, high gain thermonuclear micro-
explosion required for laser fusion power applications
(Figure 6). In these high density implosions, ρR is
sufficiently large ($\approx$ 1 g/cm^2) that thermonuclear self-
heating is important. Ion electron runaway occurs

VELOCITY $\approx$ C Log (ρ_P/ρ_{DT}) isothermal rarefaction

ION-ELECTRON RUNAWAY

$$\frac{P \ \dot{V}/M}{\rho(T_i - T_e)/T_e^{3/2}} \quad \sim \quad \frac{\dot{R}}{\rho R} \quad \frac{T_i \ T_e^{3/2}}{T_i - T_e} \quad \frac{\text{hydro}}{\text{i.e. coupling}}$$

RADIATION LOSSES

$$\frac{P \ \dot{V}/M}{M\rho T_e^{1/2}} \quad \sim \quad \frac{\dot{R}}{\rho R} \ T_e^{1/2} \quad \frac{\text{hydro}}{\text{Bremsstrahlung}}$$

DENSITY $\approx \rho_{\text{pusher}}$ isobaric

Figure 4

Exploding Pusher Implosions

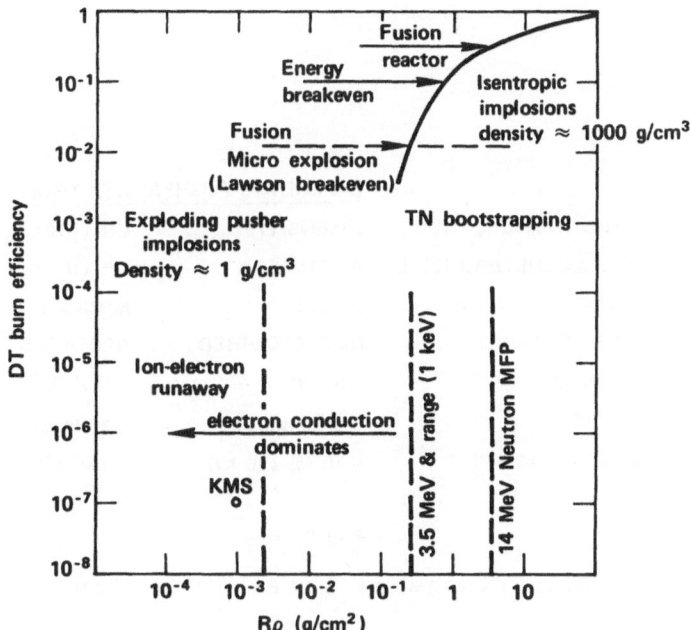

Figure 5

Laser Implosions

	EXPLODING PUSHER	ISENTROPIC
FLUID INSTABILITIES	INSENSITIVE	CRITICAL
PLASMA INSTABILITIES	INSENSITIVE	CRITICAL
ENTROPY CHANGE	LARGE	MINIMAL
PULSE SHAPING	NOT REQUIRED	REQUIRED
VELOCITY	$\sim 5 \cdot 10 \times 10^7$ cm/s	$\sim 5 \times 10^7$ cm/s
DENSITY	~ 1 g/cm^3	~ 1000 g/cm^3
BURN EFFICIENCY	$\sim 10^{-2}$% (10 kJ)	~ 10% (10 kJ)

Figure 6

Comparisons of Laser Implosions

during burn but not during implosion and temperatures
greater than several KeV must be achieved before the
thermonuclear self-heating exceeds the Bremsstrahlung
cooling. Electron conduction does not dominate energy
transport within the high ρR region as it does in the
exploding pusher fuel. The exploding pusher implosion
is not affected by fluid instabilities during the
acceleration of the pusher to high inward velocities
whereas the high density implosion is very sensitive to
fluid instabilities. Entropy changes are large in the
exploding pusher implosion and minimal in the high
density implosions. Consequently the exploding pusher
case is insensitive to plasma instabilities whereas
the high density case is highly sensitive (via entropy
changes due to superthermal electron preheat). Careful
pulse shaping is required to suppress fluid instabilities
and to minimize entropy changes so that high densities
may be achieved. With sufficiently short pulses
($\sim$ 100 ps) pulse shaping is not important in exploding
pusher implosions. Because of these fundamental differences, it is a giant step from a successful exploding
pusher laser driven implosion such as has been achieved
to a successful high density implosion which is required
for useful thermonuclear microexplosions.

What has been achieved in these laser driven implosion experiments makes laser fusion look vary attractive. The computer calculations upon which the feasibility of laser fusion has been based seem to be correct.
The problems which could possibly have made high density
laser implosions very difficult are less severe than
anticipated: laser light is being efficiently absorbed
(at least in the Livermore experiments) and the suprathermal electron tail is relatively soft. Within a few

years after inception of the accelerated laser fusion
program, and with lasers smaller than 100 Joules,
thermonuclear conditions of confinement and temperature
have been achieved which equal those achieved after
15 - 20 years of magnetic fusion research. These results
confirm the correctness of the decision by the AEC to
rapidly develop and construct 10 kJ implosion oriented
lasers. With these lasers on the horizon the outlook
for laser fusion is optimistic.

THREE QUASI-CW APPROACHES TO SHORT WAVELENGTH LASERS*

R. C. Elton

Naval Research Laboratory

Washington, D. C. 20375

ABSTRACT

Three approaches towards achieving extended-period quasi-cw amplification by stimulated emission in the vacuum-UV and x-ray spectral regions are discussed, in a somewhat logical progression towards shorter wavelengths, increased complexity, and demands. Extrapolation of visible and near-UV tuned-cavity cw lasers using higher density plasma media is first discussed for the near-to-mid VUV region. Further extension to the soft x-ray region is described, using preferential resonance charge transfer pumping. This and related intense incoherent x-ray source development could ultimately lead to successful quasi-cw $K\alpha$ inversions, as is discussed. Experiments underway to test the first two schemes are described.

---

*Supported in part by the Defense Advanced Research Projects Agency, DARPA Order 2694

I.   INTRODUCTION

The basic problems that hamper a rapid extension of lasers into the vacuum ultraviolet (VUV) and x-ray spectral regions can be summarized with a few simple relations. Since high reflectance cavities do not appear to be realistic for wavelengths shorter than ~ 1000 Å, significant gain must be achieved in a single pass; this immediately implies an increase by orders-of-magnitude in the inverted state density required for a given net gain at a particular wavelength. Hence, at truly short wavelengths we are usually speaking of amplified spontaneous emission (ASE) devices, which alone represent more of an amplifier than a tuned oscillator producing highly coherent radiation. In fact, the devices developed will probably prove most useful, at least for the near term, as amplifiers for coherent VUV radiation produced by frequency multiplication from the IR and visible regions.

For an amplifying medium of length L, the ASE gain is given by $I/I_o = \exp(\alpha L)$, where α is the gain coefficient. The product αL is often written as[1]

$$\alpha L = \frac{\lambda^2 A_{u\ell}}{4\pi^2 \Delta\nu} \; L \; \left(N_u - \frac{g_u}{g_\ell} \, N_\ell \right) \; , \qquad (1)$$

where g_u, g_ℓ and N_u, N_ℓ refer to the statistical weights and population densities of the upper and lower laser states, respectively, $\Delta\nu$ refers to the line width in frequency units, and λ refers to the wavelength of the laser transition. With the transition probability $A_{u\ell}$ for spontaneous emission scaling as $f\lambda^{-2}$, and for larger inversion (i.e., $N_u \gg N_\ell$), Eq. (1) can be written as a proportionality:

$$\alpha L \propto f N_u L / \Delta \nu, \tag{2}$$

with the oscillator strength f fairly constant along an
isoelectronic sequence. This demonstrates the need for
narrow lines and a large product $N_u L$. In plasma media
where short wavelengths occur in ionized species, $\Delta \nu$
is often dominated by Doppler or collision broadening[2]
(see Fig. 1). The lack of a wavelength-dependent factor
in Eq. (2) is somewhat deceiving, since a large N_u is
difficult to achieve against radiative depopulation
rates scaling as λ^{-2}.

Trade-offs are possible in raising $N_u L$. For example, with electron collisional pumping in a plasma
N_u scales as electron density N_e squared, so that a
high density plasma of necessarliy short length (for
pumping) is appropriate for the shortest wavelengths.
For increased density, N_u scales more as N_e due to increased collisional effects, and longer lengths can be
more appropriate. At very high densities, a longitudinal-
pump absorption length can decrease rapidly and $N_u L$ may
actually decrease with increasing density, unless transverse pumping is employed. (These scaling laws are ex-
panded for the 3p → 3s scheme below.) This illustrates
the obvious need for careful modeling for any particular
scheme, since many parameters enter. Before proceeding
to the pump power requirements, it is worth noting that
the presence of outer electrons, e.g., in innershell
laser transition schemes, decreases the net gain α_T
through photoionization losses (α_{pi}) to the main beam,
i.e., $\alpha_T \approx \alpha - \alpha_{pi}$. This is discussed further below for
the Kα innershell scheme.

A second general concern is the pump power per
unit area P/a required. This can be expressed as:

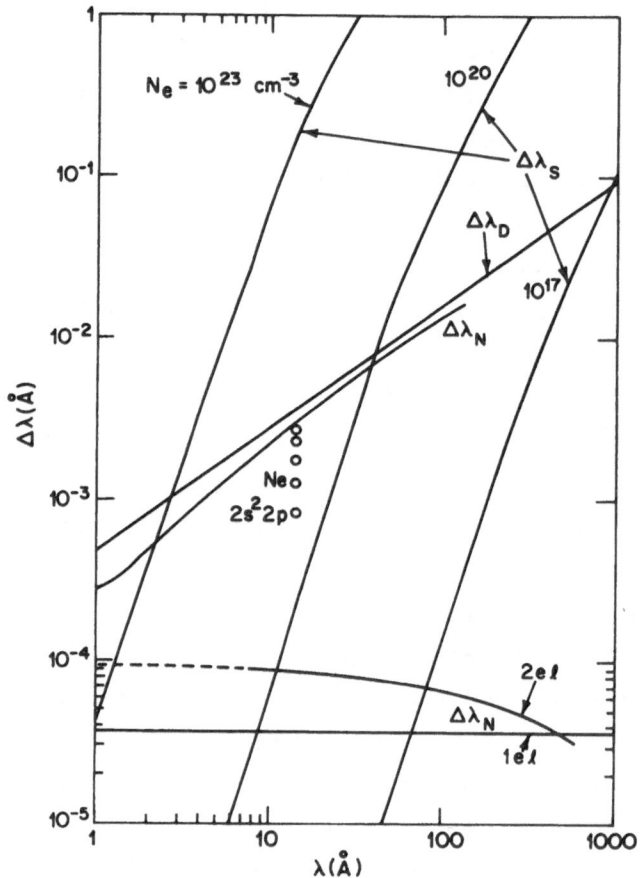

Figure 1.   Estimates of line widths for Kα type transi-
            tions versus wavelength λ, with natural
            ($\Delta\lambda_N$), Doppler ($\Delta\lambda_D$) and Stark ($\Delta\lambda_S$) effects
            included. The decrease in natural broadening
            with ionization is indicated by circles for
            neon; and hydrogenic and helium-like ionic
            species are included. (See Ref. 2.)

$$P/a = N_u Lh\nu(\Gamma+A) \propto \alpha L\nu^4 \left(1+\frac{\Gamma}{A}\right) \left(\frac{\Delta\nu}{\nu}\right) , \qquad (3)$$

using Eq. (1). Here ν is the laser frequency and Γ is
the Auger rate, which enters for innershell transitions.
For low Z and for metastable laser transitions, Γ can be
much greater than A which increases the pump requirements.
Since the presence of excess outer electrons also decreases the net gain as discussed above, there are clear
advantages to using transitions that avoid these problems
when possible, e.g., with hydrogenic and helium-like ions
(as well as alkalis for certain longer-λ transitions).
Since the same $N_u L$ product occurs in Eq. (3), similar
scaling examples to that above for αL may be applied.
It is worth noting from Eq. (3) that the ratio (P/a)
/αL, i.e., pump power density per gain product, scales
as ν^4 ($\Delta\nu/\nu \doteq \Delta\lambda/\lambda$ being approximately constant -- see
Fig. 1); thus an advance from 1 μm to 10 Å in wavelength
translates to a factor of 10^{12} in pump power density at
a particular gain product. With limited power, the need
for a small pump area a is obvious.

The third general problem that must be considered
is the lifetime of the inversion. Any transition can be
inverted for a pre-equilibrium interval limited approximately by A^{-1}, at which time lasing is "self-terminating". This time is in the nanosecond (10^{-9} sec) range
for wavelengths in the 1000 Å region, but becomes femto-
seconds (10^{-15} sec) for typical 1-2 Å dipole transitions.
Longer-lived metastable states with lower f-values and
transition probabilities require higher $N_u L$ products for
equivalent gain and comparable pumping flux, according
to Eqs. (2) and (3); the higher densities can then lead
to rapid collisional destruction of inverted metastable-
state populations so that the advantage is lost. A

sustained (cw or quasi-cw) inversion* is the most desirable mode of operation at short wavelengths and is
achieved in principle when the final laser level is depleted at a more rapid rate than it is filled; this is
particularly difficult for resonance transitions.
Three possible quasi-cw approaches will be considered
in the following.

II. SELECTED QUASI-CW APPROACHES

A. 3p → 3s

Numerous visible and near-UV lines from 3p → 3s
transitions have been reported to lase, including some
in plasmas of moderate density (using cavities).[3-5]
While the appropriate mechanisms for producing inversion
are the subject of continuing discussions, we have used
a single-ion excitation model shown schematically in
Fig. 2 for carbon-like ions. Electron collisional
pumping 2p → 3p is followed by 3p → 3s lasing, with the
final 3s level more rapidly depopulated by radiative
dipole decay to the 2p ground state, thereby maintaining
the inversion. At the densities anticipated, collisional mixing and depopulation of n = 3 states must be
included. An initial analysis for carbon-like ions extrapolated to Z as high as 42 has been published[6] and

---

*The cw-mode or stationary-inversion terminologies de-
 signate a sustained population inversion, in contrast
 to the self-terminating mode. Quasi-cw or quasistationary inversion implies that the inversion is
 limited in time by the environment or by the pumping
 pulse available, not by the basic atomic rates. The
 degree of inversion is distinct from its mere existence
 and a high value may have an additional transient dependence associated, e.g., with high initial electron
 temperatures in electron collisional pumping as discussed herein.

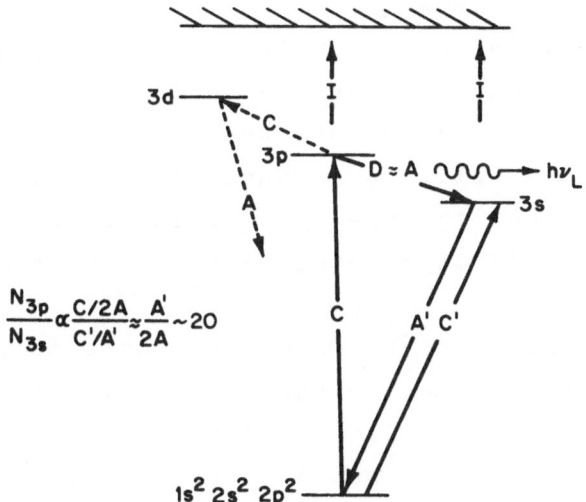

Figure 2.   Schematic energy level diagram for carbon-
            like ion species. Collisional excitation is
            designated by C and C', radiative decay by A
            and A', ionization by I, and collisional de-
            population by D. The relative 3p to 3s pop-
            ulation densities are estimated from a modi-
            fied corona model, where excitation is bal-
            anced by radiative decay and collisional de-
            excitation. Competing collisional depopu-
            lation to the 3d level is also indicated.

the resulting gain lengths are consistent with near-UV
O^{2+} observations. Results of this initial analysis
indicated a significant advantage for ASE in having a
high electron temperature T_e for collisional pumping
and a low ion temperature T_i for narrow lines. The
resulting minimum lengths for αL products of 1 (threshold)
and 5 (desired) and for $T_e = 10\ T_i$ are shown in Fig. 3
versus nuclear charge Z for the carbon-like isoelectronic
sequence. Also indicated are the electron densities
$(N_e)_{max}$ at which collisional depopulation becomes equal
to radiative decay for the $3p \rightarrow 3s$ laser transition; at
higher N_e the gain dependence becomes weaker as discussed below.

More recently, a "hot-spot" atmospheric plasma
model[7] has been applied to this problem.[8] In this
numerical model, energy is deposited in a short burst
and the electrons are assumed to be heated (by inverse
bremsstrahlung only) in times much shorter than the
electron-ion energy equipartition time. Peak T_e/T_i
ratios of 60 are predicted and, with the necessary
atomic physics included, higher gains are found for
OIII (O^{2+}) as plotted in Fig. 4 for the three densities
chosen so far. The practical and desirable duration of
such an enhanced electron temperature and associated
gain depends both upon the rate of depletion of lasing
ions through ionization as well as the rate of electron
cooling through collisions. The gain is also affected
by any rapid plasma expansion. A refined program to
properly include these effects for the particular experiment underway is now being assembled. It is of
importance to emphasize such modeling at this stage
to ascertain whether the high gains predicted so far
(Figs. 3 and 4) and associated with the enhanced electron

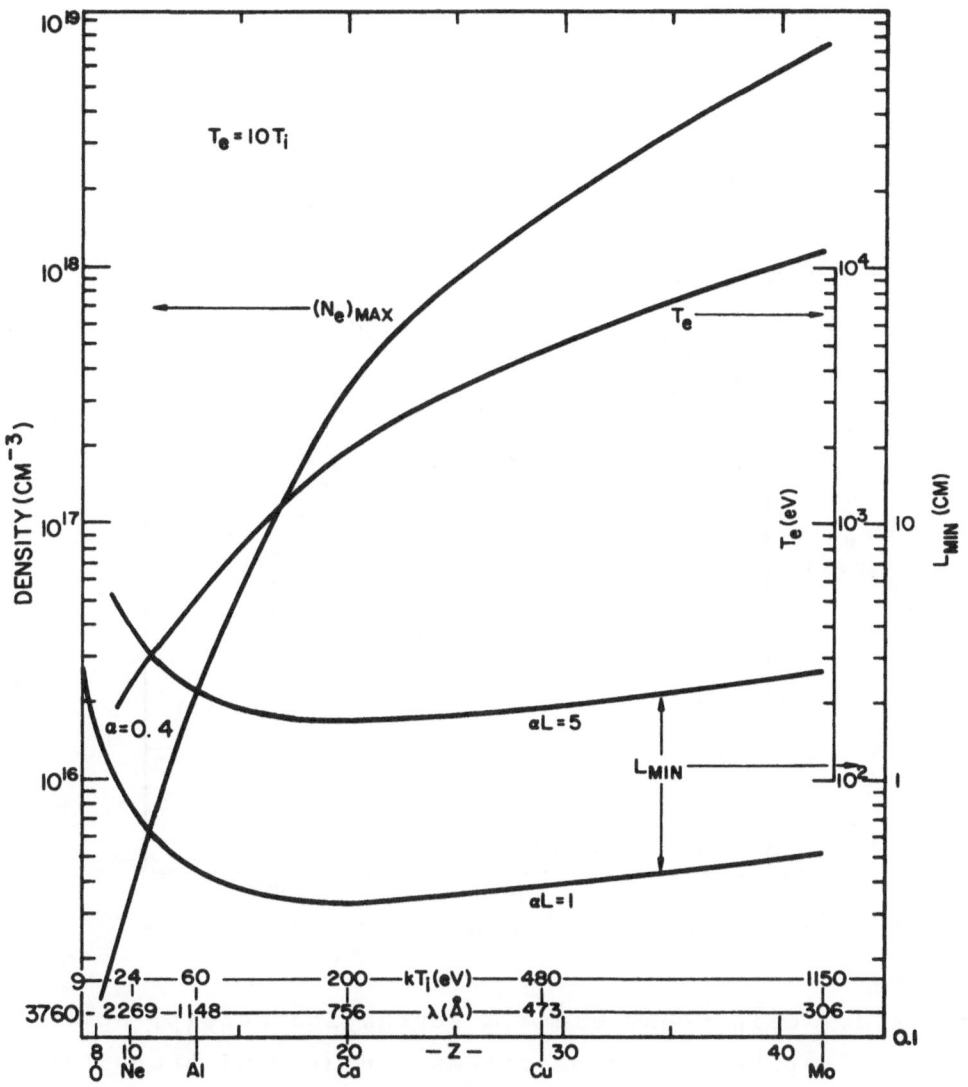

Fig. 3   Minimum Length L_{min} for amplification in carbon-
         like ions with a gain of exp (αL) versus atomic
         number Z, wavelength λ, and ion kinetic temperature
         kT_i. The electron temperature T_e is assumed equal
         to 10 T_i and is plotted. The electron density
         $(N_e)_{max}$ at which collisional mixing becomes im-
         portant is also plotted. Data extrapolated to
         O^{2+} (OIII) comparison with Fig. 4 (See also Ref.
         6).

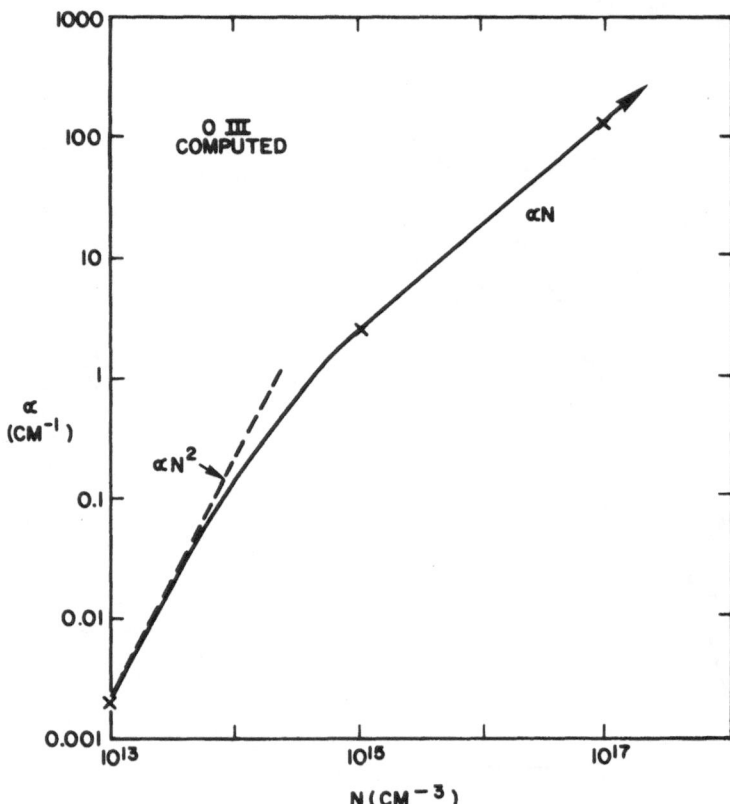

Figure 4.   Gain coefficient α versus ion density com-
            puted at three values for 3p → 3s transitions
            in the carbon-like O^{2+} (OIII) ion.   The
            varying dependence of α is understood by in-
            creased collisional mixing at higher densi-
            ties, as discussed in text.

temperature are indeed sustained for times significantly longer than A^{-1}, i.e., with a definite advantage of quasi-cw operation over self-terminating laser transitions.

The gain dependence (shown in Fig. 4) can be understood and extrapolated quite simply as follows, assuming $N_u = N_3 = N_2(N_e X_{23}/A_{33})$, where X_{23} is the collisional excitation rate coefficient from level 2 to 3. At low N_e ($> 10^{15}$ cm^{-3} for O^{2+}) and fixed length,

$$\alpha \propto N_2 \left(\frac{N_e X_{23}}{A_{33}}\right) \quad A_{33} \propto N_e^2 X_{23} \quad , \tag{4}$$

giving the N_e^2 dependence shown. For moderate densities ($\sim 10^{16}$ cm^{-3}),

$$\alpha \propto N_2 \left(\frac{N_e X_{23}}{N_e D_{33}}\right) \quad A_{33} \propto N_e A_{33} \quad , \tag{5}$$

and the density dependence becomes linear. Here D_{33} is the collisional deexcitation rate coefficient. For high densities, consider the particular case of axial laser heating of a preformed plasma. At densities $\gtrsim 10^{18}$ cm^{-3}, the axial pumping classical absorption length L_{abs} becomes comparable to the laser-medium length and scales as N_e^{-2} according to

$$L_{abs} \propto T_e^{3/2}/N_e^2 \lambda^2 \quad , \tag{6}$$

for a fixed ion charge. The gain product αL then becomes

$$\alpha L_{abs} \propto N_e^{-1} \left[A_{33} \left(\frac{X_{23}}{D_{33}}\right) \left(\frac{T_e^{3/2}}{\lambda_z^2}\right) \right] \tag{7}$$

and decreases as N^{-1}. Finally, at extremely high
densities ($\sim 10^{21}$ cm^{-3}) $N_e D_{32}$ becomes comparable to
A_{32} (the lower laser depopulation spontaneous decay
rate), collisions dominate, and equilibrium distributions
evolve with no inversion. This is all done for O^{2+} at
present, and is being extended to other ions.

In any case, the outlook is most encouraging for
reaching the mid-vacuum-UV region by extrapolation of
known laser transitions. Except at the very highest
densities, $L < L_{abs}$ and the pump power required scales
independently of L or N_e, with about 2×10^{14} W/cm^2 a
reasonable requirement.[6] Experiments are presently
underway at NRL using short pulsed (25 ps) lasers of high
quality to deliver a large P/a pumping flux axially to a
preformed linear plasma as indicated in Fig. 5 and are
discussed elsewhere in these proceedings by R. A. Andrews.
The first goal of these experiments (following initial
plasma characterization) will be to verify the existence of population inversion by relative line intensity
measurements; and then to proceed with orthogonal measurements to demonstrate net axial gain.

B. Resonance Charge Transfer Pumping

Detailed extrapolations analogous to the 3p → 3s
scheme described above have not been carried out for
transitions of higher energy, such as 4→3 or 3→2. A
preliminary analysis has indicated that for the former,
collisional mixing of the n = 4 terms will deplete the
inversion at too low a density to be practical for
significant amplification. The second possibility requires helium-like or hydrogenic ions without n = 2
electrons. With collisional excitation from the low-
lying 1s level, the large 1s-3s energy gap makes
pumping by free electron collisions in a plasma pro-

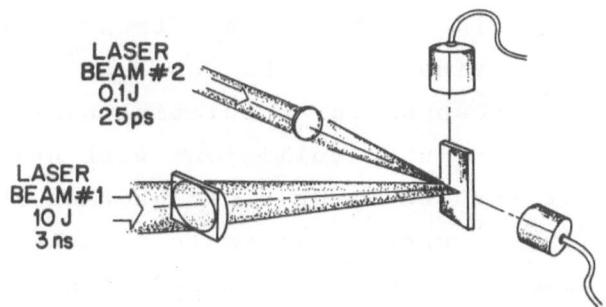

Figure 5.   Schematic of experiment designed to verify
            gain on 3p → 3s transitions pumped by
            electron collisions in a plasma.

hibitively difficult, according to the model employed
above. What is required is a less demanding (energy-
wise) mechanism for preferential population of n = 3
levels, and particularly the 3s level which could
cw-lase into 2p, in turn more rapidly depopulating to
1s. Cascading from higher levels following capture of
a free electron in a recombination transition is one
possibility considered. However, electron capture by
low-energy resonance charge transfer has by far the
largest cross section, providing the proper combination
of ion and atom can be found. The scheme is illustrated
schematically in Fig. 6 for certain transitions to be
described below. Vinogradov and Sobelman[10] originally
suggested this mechanism for populating higher states
of highly ionized neon in collisions with helium atoms,
and pointed out that the peak cross section is approximated by $\pi a_o^2 z^2$, where z is again the charge of the
ion. When this $\sim 10^{-16} z^2$ is compared to a total radiative recombination cross section[11] of about $10^{-20} z$,
the advantage of the resonance effect is obvious.

The resonance referred to here is at very low
energies (measured in eV instead of 10's of keV for
the higher energy resonance) and is described conven-
iently by the Landau-Zener formulism[12], worked out for
s-s transitions. This theory requires a classical
potential energy curve crossing between the initial
and final states of the system, and therefore requires
exothermic reactions, with an energy defect ΔE. In
Fig. 6 are plotted the resonance charge transfer cross
sections[12] versus relative particle energy for various
values of ΔE in eV, with smaller but finite defects
obviously favored. In the present analysis, we have
chosen H, He, and Ne as the atomic species, with the

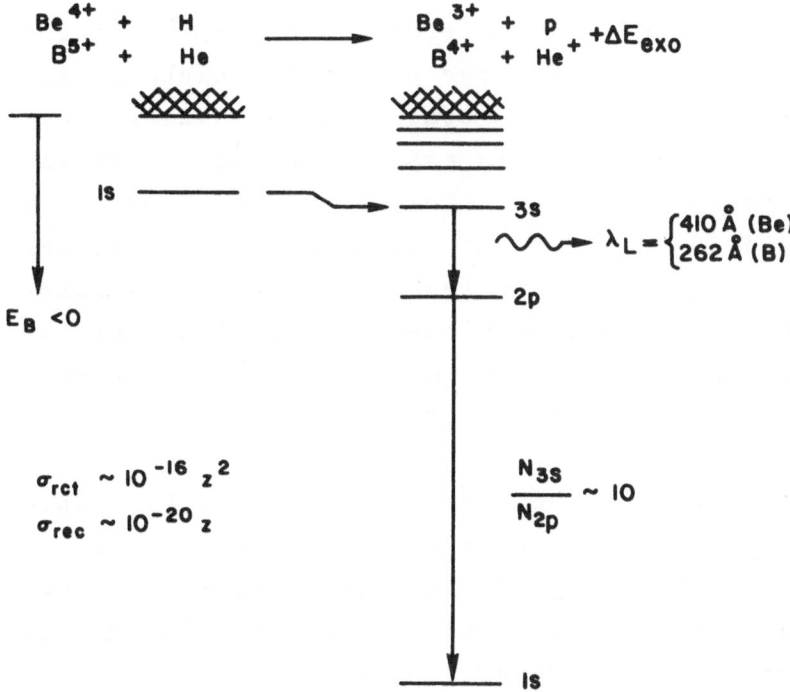

Fig. 6.   Schematic diagram of exothermic s-s resonance
          charge transfer reaction leading to a quasi-
          stationary population inversion between 3s and
          2p levels in certain helium-like or hydrogenic
          ions. Refer to Fig. 8 for other possible ion/
          atom combinations. E_B is the binding energy;
          σ_{rct} and σ_{rec} show the cross section scalings
          for resonance charge transfer and radiative
          recombination, respectively.

intention that they provide a gaseous atomic environment into which plasma ions can rapidly expand, from
a laser produced plasma for example. Hydrogen and
helium provide low-lying 1s electrons, for which the
theory is intended; neon provides six 2p electrons which
would be a useful test for p-p or p-s exchange transitions. The velocity required in utilizing Fig. 7
(and indicated there) is assumed thermal, with kT taken
as one-fourth the ionization energy required to produce
the ion desired. For initial ions, both completely
stripped and hydrogenic ions are considered, since no
n = 2 electrons are permitted for the 3-2 laser transition. The data for n+ ion stages of both classes is
sufficiently independent of element to be combined for
present purposes.

By inverting the data in Fig. 7 for each species,
we obtain the cross section as a function of ΔE shown
in Fig. 8. A survey of possible transitions between
the indicated ions and atoms shows the quantum states
into which charge transfer is likely to occur at a high
rate. For the 3s-2p transition, hydrogenic initial ions
are most promising, since the resulting 3s helium-like
ion level will not be as strongly coupled to 3p by
collisions. Here B^{4+} + H and C^{5+} + He are both promising;
for the former, hydrogen atoms do not have to be produced from gaseous molecules, as the cross section is of
similar magnitude for both the atomic and molecular
states.

Further parameters for some of the possible transitions are included in Table 1, where transitions to
the ground state are intentionally omitted since inversions of such seem most unlikely possibilities, at
least at present.[10] In this table are given the laser

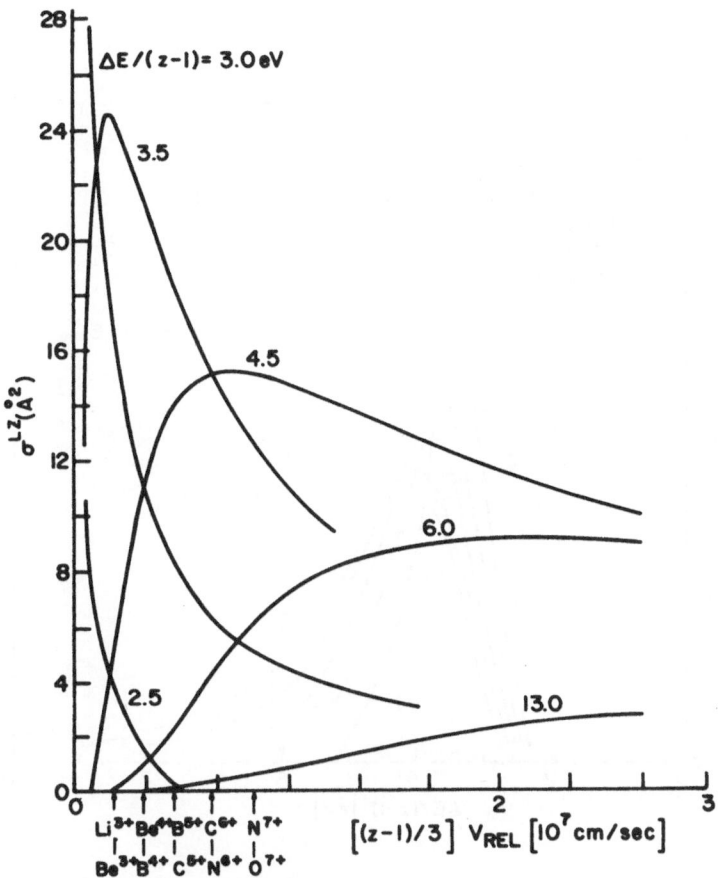

Figure 7. Resonance charge transfer cross section from
s-s Landau-Zener theory versus scaled relative velocity for the atom-ion combination.
(Data adapted from Ref. 12.) ΔE represents
the energy defect in eV for the exothermic
reaction, z the effective charge of the ion.
Velocities for ions designated are assumed
thermal, with the kinetic temperature chosen
as I.P./4 for creating the ion.

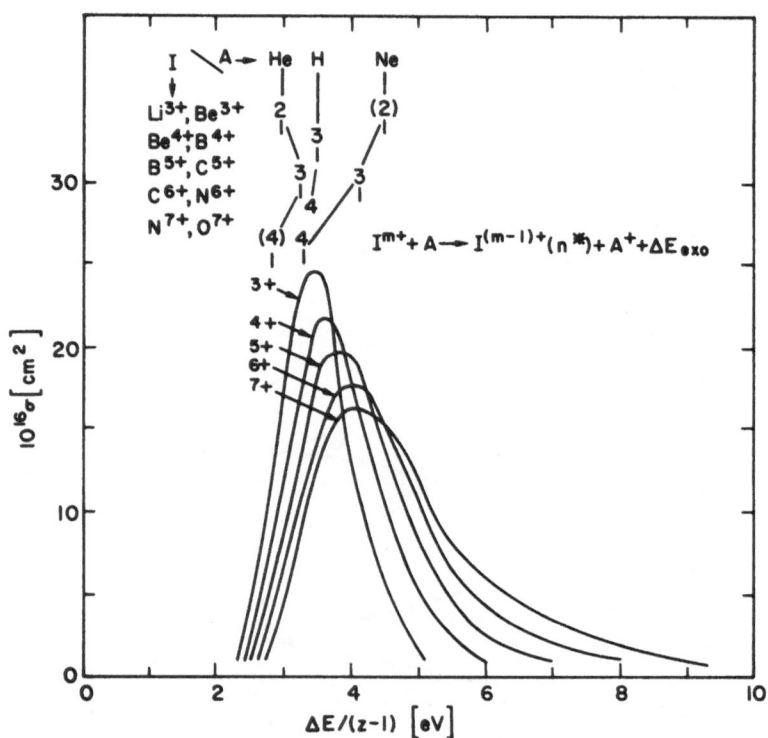

Fig. 8   Resonance charge transfer cross section ob-
         tained by inversion of Fig. 7 data.   Final
         quantum states of high capture probability for
         each ion I^{m+} and atom A combination are indi-
         cated by numerals, with parentheses to indicate
         less probable transitions.   ΔE is the exothermic
         energy defect and z the effective ion charge.

wavelengths, the "maximum" electron densities above which collisional depopulation seriously competes with radiative decay, and the maximum background-atom pressure P_A permitted for transmission of the radiation over a 3 cm length. (P_A could be increased up to about 100 Torr before laser breakdown effects become troublesome, should significant photoionization of the background gas occur; for hydrogen the protons produced by photo-ionization do not absorb). Also shown are the spon-taneous lifetimes t_u of the upper laser level, of use for self-terminating transitions (not for the 3s → 2p scheme). Finally the gain coefficient is given as de-rived from Eq. (1) assuming N_u given by $N_A N_i \sigma_{rct} v_i / A_{u\ell}$.

Table 1. Resonance Charge Transfer Laser Pumping

Δ_n	λ (Å)	$10^{-18} N_e$ (cm^{-3})	P_A (Torr)	$10^{12} t_u$ (sec)	α (cm^{-1})
(Be,B)$^{4+}$ + H	3-2 400	1	0.7	75	20
(B,C)$^{5+}$ + (He, Ne)	3-2 250	6	1	60	30
(C,N)$^{6+}$ + H	4-3 520	2	0.3	80	50
	4-2 130	2	20	90	40
(N,O)$^{7+}$ + (He, Ne)	4-3 380	6	0.3	45	30
	4-2 95	6	20	50	30

An experiment intended to test this method of pumping has been designed[9] and is shown schematically in Fig. 9. A high power glass laser beam will be focused in a line image onto various target materials placed on a rotatable disc in front of the entrance slit of a grazing incidence vacuum spectrograph. The

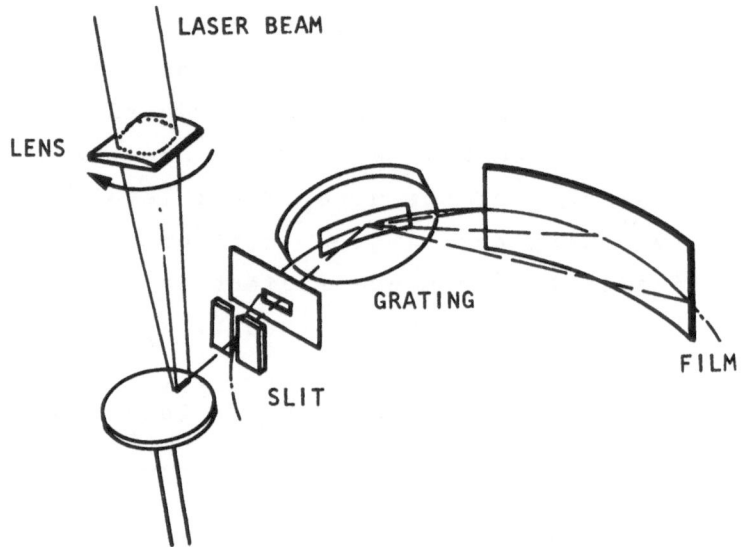

Fig. 9. Schematic diagram of the NRL resonance charge
transfer experiment, including the grazing
incidence vacuum spectrograph. The horizontal
slot provides spatial resolution along the
direction of plasma expansion from the target
surface. Rotation of the lens permits both
axial and radial viewing. The background
(atomic) gas is not indicated, nor is a planned
guiding magnetic field.

plasma formed will expand into a background gas in an
upward direction, parallel to the slit, and in a confined
slab configuration by the use of a solenoidal magnetic
field (not shown). Rotation of the focusing mirror will
permit both axial and transverse observations for indi-
cations of amplification. Spatial resolution along the
direction of expansion will be provided by a slot placed
between the entrance slit and the grating as indicated.
Experiments to verify anomalous populations under
optically-thin conditions will be carried out at lower
densities further from the target by a simple displacement of the target and lens assembly.

   C. Kα Quasi-Stationary Inversion

   As mentioned above, inversion with the ground state
by the above scheme of resonance charge transfer pumping
seems at the present unlikely, both because of the short
times involved in the self-terminating transitions and
of the pump power required.[10] However, as pointed out
by Vinogradov and Sobelman,[10] this could be a source of
intense spontaneous radiation in a single line or perhaps a series of closely spaced lines. If successfully
extended to the x-ray region, this could be a connection
to a third scheme which has received serious consideration for amplification in the x-ray region, i.e., quasi-
cw inversion of K-α innershell transitions. In a concept originally proposed by Stankevich[13], amplification
would occur by the depletion of K-shell vacancies resulting in the creation of L-shell vacancies (using
x-ray terminology). Self-termination in femtosecond
times would be avoided by Auger depletion of the L-
vacancies to higher shells at a rate higher than that
for K-vacancy decay (see sketch in Fig. 10). The relative rates R_L/R_K (reconstructed from the brief de-

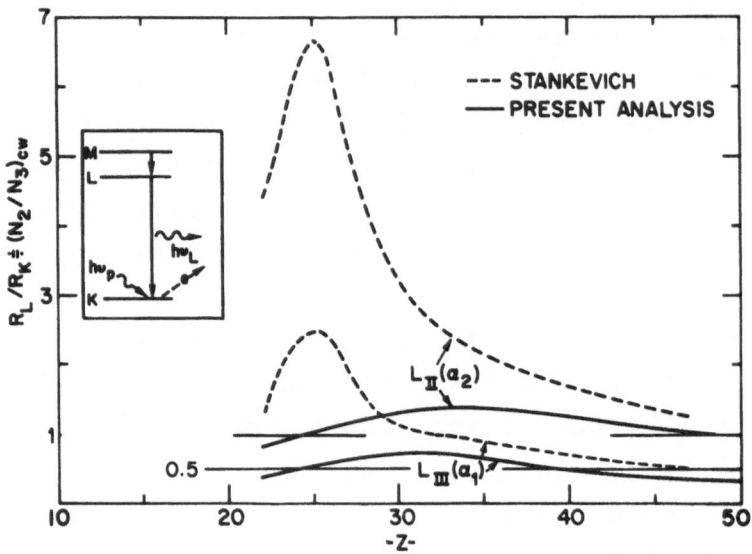

Fig. 10   Ratio of rates R_L/R_K for total transitions
out of L and K vacancy states, respectively,
versus atomic number Z. This ratio is equivalent to N_u/N_ℓ (or N_2/N_3 in Ref. 1) for
equilibrium conditions reached after long
times in cw operation. Values exceeding unity
and one-half indicate gain for the $K\alpha_2$ and
$K\alpha_1$ transitions, respectively. The model
here assumes <u>all</u> K-vacancy decay transitions
produce potential absorbers for laser radiation.
Present analysis is based on recent data; an
attempt to reproduce the results of Stankevich[13]
is shown dashed. Both $K \rightarrow L_{11}$ and $K \rightarrow L_{III}$, α_2
and α_1 respective transitions are shown. The
electron scheme is diagrammed as inset.
(See also Ref. 2.)

scription given by Stankevich) which relate to the
population inversion $N_u/N_\ell \fallingdotseq N_2/N_3$ achievable are shown
in Fig. 9. Also shown are the results of a more recent
reanalysis[2] using recent date.

What appears now to be only a marginal gain possibility in Stankevich's original concept, becomes more
promising if the further sophistication of shifts in
potentially absorbing lines accompanying K-shell Auger-
ionization transitions is introduced, as indicated in
Fig. 11 (note that recombination must continue at a
"balanced" rate). The results of this additional effect
are shown in Fig. 12. In the final analysis, the details
of which will not be repeated here except in Table 2,
sufficient pumping must be accomplished to overcome
photoionization losses to the main laser beam in the
medium, since there are by necessity many outer-shell
electrons present. This pumping must also be selective
for removal of the K-shell electron without disturbing
the outershell structure significantly. It can be shown[2]
that the magnitude and selectivity required can only be
achieved by photoionization with the "tuned" photon
flux density of magnitude N_ν, and power density F which
are indicated in Table 2, along with the resulting gain
αL for a length of 300 μm. It is suggested[2] that
blackbody sources of appropriate temperatures kT_{BB} with
peak wavelengths given by $(\lambda_m)_{BB}$ be employed, for example, with bands of intense lines in the correct wavelength region. What is needed in this approach is a
very intense x-ray pump source limited to a rather
narrow specific wavelength band concentrated onto a
small area, probably in a traveling wave pumping mode
to avoid minimum disturbance of the medium prior to
inversion. In addition to the charge transfer source

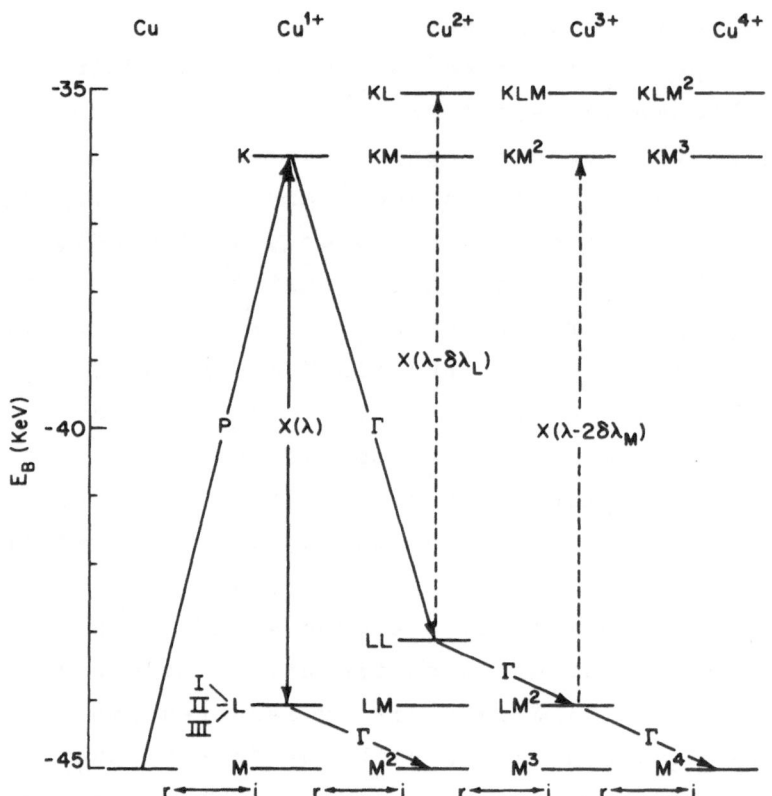

Figure 11. Vacancy diagram according to binding energies
E_B for copper. K, L and M designate shell-
vacancies. P, X, and Γ are the rates for
pumping, x-ray [emission or absorption
(dashed)] and Auger transitions, respectively. r, i indicate recombination and ionization, respectively. Auger line shifts
δλ are shown. (See also Ref. 2.)

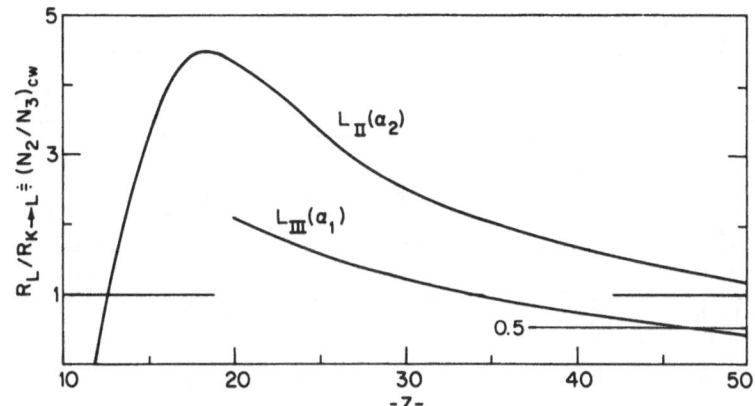

Fig. 12   Ratio of rates $R_L/R_{K \to L}$ for total transitions
out of an L vacancy state and radiative de-
cay out of a K vacancy state versus atomic
number Z. This ratio is equivalent to N_u/N_ℓ
(or N_2/N_3 in Ref. 1) for equilibrium conditions
reached after long times in cw operation.
Values exceeding unity and one-half indicate
gain for the $K\alpha_2$ and $K\alpha_1$ transitions, respectively. The model used assumes only radiative transitions produce absorbers, with Auger
transitions generating shifted ion lines.
Both $K \to L_{II}$ and $K \to L_{III}$, α_2 and α_1 respective
transitions are shown.

mentioned above, the further development of heavy
element (e.g., uranium) condensed spark pseudo-continuum
sources to the x-ray region has been suggested[2] for
pump source development; further concentration of the
x-ray flux is a separate problem.

Table 2. $K\alpha_2$ Pumping Requirements

ELEMENT	λ (Å)	$10^{-20}N_\nu$ (cm^{-3})	$10^{-3}F$ (TW/cm^2)	αL	kT_{BB} (keV)	$(\lambda_m)_{BB}$ (Å)
^{14}Si	7.1	2	4.8	70	0.5	5
^{20}Ca	3.4	4	21	30	0.7	4
^{29}Cu	1.5	20	230	15	1.2	2

III. SUMMARY

Present modeling has demonstrated that the extension of laser action can be actively pursued into
the vacuum-uv and eventually x-ray spectral regions
without high reflectance cavities, for the initial purpose of feasibility demonstration and for amplification
of coherent frequency multiplied beams. Extension of
proven visible and near-uv laser transitions into the
mid vacuum-UV region with plasmas of increased density
and decreased size seems to be a most reasonable
starting point. Extension to shorter wavelengths into
the soft x-ray region could be achieved by selective
population in, e.g., helium-like ions, where the Auger
and photoionization effects of outer electrons are absent;
here resonance charge transfer appears to be the most

promising candidate providing that a sufficient density
of ion-atom interactions can be achieved. Experiments
are well underway at NRL towards testing these two
schemes. Success here or with other intense x-ray
sources could finally lead to sufficient amplification
on Kα innershell transitions to overcome the losses
associated with the multiple outer electrons required
for quasi-cw and, eventually, to possible resonant cavity
operation.

IV. ACKNOWLEDGMENT

The cooperation of Drs. K. G. Whitney and J. Davis
in computing the 3p → 3s gain with their hot-spot program, leading to the data in Fig. 4, is greatly app-
reciated. Many illuminating discussions with Dr. R. A.
Andrews and Prof. H. R. Griem are also recalled with
appreciation.

V. REFERENCES

1. R. C. Elton, R. W. Waynant, R. A. Andrews and M. H.
    Reilly, "X-Ray and Vacuum-UV Lasers, Current Status
    and Prognosis" NRL Report 7412, May 1972.

2. R. C. Elton, "Quasistationary Inversion on K-α
    Innershell Transitions", NRL Memorandum Report
    2906, October 1974; submitted for publication.

3. P. K. Cheo and H. G. Cooper, J. Appl. Phys. 36,
    1862 (1965).

4. Y. Hashino, Y. Katsuyama and K. Fukuda, Japan J.
    Appl. Phys. 11, 907 (1972).

5. C. K. Rhodes, IEEE J. Quant. Elect. QE-10, 153
    (1974), Table of UV-Laser Transitions p. 170.

6. R. C. Elton, "Extension of 3p$\rightarrow$3s Ion Lasers into
    the Vacuum-UV Region", NRL Memorandum Report 2799,
    May 1974; also Appl. Optics, 14, 97, 1975.

7. K. G. Whitney and J. Davis, J. Appl. Phys. 45,
    5294 (1974); also K. Whitney, J. Davis and E. Oran,
    NRL Memorandum Report 2644, June 1973.

8. R. C. Elton, T. N. Lee, J. Davis, J. F. Reintjes,
    R. H. Dixon, R. C. Eckardt, K. Whitney, J. L.
    DeRosa, L. J. Palumbo and R. A. Andrews, Physica
    Fennica 9, Suppl. S1 (1974).

9. Staff Report: ARPA/NRL X-Ray Laser Program, NRL
    Memorandum Report 2910, October 1974.

10. A. V. Vinogradov and I. I. Sobel'man, Sov, Phys.
    JETP, 36, 115 (1973).

11. R. C. Elton, "Atomic Processes", (in) Methods of
    of Experimental Physics-Plasma Physics, Vol. 9A,
    eds. H. R. Griem and R. H. Lovberg, (Academic Press,
    New York, 1970).

12. H. J. Zwally and D. W. Koopman, Phys. Rev. 2, 1851
    (1970).

13. Yu. L. Stankevich, Sov. Phys. Doklady $\underline{15}$, 356 (1970).

REVIEW OF SOFT X-RAY LASERS USING CHARGE EXCHANGE†

C. D. Cantrell and Marlan O. Scully

University of California

Los Alamos Scientific Laboratory

Los Alamos, New Mexico 87544

and

University of Arizona

Optical Sciences Center

Tucson, Arizona 85721

and

K. Boyer*, R. Bousek**, J. E. Broley*, C. R.
Emigh*, F. A. Hopf**, M. Lax***, W. H.
Louisell***, W. B. McKnight*****, P. Meystre**,
D. Mueller*, J. Seely****, and R. Shnidman******

I. INTRODUCTION

This paper will review charge exchange vacuum
ultraviolet and soft x-ray lasers using ion beams[1,2].
The basic idea of the reaction is that one avoids the
rapid Auger rates characteristic of many other x-ray
laser schemes by picking up an electron in the outer

shell of a hydrogen-like, helium-like, or lithium-like
ion (see Table I).

$$(A^{n+})_{ground} + B \rightarrow (A^{(n-1)+})_{excited} + B^{+} \quad (1.1)$$

The cross section for pickup in an excited state is
larger than for pickup in the lower state for some range
of ion energies, due to a near resonance between the
state into which pickup occurs and the state of an appropriate target atom.

One of the major new results presented in this
paper will be a discussion of a coherent calculation
of the gain to be expected in swept excitation. We
show that simple estimates of the gain based on rate
equations may seriously misrepresent the actual behavior of a system subjected to traveling-wave excitation. The finite time required for the atomic
population to generate a coherent dipole moment, reduces the maximum value of the gain below the rate
equation value. Dicke superradiance increases the
energy obtained from the laser, but only after a substantial distance. Taking full account of the effect
of laser lethargy and superradiance, we derive a complete, correct gain equation for such lasers and give
numerical estimates of the gain to be expected in the
situations which concern us. We also discuss the
physical realizability of systems with the parameters
which are required according to the gain equation. In
particular, we show that gas jet targets with the appropriate density and sharpness can be constructed[3].
We discuss the experimental charge-exchange cross sections[4], and the Doppler width to be expected as the
result of the interaction between the ion beam and the
target. We discuss the production of ion beams of the
appropriate current and focused to the required spot

size, taking into account the effects of space charge
and emittance. Finally, we show that the required focus
and deflection can be achieved with a simple system. In
summary, we believe that it is possible, using current
technology, to achieve a soft x-ray laser at a wave
length of 304Å with a small signal gain of over 100,
using a duoplasmatron ion source, or a gain of 10^4, using a plasma-gun source.

II. EXCITATION METHODS

The methods by which a reaction of the kind indicated in Eq. (1.1) can be achieved, may be divided
into two broad categories: traveling-wave excitation
and quasi-steady-state excitation. A possible realization of swept excitation is shown in Fig. 1.
Traveling-wave excitation, which has been used to advantage in the vacuum ultraviolet[5], has the substantial
advantage that the energy loss due to fluoresence or
amplified spontaneous emission is minimized. This is
not the case in steady-state excitation. Further, the
entire ion beam is focused onto a small portion of the
target, thus achieving a high inversion density.
Traveling-wave methods are, of course, adaptable to
either plane or cylindrical geometries. These methods
suffer from the disadvantage that space charge imposes
a requirement of short focal length to achieve the needed tight focusing, and also that a variable focal length
lens is required. These problems are not intractable.
Space-charge neutralization may ease the limitations
imposed by electrostatic repulsion. Bunching of the
ion beam can increase the current, thus relaxing the
focusing requirement[6]. Finally, some kinematic focusing
is possible with a cylindrical or parallel-plate

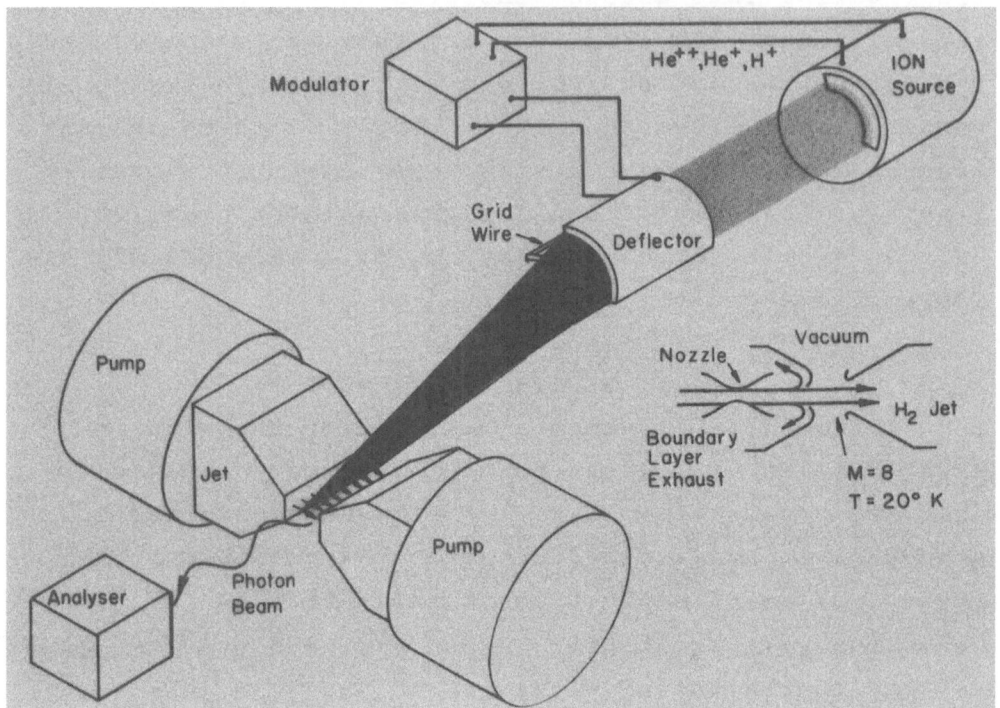

Fig. 1   Pictorial diagram of a facility now being de-
         signed at the Los Alamos Scientific Laboratory
         for studies of charge-exchange soft x-ray
         lasers. The diagram shows the production,
         focusing, and swept deflection of an ion beam,
         which intersects a high-density, sharp-boundary
         gas target.

deflection method which we call the "scissors" method,
discussed in more detail below. Quasi-steady-state
excitation is, of course, easier experimentally than
traveling wave excitation and, since one uses a line
focus rather than a point focus, there are fewer space
charge problems. A short focal lens is easily adaptable
to reasonable experimental geometries. However, as we
shall see soon, quasi-steady-state excitation offers
much lower gain for the same total ion beam current
than the traveling-wave "scissors" method. In addition,

a line-focused ion beam will result in an excitation region which is cylindrical and unfortunately optimally adapted to lose energy by amplified spontaneous emission. The use of a plasma gun[7] in a quasi-steady-state configuration offers sufficiently higher ion currents to offset many of these disadvantages.

In the bulk of this paper, we shall consider the "scissors" deflection method, so called becuase the ion beam intersects the surface of the target at a very small angle in such a way that the intersection between the ion beam and target travels at the velocity of light. Such a deflection method was originally proposed by McCorkle[8]. However, our analysis of this method differs from his. Traveling-wave deflection in the scissors method is accomplished by applying a voltage pulse traveling along two parallel plates at the velocity of light. The ion beam, which was originally traveling parallel to the plates, acquires a component of velocity perpendicular to the plates, and either exits through a slot, as in McCorkle's original scheme, or travels beyond the plates to impact the target. After exiting from the region between the plates, the beam is inclined at an angle $\theta \sim v_\perp/c$ to the plane of the plates. A possible source for confusion in analyzing this method results from the fact that the trajectories of the individual ions are inclined at a much larger angle,

$$\theta = \tan^{-1} (v_\perp /v_\parallel) \qquad (2.1)$$

These relationships are illustrated in Fig. 2.

To compare the "scissors" method of excitation with other methods, we shall estimate the density of inverted atoms to be expected in the target. This

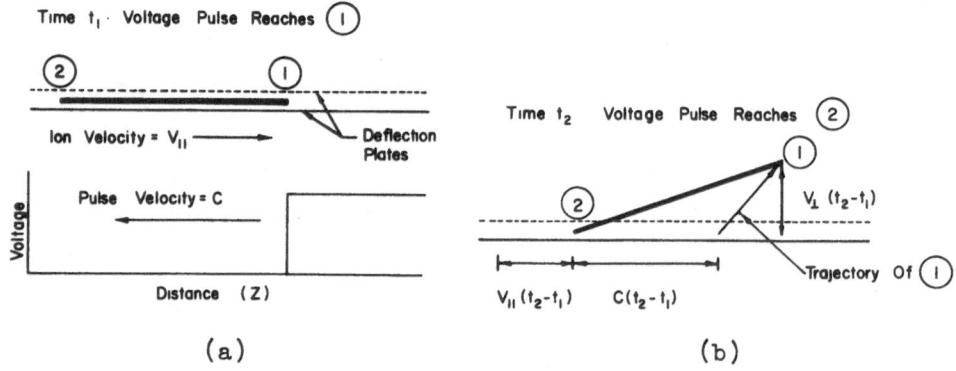

Fig. 2 Geometry of the "scissors" deflection method.
The ion beam, shown as a heavy line, is initially moving to the right with velocity v
between a pair of plane-parallel deflection
plates.
(a)   Time $t = t_1$: A TEM-wave incident from the
right has just reached the leading edge of the
beam, (1).
(b)   Time $t = t_2$: The voltage pulse has reached
the trailing edge of the beam. The individual
ion trajectories are not identical with the
axis of the ion beam.

density is simply the number contained in the initial
cylindrical volume, as shown in Fig. 3, times the fraction ΔP of the ions which wind up in the desired upper
state, divided by the total volume of the region of excitation in the target. We note that in the dimension
perpendicular to the target, the size of the excitation
region is $v_\perp \tau_s$, because the ions continue to radiate
x-rays for this distance after crossing the boundary
of the target. The inversion density is

$$\Delta N = \frac{I\ \Delta P}{Ze\ \Delta y\ v_\parallel\ v_\perp\ \tau_s} \equiv \Delta N\ \text{scissors} \qquad (2.2)$$

where I is the ion-beam current, and τ_s is the radiative
lifetime.

If $\Delta x \gg v_\perp \tau_s$, then ΔN will be reduced by a factor

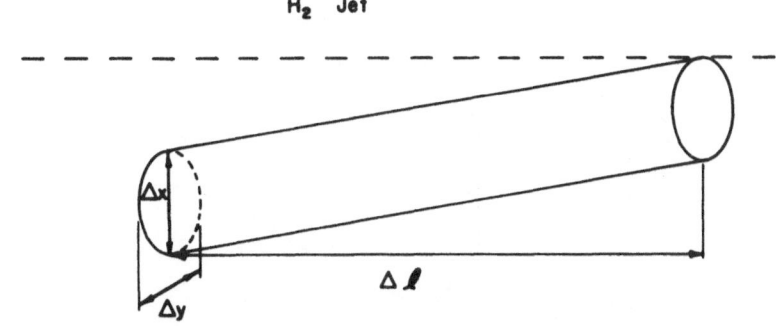

Fig. 3   Target-ion beam interaction volume in the
"scissors" scheme.

$v_{\perp} \tau_s / \Delta x$. Also, if the target density builds up gradual-
ly, then ΔN is reduced by a factor of $v_{\perp} \tau_s \lambda_p$, where λ_p
is the inverse mean free path for pickup of one electron:

$$\lambda_p = \sigma_p N_{target} \quad . \tag{2.3}$$

We compare the inversion density to be expected in
the "scissors" scheme, with that to be expected in a
scheme in which the ion beam's original velocity vector
$\vec{v}_o$ is perpendicular to the target, so that the ions impact the target primarily in a normal direction. Pic-
torially, the beam is swept like a water hose along a
wall. See Fig. 4. We see that

$$\Delta N = \frac{I}{Ze\Delta y c} \frac{\Delta P}{v_{\perp} \tau_s} \tag{2.4}$$

and therefore

$$\Delta N = \Delta N_{scissors} \cdot \frac{v_{\parallel}}{c} \quad . \tag{2.5}$$

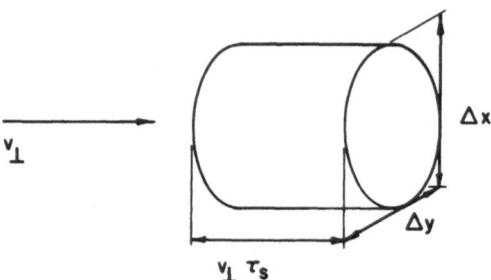

Fig. 4   Target-ion beam interaction volume in the
         "swept water hose" scheme.

Since $v_{\parallel}$ is ordinarily small compared to c, the gain
is substantially reduced in this case.   This makes such
a scheme highly unattractive, unless very large ion
currents are available.

     As an example of a quasi-steady-state excitation
method, we mention allowing the ion beam to impact the
target normally without being swept.   See Fig. 5.   In
this case the inversion density to be expected is,

$$\Delta N = \frac{I \, \tau_s \Delta P}{ZeLv_o \tau_s \Delta y} = \Delta N_{scissors} \, , \qquad (2.6)$$

which is also substantially reduced from the scissors
inversion density, since the distance $v_o \tau_s$ is expected
to be very small compared to a typical dimension of the
ion beam.   In the case we consider of a laser at 304 Å,
$v_o \tau_s$ is approximately 10^{-2} cm, while L is of order
1-10 cm.   It is possible, however, that this reduction
in gain can be offset by a great increase in ion currents
available from a device such as a plasma gun[7].   We shall
not consider quasi-steady-state methods of excitation in

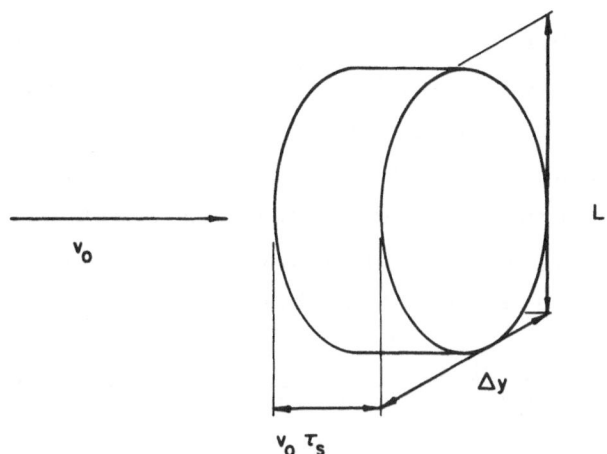

Fig. 5   Target-ion beam interaction volume in a quasi-
          steady-state "water hose" scheme, in which the
          beam is not swept along the target.

detail in this review, although this subject is currently under active investigation by our group. We shall
give a detailed discussion in a subsequent section of
the physical realizability of a scissors deflection
method. In particular, we shall show that it is possible to focus and sweep so that all the ions impact the
target at the same point, and the point of impact travels
at the velocity of light in the absence of space charge.

III.   TRAVELING-WAVE GAIN

The theory of laser amplifiers with steady state
gain is well known, having being investigated by Hopf
and Scully[9], and many subsequent authors. In this theory,
one finds significant departures from the prediction of
rate equations. The optical Bloch equations predict
phenomena such as self-induced transparency[10] which cannot be derived from the rate equations. The subject of

traveling-wave pumping has, however, been discussed very
little from a true quantum optics point of view. We
find very significant departures for our example from
the predictions of simple rate equations. In this section we will describe our analytic and numerical calculations of the traveling-wave gain in swept excitation.

The starting point for a theory of a laser amplifier is Maxwell's equations, which reduce to

$$\frac{\partial^2 E}{\partial z^2} - \frac{1}{c^2} \frac{\partial^2 E}{\partial t^2} - \mu_o \sigma \frac{\partial E}{\partial t} = \mu_o \frac{\partial^2 P}{\partial t^2} \quad , \qquad (3.1)$$

and the quantum mechanical density matrix equations of
motion. We shall assume throughout this calculation
that we are dealing with a simple two-level system. The
upper level will be called a, the lower b. The density
matrix equations of motion are

$$\frac{\partial \rho_{aa}}{\partial t} = r_a \ \delta(t-\frac{z}{c}) - \gamma_a \ \rho_{aa} - \frac{ipE(z,t)}{\hbar} \ (\rho_{ab}-\rho_{ba}) \quad ,$$

$$\frac{\partial \rho_{bb}}{\partial t} = r_b \ \delta(t-\frac{z}{c}) - \gamma_b \ \rho_{bb} + \frac{ipE(z,t)}{\hbar} \ (\rho_{ab}-\rho_{ba}) \quad ,$$

$$\frac{\partial \rho_{ab}}{\partial t} = -(i\omega+\Gamma_{ab})\rho_{ab} - \frac{ipE(z,t)}{\hbar} \ (\rho_{aa}-\rho_{bb}) \quad ,$$

$$\rho_{ba} = \rho_{ab}^{*} \quad , \qquad (3.2)$$

where γ_a and γ_b are the non-radiative decay rates of
levels a and b; $\Gamma_{ab} = \frac{1}{2}(\gamma_a+\gamma_b)+\Gamma_{ph}$ is the dephasing rate;
ν is the resonant frequency of the atomic transition;
r_α is the fraction of the atoms pumped into the level
α = a, b; and $p = e<a|x|b$. The diagonal elements of the
density matrix define a population difference equal to

$$\Delta N = N(\rho_{aa} - \rho_{bb}) \quad . \qquad (3.3)$$

The off-diagonal elements of the density matrix give rise to an atomic dipole moment which acts as the source for the electric field in Eq. (3.1). Explicitly, the macroscopic dipole moment per unit volume is

$$P(z,t) = Np\int_{-\infty}^{\infty} d\omega \; \sigma(\omega)[\rho_{ab}(z,t,\omega)+\rho_{ba}(z,t,\omega)] \; , (3.4)$$

where N is the number of atoms per unit volume, and $\sigma(\omega)$ is the distribution of atomic frequencies for Doppler broadening. This is given by

$$\sigma(\omega) = \sigma_o \; \exp-[4 \; \ell n2(\omega-\nu)^2/(\Delta\omega_D)^2] \quad , \qquad (3.5)$$

where

$$\sigma_o = \frac{2}{\Delta\omega_D} \sqrt{\frac{\ell n2}{\pi}} \quad ,$$

and $\Delta\omega_D$ is the full width at half maximum of the Doppler broadened line. We seek a so-called zero-phase solution of Maxwell's equations[9]

$$E(z,t) = E(z,t) \; \cos(kz-\nu t) \; ,$$

$$\qquad\qquad\qquad\qquad\qquad\qquad (3.6)$$

$$P(z,t) = S(z,t) \; \sin(kz-\nu t) \; .$$

subject to the well-known slowly varying amplitude and phase approximation

$$\frac{\partial E}{\partial z} \ll kE \; , \qquad\qquad \frac{\partial S}{\partial z} \ll kS \; ,$$

$$\qquad\qquad\qquad\qquad\qquad\qquad (3.7)$$

$$\frac{\partial E}{\partial t} \ll \nu E \; , \qquad\qquad \frac{\partial S}{\partial t} \ll \nu S \; .$$

In this approximation we find that the equation of motion
of the electric field is

$$\frac{\partial E}{\partial z} + \frac{1}{c}\frac{\partial E}{\partial t} + \kappa E = \frac{\nu}{2c\epsilon_o} S(z,t) \quad , \qquad (3.8)$$

where

$$S(z,t) = \frac{Np^2}{\hbar} \int_o^t dt' e^{-\Gamma_{ab}(t-t')} E(z,t')$$

$$\times [\chi_{aa}(z,t',t-t') - \chi_{bb}(z,t',t-t')] \quad .$$

In the last equation we have, for convenience, introduced the complex susceptibility

$$\chi_{\alpha\alpha}(z,t,T) = \int_{-\infty}^{\infty} d\omega \; \sigma(\omega)\rho_{\alpha\alpha}(z,t,\omega)\cos(\omega-\nu)T \quad , \quad (3.9)$$

which, as is evident from the density matrix equations
of motion, depends upon the pumping of the medium and
also upon the electric field. In the rotating-wave
approximation, we have used the fact that the off-diagonal
elements of the density matrix are given by

$$\rho_{ab}(t) = -\frac{ip}{\hbar}\int_o^t dt' E(t')[\rho_{aa}(t')-\rho_{bb}(t')]$$

$$\times \cos(kz-\nu t')e^{-(i\omega+\Gamma_{ab})(t-t')}$$

$$\cong -\frac{ip}{2\hbar} e^{i(kz-\nu t)}\int_o^t dt' E(z,t')[\rho_{aa}(t')-\rho_{bb}(t')]$$

$$\times e^{-[i(\omega-\nu)+\Gamma_{ab}](t-t')} \qquad (3.10)$$

which implies that the polarization is given by

$$P(z,t) = \frac{N p^2}{\hbar} \int_{-\infty}^{\infty} d\omega \; \sigma(\omega) \int_{0}^{t} dt' \; e^{-\Gamma_{ab}(t-t')}$$

$$\times \; [\rho_{aa}(z,t',\omega) - \rho_{bb}(z,t',\omega)]$$

$$\times \; [\sin(kz-\nu t)\cos\{(\omega-\nu)(t-t')\}$$

$$+ \; \cos(kz-\nu t)\sin\{(\omega-\nu)(t-t')\}$$

$$\times \; E(z,t') \; .$$

(3.11)

The contribution of the term proportional to $\sin((\omega-\nu)(t-t'))$ vanishes if the atomic frequency distribution σ and the population factors ρ_{aa} and ρ_{bb} are even functions of $\omega-\nu$. The equation of motion for the population factor $\rho_{\alpha\alpha}$ (where the index α can take on the value a or b) is, in the rotating-wave approximation,

$$\frac{\partial \rho_{\alpha\alpha}}{\partial t} = r_a \; \delta(t-z/c) - \gamma_\alpha \rho_{\alpha\alpha}$$

$$\mp \frac{1}{\hbar} \; p^2 E(z,t)\cos(kz-\nu t)$$

$$\times \int_{0}^{t} dt' \; e^{-\Gamma_{ab}(t-t')} E(z,t')$$

$$\times \; [\rho_{aa}(z,t') - \rho_{bb}(z,t')]$$

$$\times \; \cos(kz-\nu t + (\omega-\nu)(t-t'))]$$

$$\cong r_\alpha \; \delta(t-z/c) - \gamma_\alpha \; \rho_{\alpha\alpha}$$

$$\mp \frac{p^2 E(z,t)}{2\hbar} \int_{0}^{t} dt' \; e^{-\Gamma_{ab}(t-t')} E(z,t')$$

(3.12)

$$\times [\rho_{aa}(t') - \rho_{bb}(t')]\cos\{(\omega-\nu)(t-t')\} \; ,$$

where the upper sign is associated with level a, and the
lower sign with level b. We find an equation of motion
for the complex susceptibility by differentiating the
definition of $\chi_{\alpha\alpha}$:

$$\frac{\partial \chi_{\alpha\alpha}}{\partial t}(a,t,T) = \int_{-\infty}^{\infty} d\omega\ \sigma(\omega)\ \frac{\partial \rho_{\alpha\alpha}}{\partial t} \cos(\omega-\nu)T \quad, \quad (3.13)$$

with the result

$$\frac{\partial \chi_{\alpha\alpha}}{\partial t} = r_\alpha\ \delta(t-z/c)g(T) - \gamma_\alpha\ \chi_{\alpha\alpha}$$

$$\mp \frac{1}{2\hbar}\ p^2 E(t) \int_0^t dt'\ e^{-\Gamma_{ab}(t-t')}\ E(t')$$

$$\times \int_{-\infty}^{\infty} d\omega\ \sigma(\omega)[\rho_{aa}(z,t',\omega)-\rho_{bb}(z,t',\omega)]$$

$$\times \cos\{(\omega-\nu)(t-t')\}\cos\{(\omega-\nu)T\} \quad . \quad\quad (3.14)$$

In this equation we introduced the Fourier transform of
the atomic frequency distribution

$$g(T) = \int_{-\infty}^{\infty} d\omega\ \sigma(\omega)\ \cos\{(\omega-\nu)T\}$$

$$= \exp\ -\ [(\Delta\omega_D T)^2/16\ell n2)] \quad . \quad (3.15)$$

Our equation of motion for the complex susceptibility
alone follows:

$$\frac{\partial \chi_{\alpha\alpha}}{\partial t} = r_\alpha\ \delta(t-z/c)g(T) - \gamma_\alpha\ \chi_{\alpha\alpha}$$

$$\mp \frac{1}{4\hbar}\ p^2 E(t) \int_0^t dt'\ e^{-\Gamma_{ab}(t-t')}\ E(t')$$

$$\times \ \{\chi_{aa}(t',T+t-t')+\chi_{aa}(t',T-t+t')$$

$$-\chi_{bb}(t',T+t-t')-\chi_{bb}(t',T-t+t')\} \ . \qquad (3.16)$$

We now proceed to obtain an approximate analytical solution for the electric field and the intensity to be expected from a traveling-wave laser with swept excitation. To assist in obtaining an analytical solution we shall assume that the complex susceptibility is driven primarily by the pumping, indicated by the term

$$r_a \ \delta(t-z/c) \ g \ (T) \ .$$

This approximation amounts to neglecting the saturation of the atomic population by the optical field and is in accordance with the spirit of a calculation of the linear gain. The linearized equation of motion for the susceptibility is

$$\frac{\partial \chi_{\alpha\alpha}}{\partial t} + \gamma_\alpha \ \chi_{\alpha\alpha} = r_\alpha \ \delta(t-z/c)g(T) \ . \qquad (3.17)$$

Since the right hand side of this equation is a function only of the retarded time

$$\xi = t - z/c \qquad (3.18)$$

and the linewidth parameter T, the solution can be written down immediately:

$$\chi_{\alpha\alpha}(z,t,T) = r_\alpha \ g(T)e^{-\gamma_\alpha \xi}u(\xi) \ . \qquad (3.19)$$

We have introduced the unit step function

$$u(\xi) = \begin{cases} 1 & \text{for } \xi > 0 \\ 0 & \text{for } \xi < 0 \end{cases} . \tag{3.20}$$

Armed with a solution for the complex susceptibility χ, we can calculate the polarization which drives the electromagnetic field. We obtain the following equation for the propagation of the intensity $E^2(z,t)$:

$$\{\frac{\partial}{\partial z} + \frac{1}{c}\frac{\partial}{\partial t} + 2\kappa\} \, E^2(z,t)$$

$$= \frac{\nu N p^2}{hc\epsilon_0} \, E(z,t) \, e^{\beta^2} \, u(\xi)$$

$$\times \int_0^\xi d\tau \, E(z,t-\tau)\{r_a \, e^{-\gamma_a\xi} e^{-\alpha[\tau-\beta/\sqrt{\alpha}]^2}$$

$$-r_b \, e^{-\gamma_b\xi} \, e^{-\alpha[\tau+\beta/\sqrt{\alpha}]^2}\} \quad , \tag{3.21}$$

where we have introduced the linewidth parameters α and β

$$\beta = \sqrt{\ln 2} \, \frac{\gamma_a-\gamma_b}{\Delta\omega_D} \quad ,$$

$$\alpha = \frac{(\Delta\omega_D)^2}{16\ln2} \quad ,$$

and have set $\Gamma_{ab} = 1/2 \, (\gamma_a+\gamma_b)$. This approximation, which amounts to saying that T_2 is equal to T_1, is correct in our case, though not in all swept-excitation lasers. However, this approximation is a matter of convenience and can easily be removed. If it is further true that T_1 and T_2 are short compared to the length of the optical pulse, then we may replace

$$E(z,t-\tau) = E(z,t) \quad , \qquad (3.22)$$

to obtain the propagation equation for the intensity in the rate-equation approximation

$$\{\frac{\partial}{\partial z} + \frac{1}{c}\frac{\partial}{\partial t} + 2\kappa\} \; E^2(z,t)$$

$$= \frac{3}{2}\sqrt{\frac{\ell n2}{\pi}} \; \lambda^2 \; \frac{\Delta\omega_S}{\Delta\omega_D} \; N \; E^2(z,t)e^{\beta^2}u(\xi)$$

$$\times \; \{r_a \; e^{-\gamma_a\xi} I_a - r_b \; e^{-\gamma_b\xi} I_b\} \quad , \qquad (3.23)$$

where

$$I_a = 2\sqrt{\frac{\alpha}{\pi}} \int_0^\xi e^{-\alpha[\tau-\beta/\sqrt{\alpha}]^2} \; d\tau \quad . \qquad (3.24)$$

The quantity $\Delta\omega_s$ is the reciprocal of the spontaneous lifetime τ_s. Changing to traveling wave variables

$$(z,t) \rightarrow (z'=z, \; \xi=t-z/c) \; , \qquad (3.25)$$

$$\frac{\partial}{\partial z} + \frac{1}{c}\frac{\partial}{\partial t} \rightarrow \frac{\partial}{\partial z'} \quad , \qquad (3.26)$$

the propagation equation for the intensity becomes

$$\frac{\partial E^2(z',\xi)}{\partial z'} = [\eta F(\xi)-2\kappa]E^2(z',\xi) \; , \qquad (3.27)$$

where

$$\eta = \frac{3}{2}\sqrt{\frac{\ell n2}{\pi}} \; \lambda^2 \; \frac{\Delta\omega_s}{\Delta\omega_D} \; Ne^{\beta^2} \quad , \qquad (3.28)$$

$$F(\xi) = u(\xi)[r_a e^{-\gamma_a\xi} I_a(\xi)-r_b e^{-\gamma_b\xi} I_b(\xi)] \quad . \qquad (3.29)$$

This equation can be solved analytically with the result

$$E^2(z,\xi) = E^2(0,\xi)\exp[\eta F(\xi) - 2\kappa]z \quad . \qquad (3.30)$$

The linear gain per unit length deduced from this equation is, omitting the loss 2κ,

$$g(\xi) = \eta F(\xi) \quad ,$$

$$g(\xi) = g\, u(\xi)e^{-\gamma\xi}\, \text{erf}(\sqrt{\alpha}\xi) \quad , \qquad (3.31)$$

where $\gamma_a = \gamma_b \equiv \gamma = \Gamma_{ab}$, and $\qquad\qquad (3.32)$

and

$$g = \frac{3}{2}\sqrt{\frac{\ln 2}{\pi}}\,\lambda^2\,\frac{\Delta\omega_s}{\Delta\omega_D}\,N(r_a - r_b) \quad . \qquad (3.33)$$

If the Doppler width is so large that the exponential dies away quickly, then we can write the approximate solution

$$E^2(z,\xi) = E^2(0,\xi)\exp\{\frac{3}{2}\sqrt{\frac{\ln 2}{\pi}}\,\lambda^2\,\frac{\Delta\omega_s}{\Delta\omega_D}\,N$$

$$\times [u(\xi)(r_a e^{-\gamma_a\xi}J_a - r_b e^{-\gamma_b\xi}J_b)] \qquad (3.34)$$

$$-2\kappa\}z \quad ,$$

where

$$J_a = e^{\beta^2}[1+\text{erf}\beta] \quad , \quad J_b = e^{\beta^2}[1-\text{erf}\beta] \quad . \qquad (3.35)$$

The resulting gain coefficient at $t = z/c$ is (omitting the loss 2κ)

$$g = \frac{3}{2}\sqrt{\frac{\ln 2}{\pi}}\,\lambda^2\,\frac{\Delta\omega_s}{\Delta\omega_D}\,N\,[r_a J_a - r_b J_b] \quad , \qquad (3.36)$$

per unit length.

The gain given by Eq. (3.31) has the following noteworthy properties: For short times, $\xi \ll \gamma^{-1} = T_2$, and $\xi \ll 2 \sqrt{\ln 2} \; T_2{}^*$, the gain increases linearly with time:

$$g(\xi) \cong g \; \frac{(t - \frac{z}{c}) \Delta\omega_D}{2\sqrt{\pi \ln 2}} \quad . \tag{3.37}$$

This linear growth of the gain is due to the linear growth of the induced microscopic dipole moment. For much longer times, $\xi \gg 2\sqrt{\ln 2} \; T_2{}^*$, the gain decays exponentially with a time constant equal to T_1, the ordinary atomic lifetime:

$$g(\xi) \cong g \; e^{-\gamma(t - z/c)} \quad . \tag{3.38}$$

The competition between the initially lethargic response of the atomic medium and population decay prevents the gain from reaching the maximum value given by a simple-minded application of the rate equations. As shown in Fig. 6, this reduction can be very substantial when the Doppler width is small as is the case in the experiment we envision. We shall show subsequently, however, that superradiant effects increase the effective gain nearly to the value given by a simple rate equation, the increase being accomplished over a substantial propagation distance.

For future use we quote the gain per unit length derived from Eq. (3.34), cautioning the reader that this should be used only as an approximation to the true gain which results from many complicated, interacting factors:

$$g \cong \frac{3}{4} \sqrt{\frac{\ln 2}{\pi}} \; \lambda^2 \; \frac{\Delta\omega_s \; e^{\beta^2} (1 + \mathrm{erf}\,\beta) I \; \Delta P}{\Delta\omega_D \; e \; v_o \; \Delta x \; \Delta y} \quad . \tag{3.39}$$

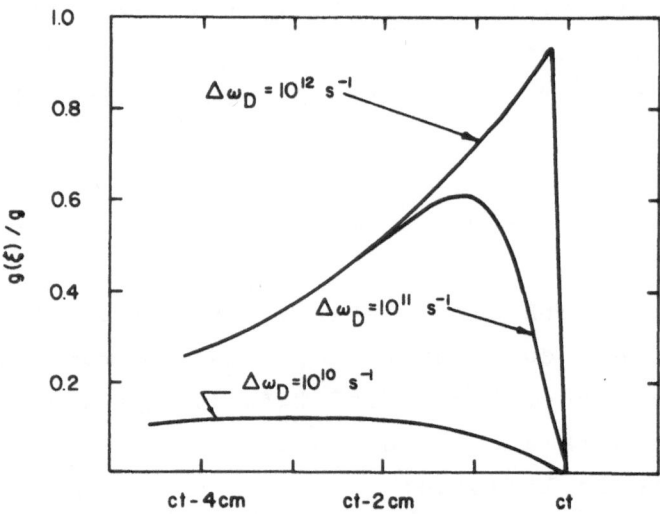

Fig. 6 The ratio $g(\xi)/g$ of the gain $g(\xi)$ according to
        Equation (3.31), to the gain g obtained assuming
        an infinitely fast buildup of the atomic polari-
        zation, is plotted against the retarded time ξ.
        The spontaneous lifetime is $\tau_s = T_1 = 10^{-10}$ s.

For a solid foil target, the spontaneous lifetime is
long compared to $(\Delta\omega_D)^{-1}$. In this case the gain is

$$g_{solid} \cong \frac{3}{4} \frac{\sqrt{\ell n 2}}{\pi} \frac{\lambda^2 \Delta\omega_s I \Delta P}{\Delta\omega_D e v_o \Delta x \Delta y} \quad . \tag{3.40}$$

However, for a gas target the spontaneous linewidth and
Doppler width are of the same order of magnitude, and
the gain is

$$g_{gas} \cong \frac{3}{4} \frac{\sqrt{\ell n 2}}{\pi} \lambda^2 \frac{\Delta\omega_s}{\Delta\omega_D} \frac{I}{e v_o} \frac{\Delta P}{x \Delta y} e^{\beta^2} (1+\text{erf}\beta) \quad . \tag{3.41}$$

To escape from the restriction of neglecting atomic
saturation and to investigate the effects of superradiance, we have performed a numerical calculation of
the field and intensity to be expected in a traveling-
wave laser for values of the parameters appropriate for

our example. The electric field obeys the equation

$$\frac{\partial E(z,\xi+z/c)}{\partial z} = - \kappa E + F(z,\xi+z/c)$$

$$+ \frac{\nu N p^2}{2c\epsilon_0 \hbar} \int_{-z/c}^{\xi} d\xi' \ e^{-\Gamma_{ab}(\xi-\xi')} \tag{3.42}$$

$$\times E(z,\xi'+z/c)\chi(z,\xi'+z/c,\xi-\xi') \ ,$$

where the noise F must be introduced phenomenologically
to account for spontaneous emission. The generalized
complex susceptibility:

$$\chi \ (z,\xi_1,\xi_2) \equiv \chi_{aa} \ (z,\xi_1,\xi_2) - \chi_{bb} \ (z,\xi_1,\xi_2) \ ,$$

obeys the equation

$$\frac{\partial \chi(z,\xi+z/c,\xi'+z/c)}{\partial \xi}$$

$$= (r_a - r_b)\delta(\xi)g(\xi'+z/c) - \gamma \chi$$

$$- \frac{1}{2\hbar} \ p^2 E(z,\xi+z/c) \int_{-z/c}^{\xi} d\xi'' \ e^{-\Gamma_{ab}(\xi-\xi'')}$$

$$\tag{3.43}$$

$$\times E(z,\xi''+z/c)\{\chi(z,\xi''+z/c,\xi'+z/c+\xi-\xi'')$$

$$+ \chi(z,\xi''+z/c,\xi'+z/c-\xi+\xi'')\} \ .$$

As shown in Fig. 7, the amplitude and phase of the pulse
at a distance z = 9 cm, roughly 1/2 the length of the
amplifier, are almost exactly what one would expect for
a fully coherent pulse. The phase varies by only a
degree or so over all of the significant amplitude.
Further downstream, at 19 cm, (see Fig. 8), we note that

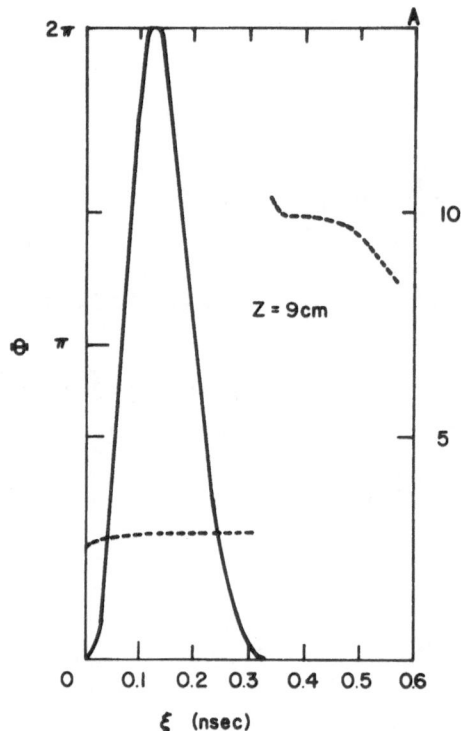

Fig. 7   The phase (left ordinate) and amplitude (right
         ordinate) of the electric field at Z = 9cm,
         plotted as a function of the retarded time ξ.
         The parameters used were: T_1 = .1 ns, T_2 = .2 ns,
         T_2^* = .1 ns. The error bars represent the ex-
         tremes obtained with six different sets of random
         numbers for the noise F.

each subpulse is coherent for essentially its entire duration. In Fig. 9 we plot the energy as a function of
distance down the amplifier. The error bars represent
the extreme values obtained for six different sets of
random numbers used to generate the noise F. We note
that at a distance of 2 cm the pulse energy is substantially below the value predicted from a simple rate
equation but that by 20 cm has nearly approached this
value. This comes about because the intense electric

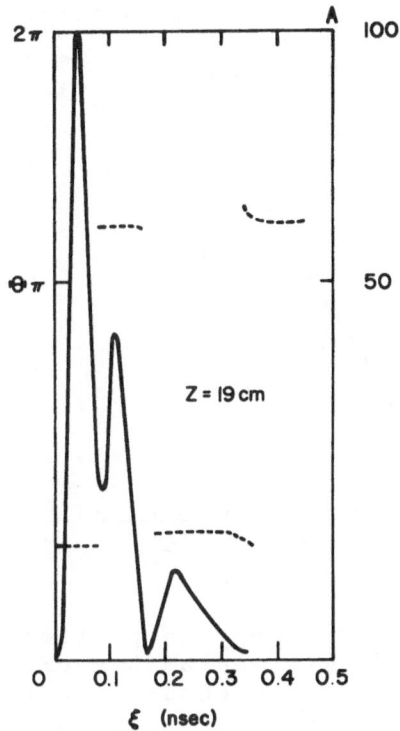

Fig. 8   The same plot as in Fig. 7, for Z = 19 cm.

field induces a dipole moment in each of the atoms of
the active medium which greatly overwhelms the dipole
moment arising from the random pumping of the medium.
The resulting rate of radiation is proportional to N^2
(where N is the number of atoms) and is thus superradiant
in character[11]. These results are consistent with our
analytical calculations to the extent that the latter
are valid. We recall that the analytical calculations
depended upon assuming that the pulse is long compared
to T_1 and T_2. However, from our figures this is mani-
festly not the case after a substantial distance has
been traveled through the active medium.

In summary, we point out that our results for the

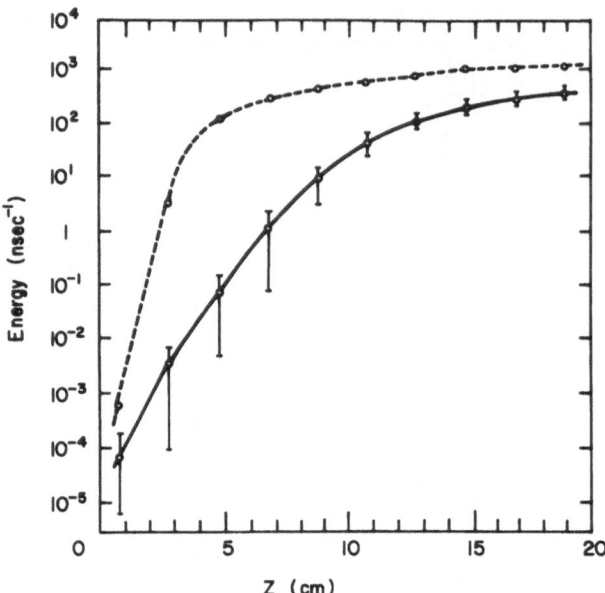

Fig. 9   The extracted energy is plotted versus the
         length z of gain medium traversed.   The dashed
         curve shows a rate-equation approximation using
         the linear gain g (see Fig. 6).   The solid curve
         shows the mean of six computer runs of a fully
         coherent pulse-propagation calculation; the
         error bars indicate the extremes obtained with
         different sets of random numbers for the random
         noise F.

gain per unit length shows very substantial differences
from the prediction of the simple rate equation.   For
this reason, we suggest that the rate equation value for
the gain be used with caution, since it substantially
overestimates the gain in certain situations.   Our
amplifier shows distinct characteristics of superradiance
which eventually overcomes the lethargy of the medium.
Finally, because the pulse soon shortens so that the
pulse length is shorter than any of the relevant atomic
lifetimes, the pulse is essentially phase coherent for

its entire length.   This fact is extremely important for possible future use of an x-ray laser of this type for holography or interferometry.

Losses

The principal radiative losses in the soft x-ray region arise from diffraction and photoelectric absorption.

The diffraction losses from a beam of diameter $\Delta\chi$ can be estimated from the diffraction angle

$$\Delta\theta \sim \frac{\lambda}{\Delta x} \quad . \tag{3.44}$$

We shall take diffraction to give rise to a distributed loss

$$\lambda_D = \ell_D^{-1} \quad , \tag{3.45}$$

where the diffraction length ℓ_D is the distance in which the beam spreads by an amount approximately equal to its original size.

$$\Delta x = \ell_D \Delta\theta \quad . \tag{3.46}$$

This is the same as requiring that the Fresnel number of the source at a distance ℓ_D be one, and leads to

$$\ell_D = \frac{(\Delta x)^2}{\lambda} \quad . \tag{3.47}$$

For our case ($\lambda = 304 \overset{\text{o}}{\text{A}}$, $\Delta x \sim 10^{-2}$ cm) we find

$$\ell_D \sim 30 \text{ cm.} \tag{3.48}$$

The photoelectric absorption constant, per cm, in
the target gas is

$$\kappa = \sigma N_{target} \quad . \tag{3.49}$$

The value of σ at a wavelength of 304 $\overset{o}{A}$ is known[19] to
be

$$\sigma = 2.8 \times 10^{-19} \text{ cm}^2 \quad , \tag{3.50}$$

leading to

$$\kappa = 1.4 \times 10^{-2} \text{ cm}^{-1} \quad ,$$

at a target density of 5×10^{16} cm^{-3}.

IV.   TARGET INTERACTIONS
Charge-Exchange Cross Sections
The energy levels of a hydrogenic atom

$$E_n = -\frac{1}{2} mc^2 \alpha^2 \left(\frac{Z}{n}\right)^2 \tag{4.1}$$

are such that a fully stripped ion with charge Z plus a
hydrogen atom in the 1s state has the same energy as a
(Z-1) charged ion in the state n = Z and a proton. However, the charge exchange between atomic hydrogen and a
Z-charged fully stripped ion, is not a resonant reaction
as one might naively conclude from the identity of the
energy at infinite separation. Both of the final particles in this reaction are positively charged and repel
each other strongly. The resulting Coulomb barrier
prevents the reaction from ocurring if the ions are too
close to each other, but also causes energy-level

crossings at slightly larger separations. The cross
section peaks at a finite energy. Since the reaction is
only quasi-resonant, it is perhaps not surprising that
the cross sections for molecular hydrogen are not greatly different from those for atomic hydrogen[12]. The
theory of such reactions was worked out long ago by
Oppenheimer, Kramers, and Brinkman[13], and more recently
modified by Schiff and others[14]. No simple calculation
gives results which agree with experiment[15]. Fortunately, the population is insensitive to the absolute value
of the cross sections. In fact, the population inversion
is approximately

$$N_{ion} \Delta P \simeq \frac{\sigma_{2P} - \sigma_{1s}}{\sigma_{total}} N_{ion} \quad . \tag{4.2}$$

As will be clear below, the absolute value of the cross
section does, of course, influence the optimal density
for a target medium. Most models for calculation of
cross sections for this type of charge exchange reaction,
are in agreement as to the value of the ratio in Eq.
(4.2). These cross sections have recently been measured
by Bayfield and Khayrallah[4] and are shown in Table 2.
Using these cross sections, and other known cross sections for the interaction between ion beam and the target,
we estimate that we can achieve a 5% inversion in the
ions which impact the target.

B. Population Dynamics

The populations of the laser species and the incident ions obey the rate equations

$$\frac{dP_{oo}}{dx} = - \sum_{\alpha} P_{oo} = - \lambda P_{oo} \quad , \tag{4.3}$$

$$\frac{dP_{\alpha o}}{dx} = \lambda_\alpha \ P_{oo}(x) - \sum_{\beta \neq \alpha} (\gamma_\alpha^\beta + \Lambda_\alpha^\beta) P_{\alpha o}(x)$$

$$(4.4)$$

$$+ \sum_{\beta \neq \alpha} \gamma_\beta^\alpha \ P_{\beta o}(x) \quad,$$

where P_{oo} is the population density of the incident ions; $P_{\alpha o}$ is the population density of the ions which have picked up one electron in the state α; λ_α is the mean free path

$$\lambda_\alpha = \sigma_\alpha \ N_{target} \quad, \qquad\qquad (4.5)$$

where N is the total density of target atoms. Also, σ_α is the cross section for pickup in the state α of the ion $A(z-1)+$; λ is the sum

$$\lambda = \sum_\alpha \lambda_\alpha \quad, \qquad\qquad (4.6)$$

and physically, is the inverse mean free path for one-electron pickup in any state. Also, λ_α^β is the inverse mean free path for a transition from the state β to the state α in the N-1 charged ions; Λ_α^β is the inverse mean free path for transitions from the α state of the N-1 charged species to the β state of the N-2 charged species. Initially, only P_{oo} differs from zero. Consequently, we consider only the most important process, namely

$$(A^{(n-1)+})_{ground} \leftrightarrow (A^{(n-1)+})_{excited} \quad . \qquad (4.7)$$

The rate for this process we shall denote by γ, taking all other rates $\gamma_{\alpha\beta}$ to be zero. Then the population of the (N-1) charged species in the state α obeys the rate equation

$$\frac{dP_{\alpha o}}{dx} = \lambda_\alpha e^{-\lambda x} - (\Lambda + \gamma) P_{\alpha o} + \gamma P_{\beta o} \quad . \qquad (4.8)$$

We find that the inversion density, expressed as a fraction of the total ion density is

$$\Delta P(x) \equiv P_a(x) - P_b(x)$$

$$= \frac{\lambda_a - \lambda_b}{\lambda - \Lambda - 2\gamma} \, [e^{-(\Lambda + 2\gamma)x} - e^{-\lambda x}] \quad . \qquad (4.9)$$

It is evident that in the limit $x \ll \gamma^{-1}$ and $x \ll (\Lambda + 2\gamma)^{-1}$ this equation reduces to

$$\Delta P \simeq x(\lambda_a - \lambda_b) \qquad . \qquad (4.10)$$

If we could choose $x = \lambda^{-1}$, this would imply that the population difference is approximately given by Eq. (4.2). Numerical values of the cross sections appropriate for a gas jet target and for a solid foil target, are shown in Table 2. Using Eq. (4.9), we find the optimum target thickness for a given density, or the optimum target density for maximizing ΔP at $x = v_o \tau_s / 2$. The inversion density calculated from Eq. (4.9) is the inversion density to be used in the gain calculation already described, that is,

$$\Delta P = r_a - r_b \quad . \qquad (4.11)$$

V.   SWEEPING AND DEFLECTION KINEMATICS

In Figure 10 we show, in block diagram, a scheme in which a beam of ions can be focused on a target and

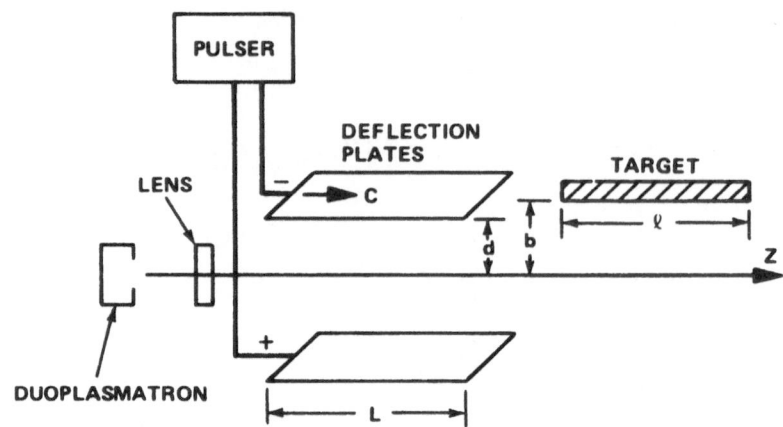

Fig. 10 Block diagram for pulsed, swept-focus ion beam.

the focus can be swept along the target at the velocity
of light. The ion source is shown as a duoplasmatron.
The ions pass through a lens which may be a symmetric
electric quadrupole triplet. In the absence of the deflection plates, the beam would come to a focus downstream below the target. By varying the voltage on the
lens, we can sweep the focus the length of the target at
the velocity of light. However, the ions travel through
a parallel plate transmission line. A finite length
voltage pulse (TEM-wave) travels downstream on the transmission line and gives the ions a transverse upward
velocity. The focused spot is then deflected upward
where it strikes the target. By changing the voltage
on the lens properly, we can move the intersection of
the ions with the target at the velocity of light.

Ion-Beam Kinematics

In Figure 11, we consider the ion beam dynamics of
a single ion, i, as it enters the transmission line with
transverse position x_{io} and transverse velocity $\dot{x}_{io}$ at
time t_i. All ions are assumed to have the same

ION BEAM DYNAMICS

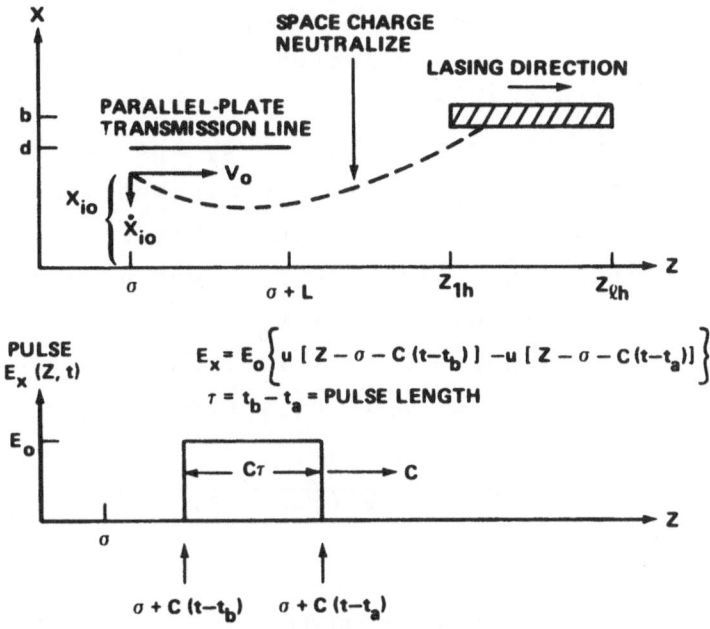

Fig. 11   The path of an ion under the influence of a
          finite-length pulse traveling downstream on a
          transmission line.

longitudinal velocity, v_o. A pulse is launched downstream along the transmission line at time t_a and ends
at time t_b so that

$$E_x = E_o \{u[z-\sigma - c(t-t_b)] - u[z-\sigma-c(t-t_a)]\} \quad (5.1)$$

where $u(x)$ is the unit step function. The pulse length
is

$$t = t_b - t_a . \quad (5.2)$$

While the ion "sees" the pulse, it travels in a parabolic

trajectory. The time an ion "sees" the pulse is
$c\tau/(c-v_o)$, the pulse width divided by the pulse velocity
relative to the ion. After the ion leaves the pulse, it
follows a straight line trajectory to the target.

Consider first ions which are traveling along the
axis. In Figure 12 we show as ion 1, the ion which was
located at $z = \sigma$ (left end of transmission line) at the
time when the pulse was turned on. This ion will see
the entire pulse, and will have followed a parabolic
trajectory to the point $z = \sigma + v_o\tau\,(1-\beta)$, where
$\beta \equiv v_o/c$, when the pulse leaves it. It will then head
toward the target at an angle θ, where

$$\tan\,\theta = \frac{\eta E_o\tau}{v_o(1-\beta)} \quad , \tag{5.3}$$

PULSE LENGTH

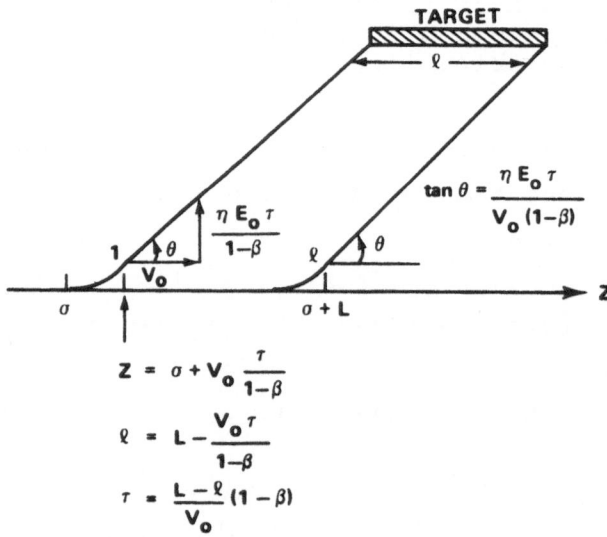

Fig. 12   The path of two ions originally on axis after
          seeing pulse for time $\tau/(1-\beta)$. Each such ion
          heads toward target at angle θ.

as seen in Figure 12. The last ion, ℓ, to see the pulse
for the full time $\tau/(1-\beta)$ will be at $z = \sigma + L$ as the
pulse reaches the end of the transmission line. It too
will head for the target at angle θ. From the figure
we see that the target length, ℓ, over which ions will
see the pulse for the same time and have angle θ is

$$\ell = L - \frac{v_o \tau}{1-\beta} \quad , \tag{5.4}$$

so that the pulse length must be

$$\tau = \frac{(L-\ell)}{v_o} (1-\beta) \quad . \tag{5.5}$$

Although each on-axis ion which sees the pulse for
time $\tau/(1-\beta)$ has angle θ, the beam as a whole will be
tilted at a very small angle ϕ as shown in Figure 13,
after the pulse. This angle may be calculated quite
simply from Figure 13. In the upper part of the figure,

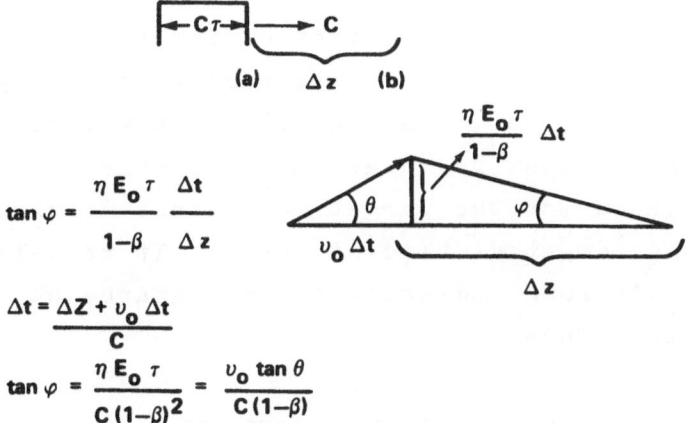

Fig. 13 Diagram to show that the entire beam is tilted
at an angle ϕ after interaction with the deflection pulse.

we consider two ions (a) and (b) located a distance Δz
apart just as the pulse hits ion (a). Both ions travel
with velocity v_o along the z axis. If Δt is the time
required for the pulse to reach ion (b), we have

$$\Delta t = \frac{\Delta z + v_o \Delta t}{c} \quad , \tag{5.6a}$$

or

$$\frac{\Delta z}{\Delta t} = c(1-\beta) \quad . \tag{5.6b}$$

In this time ion (a) has traveled upward a distance
$\eta E_o \tau \Delta t/(1-\beta)$ where η is the ion charge to mass ratio.
Thus

$$\tan \phi \; \frac{\eta E_o \tau}{1-\beta} \; \frac{\Delta t}{\Delta z} = \frac{\eta E_o \tau}{c(1-\beta)^2} = \beta \; \frac{\tan \theta}{(1-\beta)} \quad , \tag{5.7}$$

where we used (5.3). Since $\beta = v_o/c \ll 1$, we see that
$\phi \ll \theta$. Thus, a beam originally parallel to the z axis
is tilted at the small angle ϕ after the pulse passes.

We next show that the point of intersection of the
tilted beam with the target sweeps along the target at
the velocity of light. In Fig. 14 we show a portion of
the beam as it hits the target at some time t. An ion
which is headed for the target a distance Δz downstream,
will hit the target at time Δt later. It travels a
distance $v_o \Delta t$ longitudinally and a distance $\eta E_o \tau \Delta t/(1-\beta)$
transversely. Thus

$$\tan \phi = \frac{\eta E_o \tau \Delta t}{(1-\beta)(\Delta z - v_o \Delta t)} \quad . \tag{5.8}$$

If we compare this with Eq. (5.7), we see that the point
of intersection sweeps at velocity c, i.e., $\Delta z/\Delta t = c$.

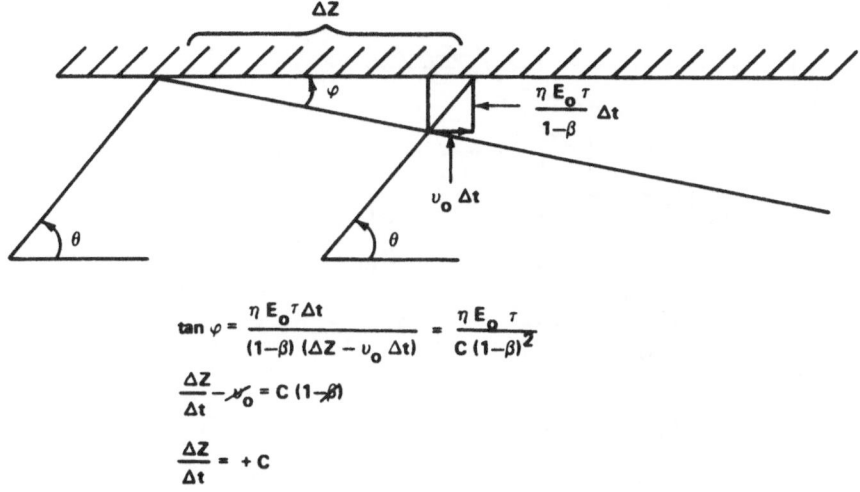

$$\tan \varphi = \frac{\eta \, E_0 \, \tau \Delta t}{(1-\beta)(\Delta Z - \upsilon_0 \, \Delta t)} = \frac{\eta \, E_0 \, \tau}{C \, (1-\beta)^2}$$

$$\frac{\Delta Z}{\Delta t} - \upsilon_0 = C \, (1-\beta)$$

$$\frac{\Delta Z}{\Delta t} = +C$$

Fig. 14   Diagram to show that the intersection of the
          ion beam with the target travels at the velocity
          of light.

Therefore, all ions originally along the z-axis
which see the pulse for the same time will intercept the
target at the velocity of light. One should not visual-
ize the process as a waterhose which is moved along at
the velocity of light, but rather as the entire beam
being tilted at a small angle and then intersecting the
target in such a way that the point of intersection
travels at velocity c.

Swept Focus

Let us next consider what happens to ions un-
fortunate enough to be off axis as shown in Fig. 15.
Here we show two ions (a) and (b) off axis at the input
$z = \sigma$ at the same time t_i. We would like these to hit
the target at the same time and position as the axis ion

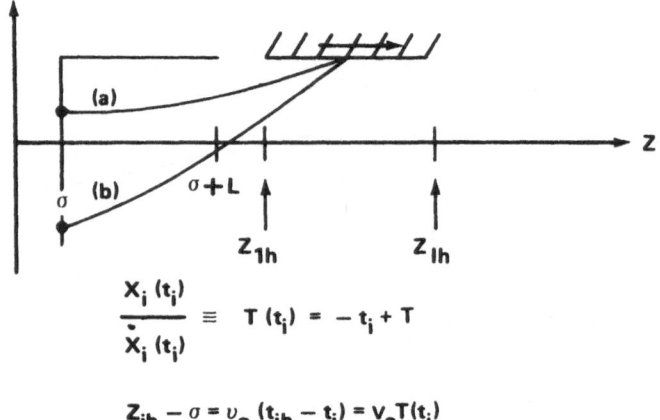

$$\frac{x_i\,(t_i)}{\dot{x}_i\,(t_i)} \equiv \; T\,(t_i) = -\,t_i + T$$

$$Z_{ih} - \sigma = v_o\,(t_{ih} - t_i) = v_o T(t_i)$$

Fig. 15   Conditions all ions must satisfy if the beam
            is to be focused on the target.

in order to have a focused beam which will "sweep" along
the target at velocity c. In reference 16, it is shown
that if each ion that enters at time t_i satisfies the
relation

$$\frac{x_i(t_i)}{\dot{x}_i(t_i)} \equiv \; - \; T(t_i) = + \; t_i \; - \; T \; , \qquad (5.9)$$

where T is a constant, then they will all hit the target
at the same position given by

$$z_{ih} \; - \; \sigma = v_o(T-t_i) \; . \qquad (5.10)$$

In order to satisfy (5.9), the ions must be headed for a
focus downstream at position z_{ih} and the focus must be
swept in order that (5.9) be satisfied for successive
ions.

For simplicity in Figure 16, we show a single electric quadrupole lens which focuses in the x-direction. In practice, a symmetric triplet would be used to provide focusing in both transverse directions. The lens voltage is $V_o(t)$, the dimension $\underline{a}$ is shown on the figure and $\underline{s}$ is the length of the lens. An ion enters the lens at t_{oi}, exits at t_I, drifts in a straight line, and enters the transmission line at time t_i. It is headed toward a focus at z_f. While the ion is in the lens it obeys the equation of motion

$$\ddot{x}(t) = -\frac{2\eta\,V_o}{a^2}(+)\,x \equiv -\omega^2(t)x \quad . \tag{5.11}$$

However, $V_o(t)$ is unknown. We require at the output

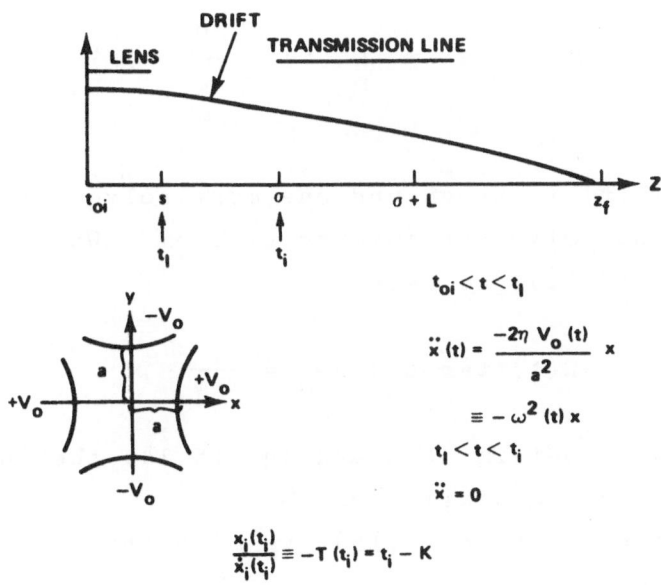

Fig. 16   Quadrupole lens to be used to sweep the ion beam focus along target at the velocity of light.

$$\frac{x(t_I)}{\dot{x}(t_I)} = t_I - K \ , \tag{5.12}$$

where we have translated (5.9) through the drift region to the lens output.

To attempt to find a voltage variation which satisfies (5.11) for all t, we look for a solution of Eq. (5.11) of the form

$$x = \frac{B\cos \phi(t)}{\sqrt{\Omega(t)}} \ , \tag{5.13}$$

where

$$\phi(t) = \int_{t_{oi}}^{t} \Omega(t')dt' \quad . \tag{5.14}$$

Then Equation (5.11) reduces to

$$\Omega \frac{d^2}{dt^2} \Omega^{-\frac{1}{2}} - \Omega^2 = -\omega^2(t) \equiv \frac{2\eta V_o}{a^2} (+)x \quad . \tag{5.15}$$

Normally, $\omega^2(t)$ is known and one must solve for Ω. Here, we must solve the inverse problem. When we put (5.13) into (5.12), we have

$$\Omega(t) \tan \phi + \frac{1}{2} \frac{\dot{\Omega}}{\Omega} = \frac{1}{T-t} \quad . \tag{5.16}$$

We use this to determine Ω and Eq. (5.14) then determines the correct voltage variation.

To solve Equation (5.15), we make the adiabatic approximation

$$\frac{1}{\omega^2} \dot{\omega} \ll 1 \ , \tag{5.17}$$

and the approximation that Ω does not change much in going through the lens, viz.,

$$\phi \simeq \omega s/v_o \ll 1 . \qquad (5.18)$$

The solution of (15) to second order in the WKB approximation is then found to be

$$\omega^2 (t) = \frac{3}{4} \frac{1}{T-t} = \frac{2\eta V_o(t)}{a^2} . \qquad (5.19)$$

It then follows easily that the two approximations are equivalent. Furthermore, Equation (5.17) is also just the thin lens approximation[17] as may be seen from Figure 17.

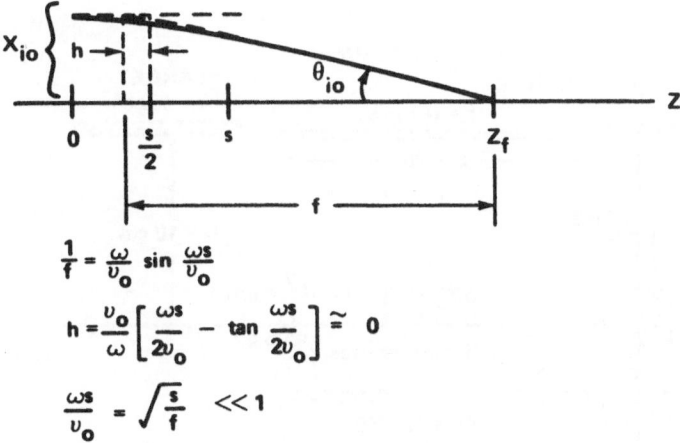

$$\frac{1}{f} = \frac{\omega}{v_o} \sin \frac{\omega s}{v_o}$$

$$h = \frac{v_o}{\omega} \left[\frac{\omega s}{2v_o} - \tan \frac{\omega s}{2v_o} \right] \simeq 0$$

$$\frac{\omega s}{v_o} = \sqrt{\frac{s}{f}} \ll 1$$

Fig. 17   Ion path through lens. For the lens to be thin the principal pulse must lie at lens center so that h = 0.

In Figure 18, we show a set of operating parameters which seem easily achievable.

In this analysis we have neglected space charge effects with the understanding that the ion density is sufficiently low in the transmission line. In the target region, we could space-charge neutralize the beam and allow it to focus. In a subsequent section, we consider an alternate scheme to neutralize space charge in a different geometrical arrangement.

VI. DOPPLER BROADENING

A. Broadening Due to Charge Exchange

The linear gain of the charge-exchange x-ray laser is inversely proportional to the Doppler width $\Delta\omega_D$. We have calculated the Doppler broadening using a simple model for the charge-exchange process, shown in Figure 19. A 25 keV alpha particle approaches a neutral hydrogen atom or molecule with impact parameter s. Charge exchange occurs at the point of closest approach,

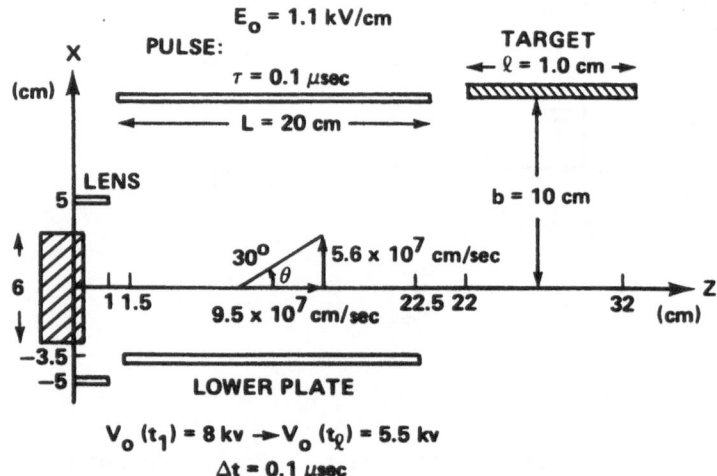

Fig. 18   Typical operating parameter for swept-focus ion beam.

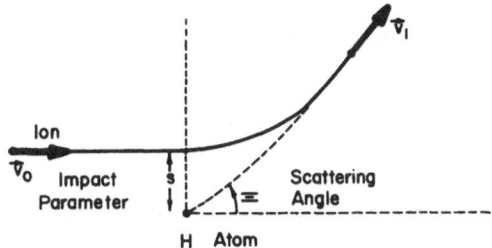

Fig. 19   The charge-exchange scattering process.

and then the He^+(or H_2^+) ions separate along a hyperbolic path.   During the scattering encounter, there is no force between the alpha particle and the neutral hydrogen.   In this simplified model, the scattering angle θ is half the usual Rutherford scattering angle.   The Doppler broadening is given by

$$\Delta\omega_D = (\omega/c) \ [<v_\parallel^2> - <v_\parallel>^2]^{\frac{1}{2}} \ , \qquad (6.1)$$

where $<v_\parallel>$ is the average (over the scattering angle θ) of $v_\parallel$, the component of velocity of the He^+ ion parallel to the laser axis.   Thus, the Doppler broadening is due to a spread in values of v corresponding to a spread in values of scattering angle.   In this simple model, we ignore the details of the charge-exchange process, and assume that the scattering angle is completely determined by the impact parameter.

After a straightforward calculation[18], the center-of-mass scattering angle θ is given by

$$\sin \theta = \epsilon^{-1} \ , \qquad (6.2)$$

where the eccentricity of the hyperbolic orbit is

$$\varepsilon = [1 + (rs)^2]^{\frac{1}{2}} \quad ,$$

$$r = 2E/(ZZ'e^2) \quad , \tag{6.3}$$

and s is the impact parameter. The scattering angle in
the lab frame Ξ is related to the scattering angle in
the center-of-mass frame θ by

$$\tan \Xi = \frac{\sin \theta}{4 + \cos \theta} \quad . \tag{6.4}$$

In writing Equations (6.3) and (6.4), we have assumed
that the hydrogen atom (or molecule) is at rest before
the encounter, and we have neglected the Coulomb energy
of repulsion compared to 25 keV, the energy of the alpha
particle. Using conservation of energy and momentum,
the velocity of the He$^+$ ion after the encounter is

$$v_1(\Xi) = \frac{4}{5} v_o (\cos \Xi + \sqrt{1/16 - \sin^2 \Xi}) \quad , \tag{6.5}$$

where $v_o = 10^8$ cm/sec is the incident velocity of the
σ particle. As shown in Figure 20, we assume that the
alpha particle is incident at angle β to the laser

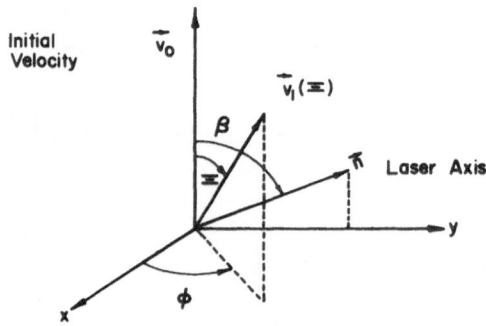

Fig. 20   The alpha particles are incident at an angle β
to the laser axis $\hat{n}$.

axis $\vec{n}$. Then the projection of $\vec{v}(\Xi)$ onto the laser axis
is

$$v_{\|} = v_1(\Xi)(\sin \phi \sin \Xi \sin \beta + \cos \Xi \cos \beta) . \quad (6.6)$$

In equations (6.2) - (6.4), we may average over s rather
than Ξ. Thus

$$<v_{\|}> \int_0^{2\pi} \int_0^{\infty} v \, p(s,\phi) \, ds \, d\phi \quad (6.7)$$

where $p(s,\phi)$ is the probability distribution. We assume
that all azimuthal angles ϕ are equally likely, and that
charge exchange occurs only if the impact parameter is
less than a cutoff value s_c, where πs_c^2 is the total
measured cross-section. Thus

$$p(s,\phi) ds \, d\phi = \begin{cases} \dfrac{2\pi \, s \, ds}{\pi s_c^2} \dfrac{d\phi}{2\pi} & \text{for } s \leq s_c \\[2mm] 0 & \text{for } s > s_c . \end{cases} \quad (6.8)$$

From the experimentally measured[4,15] charge-exchange
cross section at 25 keV, the cutoff impact parameter is
approximately 2.26 $\overset{o}{A}$. Substituting Equations (6.5),
(6.6), and (6.8) into (6.7), we obtain

$$<v_{\|}> = \frac{8v_0 \cos \beta}{5 \, s_c^2} \, I_1 \quad , \quad (6.9)$$

where $I_1 = \displaystyle\int_0^{s_c} \cos \Xi \, (\cos \Xi + \sqrt{1/16 - \sin^2 \Xi}) s \, ds$.(6.10)

The integral in Eq. (6.10) must be evaluated numerically.
In a similar manner, we find

$$<v_{\parallel}^2> = (\frac{4v_o \sin \beta}{5 \, s_c})^2 \, I_2 + 2 \, (\frac{4v_o \cos \beta}{5 \, s_c})^2 \, I_3, \quad (6.11)$$

where $I_2 = \int_0^{s_c} \sin^2 \Xi \, (\cos \Xi + \sqrt{1/16 - \sin^2 \Xi})^2 \, s \, ds$,

$$(6.12)$$

$$I_3 = \int_0^{s_c} \cos^2 \Xi \, (\cos \Xi + \sqrt{1/16 - \sin^2 \Xi})^2 \, s \, ds \, .$$

$$(6.13)$$

Substituting Equations (6.9) and (6.11) into (6.1), the Doppler broadening is found to be

$$\Delta\omega_D = (8.36 \times 10^9 \, \text{sec}^{-1})[1 + 6.15 \sin^2 \beta]^{\frac{1}{2}} \, , \quad (6.14)$$

where we used $\omega = 2\pi \, c/\lambda$, $\lambda = 304 \, \overset{o}{\text{A}}$, $v_o = 10^8$ cm/sec, $s_c = 2.26 \, \overset{o}{\text{A}}$, $I_1 = 3.19$, $I_2 = 8.24 \times 10^{-7}$, and $I_3 = 3.99$. As shown in Figure 21, the Doppler broadening decreases with decreasing angle of incidence. This decrease may be understood by referring to Figures 22 and 23, where we show schematically the scattering process for the two cases $\beta = 0$ and $\beta = 90°$.

B. Broadening Due to Ion Source Temperature

It is perhaps not initially obvious that the major component of the Doppler width of the ion beam in the target is due to scattering and charge-exchange colli- sions. Every ion beam from a real ion source has a longitudinal velocity spread resulting from the tempera- ture of the plasma in the ion source. However, this velocity spread is, in fact, decreased through uniform acceleration to the velocity required for optimization of the cross sections. This may be seen simply as follows: The formula for the velocity resulting from

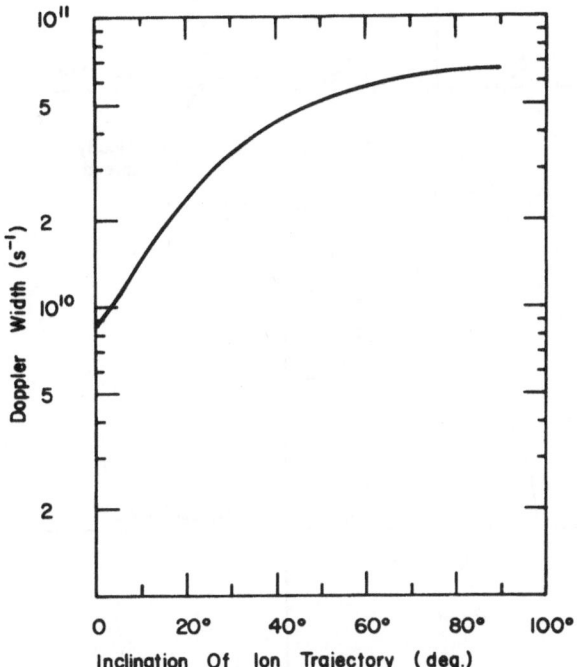

Fig. 21   Plot of the Doppler FWHM, $\Delta\omega_D$, versus the particle angle of incidence β, from Equation (6.14).

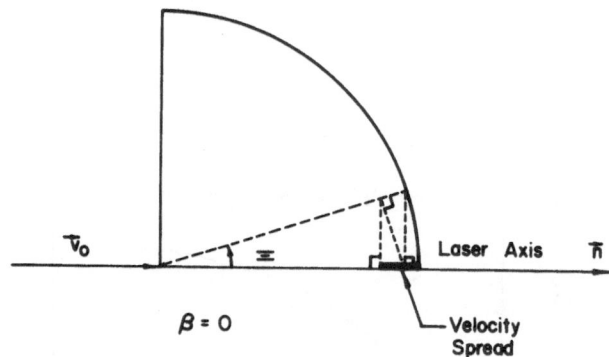

Fig. 22

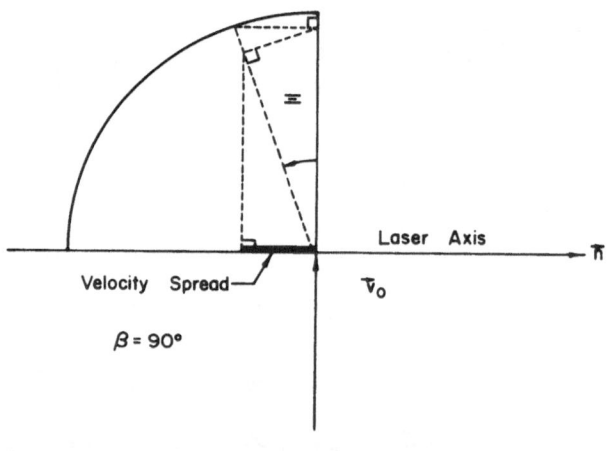

Fig. 23

Fig. 22 and 23   The velocity spread projected onto the
                laser axis is smaller for β = 0 than for
                β = 90°.

a uniform acceleration $\underline{a}$ applied through a distance $\underline{s}$
is

$$\langle v^2 \rangle = \langle v_o^2 \rangle + 2as = v_{th}^2 + 2as \quad , \qquad (6.15)$$

where

$$v_{th} = \sqrt{\frac{kT}{m}} \quad . \qquad (6.16)$$

The root-mean-square velocity is, therefore

$$V_{RMS} = \sqrt{2as} \left[1 + \frac{v_{th}^2}{2as}\right]^{\frac{1}{2}} \cong \sqrt{2as} + \frac{v_{th}^2}{2\sqrt{2as}} \quad . \qquad (6.17)$$

One anticipates that the final velocity will range between $\sqrt{2as}$ and v_{rms}. Therefore, the spread in the
velocity distribution should be

$$\Delta v \sim \frac{v_{th}^2}{2\sqrt{2as}} \ll v_{th} \quad . \qquad (6.18)$$

A more sophisticated approach to answering this
important question begins with the Boltzmann equation
for the probability density $P(\vec{r}, \vec{v}, t)$ in position and
velocity:

$$\frac{\partial P}{\partial t} + \vec{v}.\nabla_r P + \vec{a}.\nabla_v P = 0 \quad . \qquad (6.19)$$

We seek the steady-state distribution disregarding
collisions, space-charge forces, etc.:

$$\frac{\partial P}{\partial t} = 0 \quad . \qquad (6.20)$$

In one dimension the Boltzmann equation, in steady state,
reduces to

$$v \frac{\partial P}{\partial z} + a \frac{\partial P}{\partial v} = 0 \ . \qquad (6.21)$$

We apply the method of Lagrange[20] to discover that the probability distribution is constant along a trajectory of z vs. v such that

$$\frac{dz}{v} = \frac{dv}{a} \ , \text{ i.e. } z - z_o = \frac{1}{2a} (v^2 - v_o{}^2) \ . \qquad (6.22)$$

We conclude that

$$P(z,v) = P(z_o, \sqrt{v^2 - 2a(z - z_o)}) \ . \qquad (6.23)$$

This equation tells us that apart from normalization, the velocity distribution probability downstream following acceleration is the same function of the final v^2 as the original probability distribution is of $v^2 - 2as$. This is shown in Figure 24. From this figure the

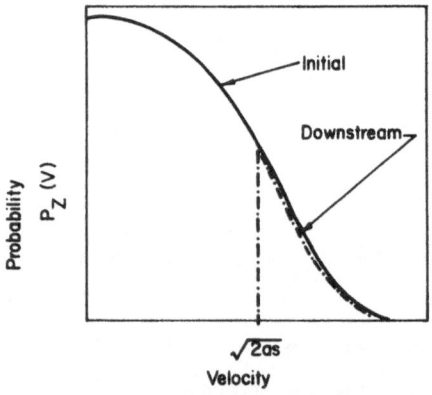

Unnormalized Velocity
Probability Distributions

Fig. 24   Apart from normalization, the velocity probabili
ty distribution after acceleration is the same
function of v^2 as the original distribution is
of v^2 - 2as. This narrows the spread of the
distribution downstream.

reduction in the width in velocity space following acceleration is quite evident.

To calculate a probability distribution for the velocity alone, we note that the conditional probability density for the velocity given the position z is

$$P_z(v) = [\int_{-\infty}^{\infty} P(z,v)dv]^{-1} P(z,v) \quad . \qquad (6.24)$$

Initially at $z = z_0$, the conditional probability is

$$P_{z_0}(v_0) = \begin{cases} \dfrac{2}{v_{th}\sqrt{\pi}} \; e^{-v_0^2/v_{th}^2} & \text{for } v_0 > 0 \; , \\[2ex] 0 & \text{for } v_0 < 0 \; , \end{cases} \qquad (6.25)$$

since only those ions traveling in the correct direction, to the right in our coordinate system, are of interest. Downstream following acceleration the conditional probability is

$$P_z(v) = \begin{cases} [\frac{\sqrt{\pi}}{2} v_{th} \; \text{erfc} \cdot (\frac{\sqrt{2as}}{v_{th}})]^{-1} \; e^{-v^2/v_{th}^2} & \text{for } v > \sqrt{2as} \\[2ex] 0 \qquad \text{for} \quad v < \sqrt{2as} \; . \end{cases} \qquad (6.26)$$

It is convenient to subtract off the velocity which would result from acceleration of an ion initially of zero velocity. In terms of the coordinate $v' = v - 2as$ the conditional probability density is (for $v' > 0$)

$$P_z(v') = \frac{2e^{-2as/v_{th}^2}}{v_{th}\sqrt{\pi} \; \text{erfc}(\sqrt{2as}/v_{th})}$$

$$(6.27)$$

$$\times \exp-[(v'^2 + 2v'\sqrt{2as})/v_{th}^2] \quad .$$

This is a product of two functions: A gaussian with width v_{th} and an exponential with width $v_{th}^2/2\sqrt{2}as$ which is much smaller than v_{th}. We anticipate that the root-mean-square width in velocity $<(\Delta v')^2>^{1/2}$, to the lowest order in the small parameter $v_{th}/\sqrt{2}as$, is

$$<(\Delta v')^2>^{1/2} \cong \frac{v_{th}^2}{2\sqrt{2}as} \quad . \tag{6.28}$$

This expectation is confirmed by further calculations.

To verify that this distribution has no pathologies, we calculate the characteristic function of the velocity distribution (6.27). We recall that the characteristic function is defined as

$$\chi_z(\eta) = \int_0^\infty d\eta \ e^{-v'\eta} P_z(v') \quad , \tag{6.29}$$

and that the moments of the probability distributions may be calculated by successive differentation:

$$\mu_m \equiv <(v')^m> = \int_0^\infty P_z(v')(v')^m dv'$$
$$= (-\frac{d}{d\eta})^m \chi_z(\eta) \quad . \tag{6.30}$$

The moment we seek is

$$<(\Delta v')^2> = \mu_2 - \mu_1^2 \quad . \tag{6.31}$$

Explicit calculation shows that the characteristic function is

$$\chi_z(\eta) = \frac{e^{\theta^2} \ \text{erfc}\theta}{e^{\beta^2} \ \text{erfc}\beta} \quad , \tag{6.32}$$

where

$$\beta = \frac{\sqrt{2as}}{v_{th}} \quad , \tag{6.33}$$

$$\theta = \beta + \eta \frac{v_{th}}{2} \quad . \tag{6.34}$$

The moments of the velocity distribution, Equation (6.30), where

$$\chi_z(\eta) = \phi(\theta(\eta)) \quad , \tag{6.35}$$

must be calculated by successive differentation using a formula due to diBruno[21].

$$\frac{d^m \phi(\theta(\eta))}{d\eta^m} = \sum_{\ell=1}^{m} \sum_{\{a\}} c(\ell; a_1, \ldots, a_m)$$
$$\times \frac{d^\ell \phi}{d\theta^\ell} \left(\frac{d\theta}{d\eta}\right)^{a_1} \left(\frac{d^2\theta}{d\eta^2}\right)^{a_2} \ldots \left(\frac{d^m\theta}{d\eta^m}\right)^{a_m} \quad , \tag{6.36}$$

where $\sum_{\{a\}}$ runs over all non-negative integers $a_1, \ldots, a_m$ such that

$$\sum_{j=1}^{m} a_j = \ell \quad , \tag{6.37}$$

$$\sum_{j=1}^{m} j a_j = m \quad , \tag{6.38}$$

and

$$c(\ell; a_1, \ldots, a_m)$$
$$= \frac{m!}{\prod\limits_{j=1}^{m} a_j! (j!)^{a_j}} \quad . \tag{6.39}$$

Because of Equation (6.34), the derivativies $\frac{d^n\theta}{d\eta^n}$ vanish for $n > 2$. Thus, the derivatives of the characteristic function reduce to

$$\frac{d^m\chi_z(\eta)}{d\eta^m} = \frac{d^m\phi}{d\theta^m} \left(\frac{d\theta}{d\eta}\right)^m$$

$$= \frac{d^m\phi}{d\theta^m} \left(\frac{1}{2}v_{th}\right)^m \quad , \tag{6.40}$$

where $\phi(\theta)$ is given by Equation (6.32). Since the parameter β is much larger than one and we are interested in the moments of the velocity distribution only to lowest order in β^{-1}, we make an asymptotic expansion in powers of β^{-1}:

$$\phi(\theta) \sim \sum_n b_n(\theta)\beta^{-n} \quad . \tag{6.41}$$

When the derivatives possess the asymptotic expansion

$$\frac{d^m\phi}{d\theta^m} \sim \sum_n \frac{d^m b_n(\theta)}{d\theta^m} \beta^{-n} \quad , \tag{6.42}$$

and the derivatives of the characteristic function are

$$\frac{d^m\chi_z(\eta)}{d\eta^m} \sim \left(\frac{1}{2}v_{th}\right)^m \left(\frac{d}{d\theta}\right)^m \sum_n b_n(\theta)\beta^{-n} \quad . \tag{6.43}$$

This implies that we may calculate Equation (6.40) by first obtaining the asymptotic expansion of the function $\phi(\theta)$, then differentiating. From the well known asymptotic expansion[21]

$$e^{z^2} \text{erfc } z \sim \frac{1}{\sqrt{\pi} z} \left[1 + \sum_n (-1)^n \frac{(2n-1)!!}{(2z^2)^n} \right. \quad , \tag{6.44}$$

we obtain

$$\phi = \frac{e^{\theta^2} \text{erfc}\theta}{e^{\beta^2} \text{erfc}\beta} \sim \frac{\beta}{\theta} (1 + \frac{1}{2\beta^2} - \frac{1}{2\theta^2}) \quad . \qquad (6.45)$$

The resulting moments are

$$\mu_1 = - \frac{d\phi}{d\theta}\Big|_{\theta=\beta} \cdot \frac{1}{2} v_{th}$$

$$\sim \frac{\beta}{\theta^2} \cdot \frac{1}{2} v_{th} = \frac{v_{th}^2}{2\sqrt{2}as} \quad , \qquad (6.46)$$

$$\mu_2 = \frac{d^2\phi}{d\theta^2}\Big|_{\theta=\beta} (\frac{1}{2}v_{th})^2$$

$$\sim \frac{2\beta}{\theta^3} (\frac{1}{2}v_{th})^2 \sim \frac{v_{th}^4}{2.2as} \quad . \qquad (6.47)$$

The variance of the excess velocity v' is

$$\sigma^2 = \mu_2 - \mu_1^2 = <(\Delta v')^2> \sim \frac{v_{th}^4}{2.2as} \quad , \qquad (6.48)$$

so that the RMS velocity fluctuation is

$$<(\Delta v')^2>^{\frac{1}{2}} \sim \frac{v_{th}^2}{2\sqrt{2}as} << v_{th} \quad , \qquad (6.49)$$

to lowest order in the small parameter $v_{th}/\sqrt{2as}$.

VII. SPACE CHARGE AND EMITTANCE

Until this point, we have calculated as if the positively charged particles in the ion beam were insensitive to their mutual electrostatic repulsion. In fact, space charge is a very important influence on the dynamics of the ion beams and, together with emittance, is a fundamental limitation on the tightness to which a beam can be focused in a practical situation. The gain we expect in our x-ray laser is inversely proportional

to the area of the region to which we can focus the ion
beam. It is important to know what physical limitations
on focusing these two effects imply.

Simple Theory

Most analytical theories of space charge in ion
beams make use of a cylindrical approximation in which
the electric field experienced by an ion at a distance r
from the axis of an ion beam carrying current I is

$$E_r = \frac{I}{2\pi\epsilon_o v_z r} \quad .\tag{7.1}$$

This implies a particle equation of motion

$$m\ddot{r} = \frac{qI}{2\pi\epsilon_o v_z r} \quad ,\tag{7.2}$$

which may be scaled by a transformation

$$r = \sigma r_o \quad ,\tag{7.3}$$
$$\tau = \omega_p t \quad ,$$

where

$$\omega_p = [\frac{qI}{2\pi\epsilon_o m v_z r_o^2}]^{\frac{1}{2}} \quad ,\tag{7.4}$$

to give the equation

$$\frac{d^2\sigma}{d\tau^2} = \frac{1}{\sigma} \quad .\tag{7.5}$$

This equation has the solution

$$\int_o^{\sqrt{\ln\sigma}} e^{y^2} dy = \frac{\omega_p t}{\sqrt{2}} \quad .\tag{7.6}$$

The results of this analysis can be displayed in a
universal beam spread chart, as shown in Figure 25.

Statistical Theory

While this standard analysis gives much insight into charge particle beam dynamics, Emigh has found that
a statistical theory of beam transport which includes
emittance and focusing as well as space charge can be
constructed[22]. This theory (which is summarized in the
Appendix) makes use of the root-mean-square beam radius
and transverse velocity

$$R_x(t)^2 = \frac{4}{N} \sum_{j=1}^{N} x_j(t)^2 \quad , \tag{7.7}$$

$$V_x(t)^2 = \frac{4}{N} \sum_{j=1}^{N} \dot{x}_j(t)^2 \quad . \tag{7.8}$$

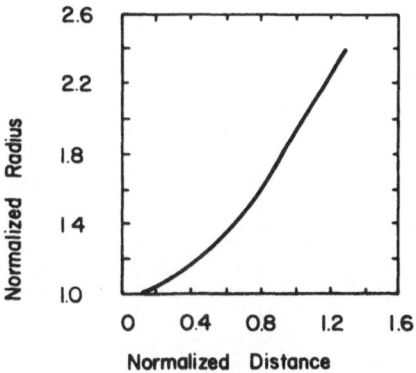

Fig. 25   Universal beam-spreading diagram.   The ratio
of final to initial radius, r/r_o, is shown as
a function of the normalized distance,
$[qI/4\pi\epsilon_o m v_z^3]^{\frac{1}{2}} z/r_o$.

It is possible to obtain equations of motion including space charge, emittance, and linear focusing forces for these quantities.

It is useful to define a generalized emittance

$$E_x^2 = R_x^2 \, v_x^2 - (R_x \dot{R}_x)^2 \quad , \qquad (7.9)$$

which turns out to be constant for space charge forces and linear focusing. This differs from the normal definition of emittance

$$E_x = \Delta x \, \Delta v_x \quad , \qquad (7.10)$$

where Δx and v_x are illustrated in Figure 26. The generalized emittance as defined here, when squared consists of the sum of the square of the usual emittance and an additional term which is required to make the generalized emittance constant under space charge forces and linear focusing forces. Since the generalized

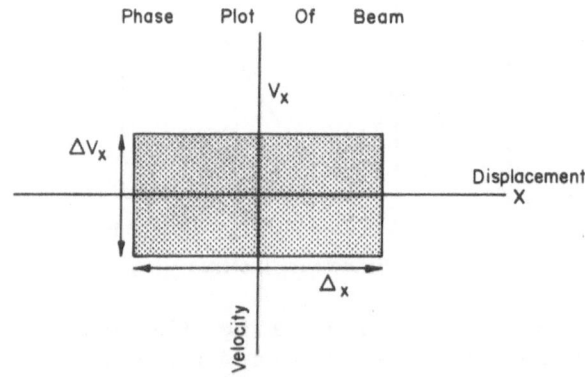

Fig. 26   An idealized phase plot of an ion beam in one transverse dimension x, illustrating the conventional definition of emittance, $E_x = \Delta x \Delta v_x$.

emittance has much to recommend it, we use it rather than the ordinary definition. The use of root-mean-square rather than maximum quantities implies that the theory is valid for non-uniform beam distributions, an important point for practical applications. The root-mean-square beam dimension $R_x(z)$ obeys the equation

$$\frac{d^2 R_x}{dz^2} = \sqrt{\frac{m}{2q}}\ \frac{I}{2\pi\varepsilon_o \phi^{3/2}}\ \frac{R_x}{(R_x^2 + R_y^2)}$$

(7.11)

$$+\ \frac{E_x^2}{\frac{2q\phi}{m} R_x^3}\ +\ \frac{K_x\, R_x}{2\phi}$$

where the average translational energy is

$$q\phi = \frac{1}{2}\, m\, v_z^{\,2}\quad .$$

(7.12)

The first term in this equation represents the dispersive effect of space charge forces; the second term represents the effect of emittance; and the third term represents the effect of linear focusing forces. When all three terms in the equation are present, a numerical solution is required. Such calculations for a He^{++} beam with a current of 30 mA, an energy of 25 keV, a transverse temperature of a few eV, and an initial radius of 1 cm, show that the beam can be focused to a diameter of order 10^{-2} cm.

It is possible to obtain much insight into the dynamics of the beam by considering each process separately. We shall take the effect of a linear focusing force to be a net inward radial impulse of the beam. Using this as an initial condition, we can then investigate the effects of space charge and of emittance separately.

Considering space charge first, we see that the electrostatic potential due to the charged particles at a distance r from the center of the beam, is

$$\phi(r) = -\frac{I}{2\pi\epsilon_o v_z} \ln r \quad .$$

(7.13)

By conservation of energy, the radial velocity obeys the equation

$$\frac{1}{2} m \dot{r}^2 + q\phi(r) = \text{constant}$$

$$= q\phi(r_m) \quad .$$

(7.14)

At the focus ($r=r_m$) the radial velocity vanishes and

$$\frac{r_m}{r_o} = \exp - \left\{ \frac{1}{\alpha} (\frac{r_o}{f})^2 \right\} \quad ,$$

(7.15)

where $\underline{f}$ is the focal length of the electrostatic lens, and

$$\alpha = \frac{qI}{\pi m \epsilon_o v_z^3} \quad .$$

(7.16)

The effect of space charge for a relevant set of parameters is shown in the steeply rising curve in Figure 27. We note that this curve is nearly vertical for values of the parameter f/r_o larger than about 2. This means that we are confined by space charge to using small f-numbers.

We can derive the restriction on focusing due to emittance alone by considering suitably averaged single-particle equations of motion. In the absence of focusing

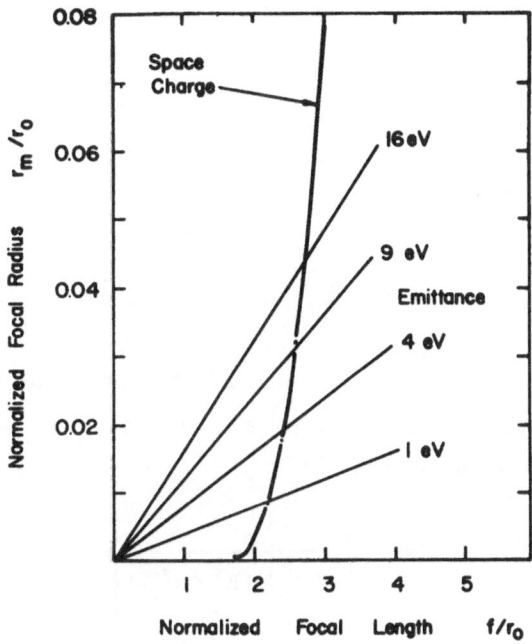

Fig. 27   Composite graph of the normalized focal radius
           r_m/r_o versus the normalized focal length f/r_o,
           for a He^{++} beam with velocity $v_z = 10^8$ cm s^{-1},
           and current I = 25 mA, showing the effects of
           space charge. The straight lines showing the
           emittance limit are plotted for four different
           transverse temperatures of the ion beam.   In
           the region where the curves overlap a numerical
           solution of Equation (7.11) must be performed.

the X or Y coordinate of an individual typical particle,
will obey the equation

$$x_j = v_{xj} t + x_{oj} \quad .$$

(7.17)

With focusing, the velocity after leaving the lens will
be

$$v_{xj} = v_{oj} - kx_{oj} \quad , \tag{7.18}$$

where

$$k = \frac{v_z}{f} \quad ,$$

and $\underline{f}$ is the focal length. We find (see the Appendix)
that the mean-square position at the focus, where the
radial velocity vanishes, is

$$\langle x_m^2 \rangle = \langle x_o^2 \rangle \left\{ \frac{\langle v_o^2 \rangle}{\langle v_o^2 \rangle + k^2 \langle x_o^2 \rangle} \right\} \quad . \tag{7.19}$$

In the limit of tight focusing ($\langle x_m^2 \rangle \ll \langle x_o^2 \rangle$), the mean-
square beam radius is

$$\langle x_m^2 \rangle \cong \frac{\langle v_o^2 \rangle}{\langle v_z^2 \rangle} f^2 \quad , \tag{7.20}$$

so that

$$\frac{r_m}{r_o} \cong \left[\frac{\langle v_o^2 \rangle}{\langle v_z^2 \rangle} \right]^{\frac{1}{2}} \frac{f}{r_o} \quad . \tag{7.21}$$

This implies that the focal spot size is roughly independent of the original beam radius, in the limit of
tight focusing. In deriving these equations, we have
assumed that prior to entering the lens there was no
net x-component of velocity and that the distribution of
velocities upon entering the lens was independent of the
radial position. The linear relation between focused
beam size and f number is also shown on Figure 27, together with the curve due to space charge alone. In the

region where the emittance curves lie above the space
charge curve, and in the region where the space curve
lies far above the emittance lines, these approximate
solutions are equal to the solutions obtained by an
integration of the full equation of motion, Equation
(7.11). In the intermediate region, of course, the full
equation must be used.

Harris Flow

In order to neutralize the effects of the space
charge and obtain a focused ion beam which intersects a
target at the velocity of light, we consider the scheme
shown in Figure 28. Here we have a shaped ion source
which puts out a hollow cylindrical beam of inner radius
a and outer radius b such that the space charge in the

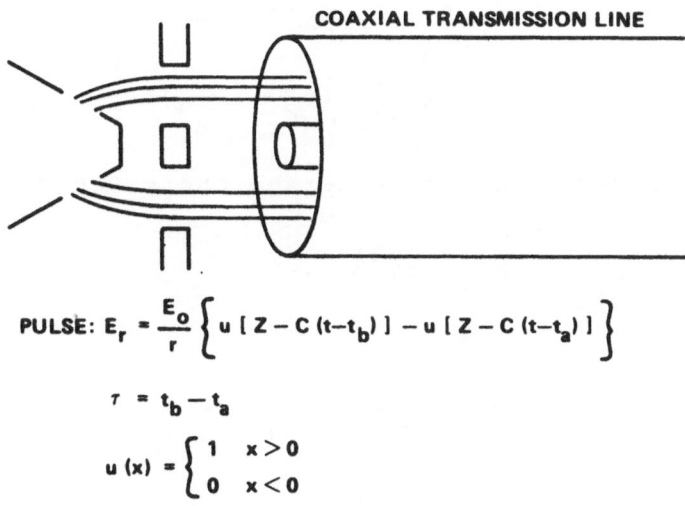

$$\text{PULSE: } E_r = \frac{E_o}{r}\left\{ u\left[Z - C\left(t - t_b\right)\right] - u\left[Z - C\left(t - t_a\right)\right]\right\}$$

$$\tau = t_b - t_a$$

$$u(x) = \begin{cases} 1 & x > 0 \\ 0 & x < 0 \end{cases}$$

Fig. 28   A hollow cylindrical beam in Harris flow. A
         pulse with the electric field E_e deflects ions
         to the center, where the target is located.

beam varies with radius as

$$\rho = kr^{-4} . \tag{7.22}$$

The beam then goes through a radial magnetic field of
magnitude and length such that the beam acquires an
azimuthal velocity which varies as

$$\dot{\theta}^2(r) = \frac{\eta k}{2a^2\epsilon_o} r^{-4} . \tag{7.23}$$

In addition, a d.c. electric field given by $-V_a/r$ is
applied between the electrodes of a coaxial cable. The
radial space charge force for a beam described by (7.22)
is

$$E_{s.c.} = \frac{k}{2\epsilon_o} [\frac{1}{a^2 r} - \frac{1}{r^3}] . \tag{7.24}$$

The radial equation of motion is thus

$$\ddot{r} - \dot{\theta}^2 r = \eta \left\{ \frac{k}{2\epsilon_o} (\frac{1}{a^2 r} - \frac{1}{r^3}) - \frac{V_a}{r} \right\} . \tag{7.25}$$

The angular momentum is conserved and given by

$$p_\theta = mr^2\dot{\theta} . \tag{7.26}$$

If we let

$$V_a = \frac{k}{2a^2\epsilon_o} , \tag{7.27}$$

and let $\dot{\theta}^2$ be given by (7.23) or (7.25), then the space
charge forces are cancelled across the entire beam, as
seen by inspection of Equation (7.25). Such flow is
called Harris flow[23].

The constant k is determined by the current and longitudinal beam velocity:

$$k = \frac{I}{\pi v_o (a^{-2} - b^{-2})} \quad . \tag{7.28}$$

The radial magnetic field required varies as

$$B_r = \frac{B_o a}{r} \quad . \tag{7.29}$$

If ℓ is the longitudinal extent of the magnetic field, then it is found that,

$$(B_o a \ell)^2 = k/2 \epsilon_o \eta \quad , \tag{7.30}$$

will give the beam the azimuthal velocity of Equation (7.23).

For an alpha-particle beam of 20mA, $v_o = 10^8$ cm/sec, a = 5 cm, b = 10 cm, then V_A = 480 volts, and B ℓ = 3.2×10^3 gauss-cm. At r = a, the beam rotational velocity is 3.0×10^6 sec^{-1}.

As shown in Figure 28, a pulse in the form of a TEM-mode can propagate downstream and cause the ions to intersect a target just inside the central (porous) conductor at the velocity of light.

VIII. EXPERIMENTAL REALIZATION

A. Gas-Jet Target

As we have seen in the section on Doppler broadening, the Doppler width to be expected with a solid foil target greatly exceeds the Doppler width to be expected with a gas jet target. This increase much more than offsets the slightly larger inversion which can be obtained with the solid foil target. We turn our attention to

the practical realization of a gas jet with the required density and with a sufficiently sharp profile
grading from vacuum to the desired density. Figure 29
shows a schematic of a gas jet with the desired characteristics which has been operated at the Los Alamos Scientific Laboratory[4]. Figure 30 gives the dimensions of the
Mach 8 nozzle which was used in this experiment. Figure
31 is a trace through the gas jet using an ultraviolet
photon beam. The upper trace is essentially in the
throat of the nozzle; the lower trace was taken 4 mm
below the throat. We note that the target has the required steep edge rising from essentially vacuum to a
density of approximately 10^{17} atoms per cubic centimeter,
within a small fraction of a mm. The small gradation

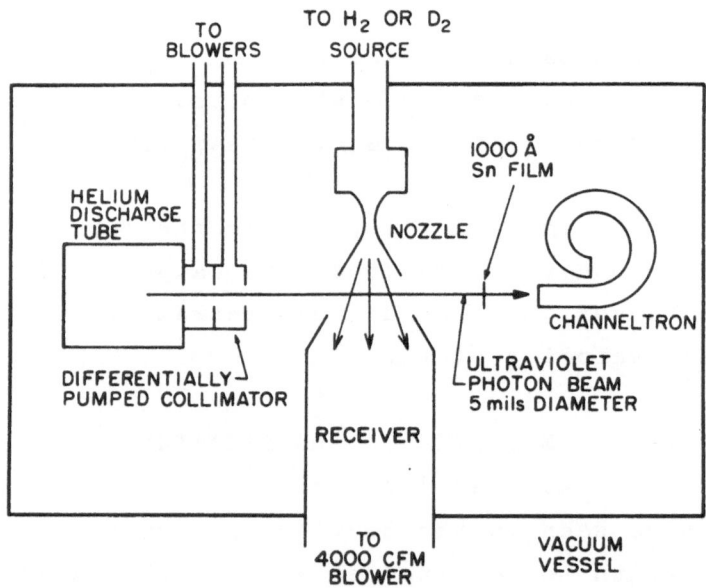

Fig. 29   Schematic diagram of the production of a gas-
          jet target (Ref. 4). This apparatus has been
          successfully operated at the Los Alamos
          Laboratory.

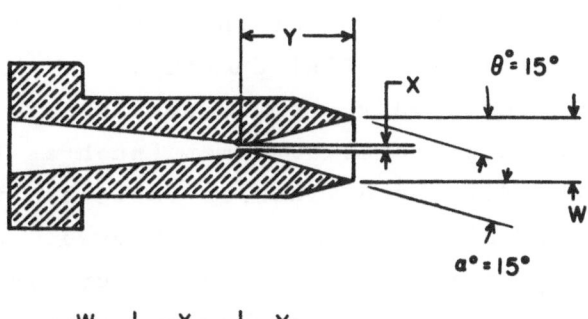

W	X	Y
0.42	0.010	0.76
0.41	0.0076	0.76
0.41	0.0051	0.76
0.82	0.0025	1.5

Centimeters

Fig. 30   Dimensions of the nozzle used to produce the gas-jet target.

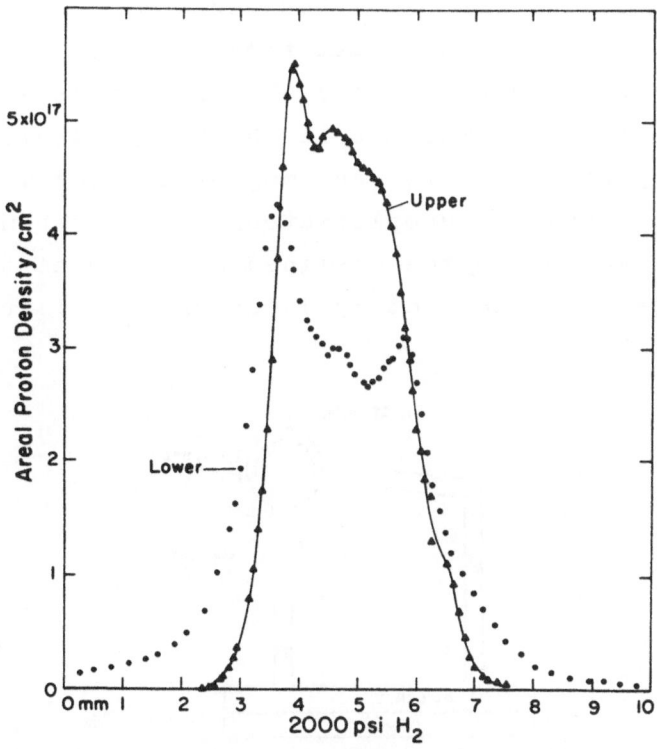

Fig. 31   Density trace of the gas jet.   The upper trace was taken nearly in the throat of the nozzle; the lower trace was taken 4mm below the throat.

at low density at the edge of the target can be removed
by skimming and differential pumping.

In experiments where atomic or diatomic cesium
vapor would be a preferable target (such as the reactions in the first two lines of Table I), the apparatus sketched in Figure 32 may be employed. Cesium vapor
in a heat pipe at an elevated temperature is allowed to
expand through a nozzle. The boundary layer may be removed by skimming (as shown schematically in Figure 1).
The cesium jet may be collected by a room-temperature
condenser, which acts as a "cryogenic" pump for cesium.
At this writing, such a source is being designed at the
Los Alamos Scientific Laboratory.

B.   Ion Sources
Duoplasmatron

Sources which accelerate electrons into a neutral
gas to create a plasma, and in which the electrons are
confined by both electric and magnetic forces, are con-
ventionally known as duoplasmatrons[24]. A schematic
drawing of an ion source designed and used at the Los
Alamos Scientific Laboratory is shown, for illustrative

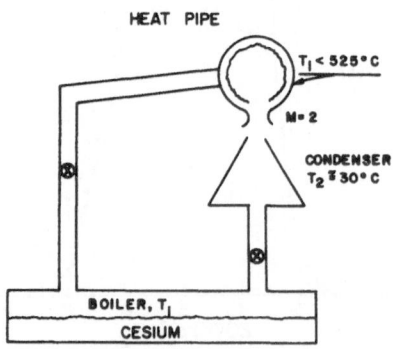

Fig. 32 Schematic of cesium gas-jet production.

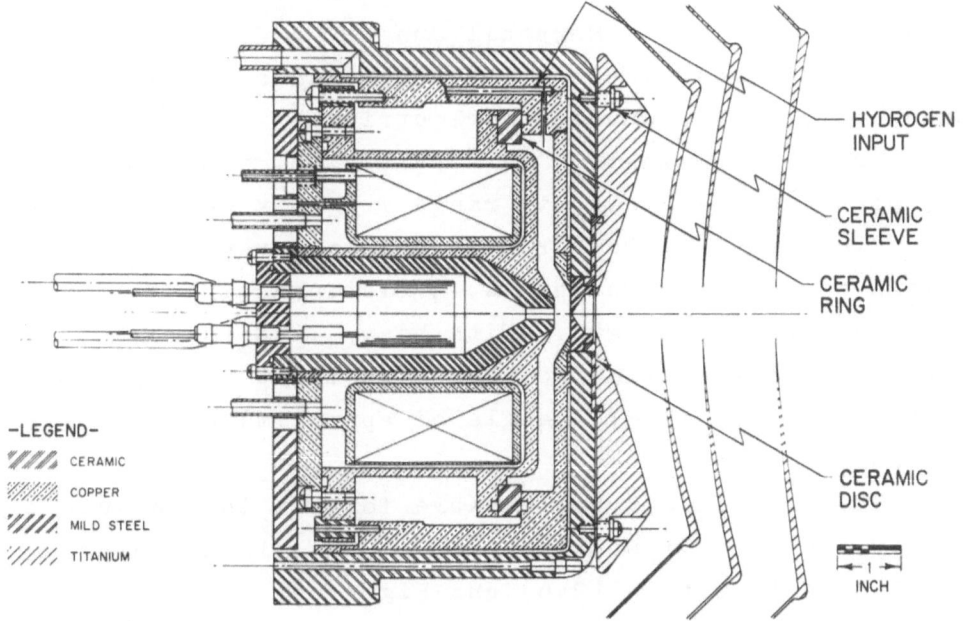

HYDROGEN
INPUT

CERAMIC
SLEEVE

CERAMIC
RING

CERAMIC
DISC

-LEGEND-

▨ CERAMIC

▨ COPPER

▨ MILD STEEL

▨ TITANIUM

1 INCH

Fig. 33   Diagram of a duoplasmatron.

purposes, in Figure 33. This particular source produces
currents of $\sim$ 40mA of H^+ or He^+, which are adequate for
the reactions listed in the first two lines of Table I.

Duoplasmatrons operated under the proper conditions
using He gas can produce a few percent of the total ion
current as He^{++} [25]. Other sources of this general type
have been operated at several A total current[26], and a
multiampere duoplasmatron is currently under development
at Los Alamos. A multiampere duoplasmatron or duo-
PIGatron would produce at least 30 mA of He^{++} if the
proportion of He^{++} scales to larger devices. With the
other parameters mentioned in Section IV, this implies
a gain $g_o \ell \gtrsim 50$ for a H_2 gas-jet target.

Marshall Gun

A device developed at the Los Alamos Scientific Laboratory[7] has produced an electrically neutral plasma containing approximately 10^{22} particles, in 10^{-4} s [27]. The plasma stream velocity range from 5×10^{6} cm-s^{-1} to 2.5×10^{7} cm-s^{-1}. The cross-sectional area of the plasma stream is a few cm^{2} at a distance of 1 m from the aperture of the gun. With heavy ions (Ne), the stream is well collimated, but with light ions the beam is emitted into a solid angle of approximately one steradian.

This source may be suitable for use in a quasi-steady-state configuration such as was discussed in Section II (Equation (2.6) and Figure 5). We estimate that $\Delta\omega_{s}/\Delta\omega_{D} \sim 10^{-2}$; with the current mentioned above, the gain is approximately 10^{4}.

IX.   SUMMARY

From Equations (3.39) and (3.40), we can calculate the gain at 304 $\overset{o}{A}$ for a solid hydrogen foil target and a hydrogen gas-jet target using a focused ion beam. We take $\Delta x = v_{o}\tau_{s} = \Delta y = 10^{-2}$ cm, $\frac{1}{2} mv_{o}^{2} = 40$ keV, I = 30 mA, and the length of the focal region $\ell = 10$ cm for our focused He^{++} beam. For a solid-foil target 25 $\overset{o}{A}$ thick, the Doppler width is $\Delta\omega_{D} = 10^{12}$ s^{-1}, and the fractional inversion $\Delta P = 0.1$, giving

$$(g_{o}\ell)_{foil} = 1.1 \quad . \tag{9.1}$$

For a gas-jet target with a density and an angle of incidence $\beta = 30^{\circ}$, $N_{target} = 5 \times 10^{16}$ cm^{-3}. The Doppler width is $\Delta\omega_{D} = 3 \times 10^{10}$ s^{-1} and the fractional inversion $\Delta P = 0.05$, so that

$$(g_o \ell)_{gas} \gtrsim 50 \ . \tag{9.2}$$

The gain obtained using a quasi-steady-state source such as a Marshall gun, and a gas target, can be calculated from Equations (2.6) and (3.39). We find that

$$(g_o \ell)_{Marshall} \sim 10^4 \ . \tag{9.3}$$

These numbers speak for themselves.

We have shown that a charge-exchange soft x-ray laser is a real possibility within the grasp of current technology. The charge-exchange cross sections are known and favorable. Ion beams of the required current, focused sufficiently tightly, can be obtained using known techniques. It is possible to sweep, deflect, and focus the ion beam so that all particles are focused to the same point, and so that the focus travels at the velocity of light. Analytical tools exist to analyze other sweeping and deflection schemes, for example using cylindrical geometry and Harris flow. Unfocused ion sources such as the Marshall gun give very large currents, so that the resulting gain is high despite a lack of sweeping and focusing. Of general interest is our observation that the output of a swept-excitation laser can be fully phase-coherent throughout the pulse, if T_2^* is of the same order of magnitude as T_1 and T_2. Spatial coherence is guaranteed in tightly focused schemes where one end of the gain region subtends roughly one Fresnel zone as seen from the other end. The output of a laser of the type we have considered should be useful for holography and interferometry in the soft x-ray region.

X.   ACKNOWLEDGEMENTS

We thank J.E. Bayfield, M. Ciftan, B.L. Donnally, H. Hill, G.A. Khayrallah, W.E. Lamb, Jr., R.A. McCorkle, J.D. McCullen, D. Rogovin, and W. Wing for useful dis- cussions. We are particularly indebted to Professor Bayfield and Dr. Khayrallah for communicating their results for charge-exchange cross sections prior to publication. We also thank B. Carpenter, Rose M. Borrego, and Robert E. Stapleton for essential assistance.

TABLE I.

POTENTIAL   (2p) - (1s) LASERS

Lasing Species	Wavelength (10^{-8} cm)	Lifetime (10^{-10}s)
H	1216	16
He	584	5.6
He^{+}	304	1.0
Li^{+}	199	0.39
Li^{++}	135	0.20
Be^{++}	100	8.2×10^{-2}
B^{3+}	60	2.7×10^{-2}
C^{4+}	40	1.13×10^{-2}

TABLE II.

	Scully et al.	$\sigma(H)$ measured	$\sigma(H2_o)$ measured
σ_{2P}	100	36	15
σ_{1s}	1	1.2	0.5
σ_{total}	500	150	60
$\sigma_{\alpha\beta}$	1	1	1
$\sigma_{\alpha o}$	8.5	8.5	20

Table II. Calculated[28] and experimental[4] values of the charge-exchange cross sections in molecular and atomic hydrogen, in units of 10^{-17} cm^2.

TABLE III.

	Scully et al.	From $\sigma(h)$ experimental cross sections	From $\sigma(H_2)$ experimental cross sections
ΔP_{solid}	.10	-	-
ΔP_{gas}	.05	.043	.055

Table III. Values of the fractional population inversion ΔP obtained from theoretical[28] and experimental[4] cross sections for He^{++} on solid, atomic-gas, and molecular-gas hydrogen.

TABLE IV.

	Solid	Gas
Λ	3.6×10^6 cm^{-1}	4.3 cm^{-1}
γ	4×10^5 cm^{-1}	0.5 cm^{-1}

Table IV. Values of the inverse mean free path Λ for second-electron pickup, and of the inverse mean free path γ for one-electron transitions in He$^+$, for a solid hydrogen foil 25 Å thick with a density of 4×10^{22} atoms cm^{-3} and a H_2 gas jet with density 5×10^{16} molecules cm^{-3}. The cross section for second-electron pickup is 8.5×10^{-17} cm^2/atom (Ref. 19), while the cross section for one-electron transitions in He$^+$ was estimated as 10^{-17} cm^2/atom (Ref. 2, Equation (5.11)).

FOOTNOTES

† Work performed under the auspices of the United States Energy Research and Development Administration; The United States Army Research Office; and The Advanced Research Projects Agency.

* University of California, Los Alamos Scientific Laboratory, Los Alamos, NM 87544

** University of Arizona, Optical Sciences Center Tucson, AZ 87521

*** Department of Physics, City College of the City University of New York, New York, NY 10031

**** Department of Physics, University of Southern California, Los Angeles, CA 90007

***** Department of Physics, University of Alabama, Huntsville, AL 35804

****** Applied Mathematics Laboratory, U.S. Army Ballistic Research Laboratory, Aberdeen Proving Ground, Aberdeen, MD 21005.

REFERENCES

1. M.O. Scully, W.H. Louisell and W.B. McKnight, Optics Comm. 9, 246 (1973).

2. W.H. Louisell, M.O. Scully and W.B. McKnight, Phys. Rev. A (to be published).

3. J.E. Brolley, IEEE Trans. Nucl. Sci. (USA) 20, 475 (1973).

4. J.E. Bayfield and G.A. Khayrallah, private communication.

5. R.W. Waynant, J.D. Shipman, Jr., R.C. Elton, and A.W. Ali, Appl. Phys. Letters 17, 383(1970); R.T. Hodgson, Phys. Rev. Letters 25, 494 (1970).

6. C.R. Emigh, Proc. 1968 Proton Linear Accelerator Conference, p. 338.

7. J. Marshall, Phys. Fluids 3, 135 (1960).

8. R.A. McCorkle, Phys. Rev. Letters 29, 982 (1972) and 29, 1428 (1972).

9. W.E. Lamb, Jr., Phys. Rev. 134, A 1429 (1964); F.A. Hopf and M.O. Scully, Phys. Rev. 179, 399 (1969).

10. S.L. McCall and E.L. Hahn, Phys. Rev. Letters 18, 908 (1967).

11. R.H. Dicke, Phys. Rev. 93, 99 (1954).

12. T.F. Tuan and E. Gerjuoy, Phys. Rev. 117, 756 (1960).

13. J.R. Oppenheimer, Phys. Rev. 31, 349 (1928); H.C. Brinkman and H.A. Kramers, Proc. Acad. Sci. Amsterdam 33, 973 (1930).

14. H. Schiff, Can. J. Phys. 32, 393 (1954); M.B. McElroy, Proc. Roy. (London) A 272, 542 (1963); D.R. Bates and N. Lynn, Proc. Roy. Soc. A 253, 141 (1959).

15. J.E. Bayfield and G.A. Khayrallah, Phys. Rev. A (to be published); J.E. Bayfield, Proceedings of the Fourth International Conference on Atomic Physics,

Heidelberg, July 1974.

16. W.H. Louisell, J.F. Seely, R. Shnidman, and M. Lax, (to be published).

17. K.G. Steffen, "High Energy Beam Optics" (Inter-science Publishers, New York, 1965).

18. H. Goldstein, "Classical Mechanics" (Reading, Addison-Wesley Publishing Company, 1950).

19. W.L. Fite, A.C.H. Smith, and R.F. Stebbins, Proc. Roy. Soc. (London) A 268, 527 (1962).

20. J.L. Lagrange, Nouv. Mem. de l'Acad. de Berlin, 1779 [Oeuvres (Paris, Gauthier-Villars, 1869), Vol. 4, pp. 585-634].

21. M. Abramowitz and I.A. Stegun, "Handbook of Mathematical Functions" (National Bureau of Standards, U.S. Department of Commerce, 1964).

22. C.R. Emigh, Proc. 1972 Proton Linear Accelerator Conference, p. 182.

23. P. Kirstein, G. Kino, and W. Waters, "Space Charge Flow" (New York, McGraw-Hill Book Company, Inc., 1967).

24. M. von Ardenne et al, "Tabellen der Elektronen-physik, Ionenphysik, und Übermikroskopie" (VEB Deutscher Verlag der Wissenschaften, Berlin, 1956).

25. D. Mueller (unpublished).

26. O.B. Morgan, G.G. Kelley, and R.C. Davis, Rev. Sci. Instr. 38, 467 (1967); R.C. Davis, O.B. Morgan, L. D. Stewart, and W.L. Stirling, Rev. Sci. Instr. 43, 278 (1972).

27. J. Marshall, private communication.

28. M.O. Scully, D. Rogovin, and C.D. Cantrell (un-published).

APPENDIX: STATISTICAL THEORY OF BEAM TRANSPORT
A. COMPLETE THEORY[22]

We define the RMS beam radius and radial velocity as follows:

$$R_x(t)^2 = \frac{4}{N} \sum_{j=1}^{N} x_j^2(t) \quad , \tag{A.1}$$

$$V_x(t)^2 = \frac{4}{N} \sum_{j=1}^{N} \dot{x}_j^2(t) \quad . \tag{A.2}$$

Differentiating Equation (A.1) we obtain

$$R_x \dot{R}_x = \frac{4}{N} \sum_{j=1}^{N} x_j \dot{x}_j \quad . \tag{A.3}$$

Differentiate again:

$$R_x \ddot{R}_x + \dot{R}_x^2 = \frac{4}{N} \sum_{j=1}^{N} x_j \ddot{x}_j + \frac{4}{N} \sum_{j=1}^{N} \dot{x}_j^2 \quad . \tag{A.4}$$

Use Equation (A.2) and divide both sides by R_x:

$$\ddot{R}_x - \frac{R_x^2(V_x^2 - \dot{R}_x^2)}{R_x^3} = \frac{4}{N R_x} \sum_{j=1}^{N} x_j \ddot{x}_j \equiv A \quad . \tag{A.5}$$

Space Charge

The electrostatic force on a single particle gives rise to an acceleration

$$\ddot{x}_j = - \frac{q}{m} \frac{\partial \phi_j}{\partial x_j} \quad , \tag{A.6}$$

where the electrostatical potential is

$$\phi_j = \frac{q}{4\pi\epsilon_o} \sum_{i \neq j} [(x_j - x_i)^2 + (y_j - y_i)^2 + (z_j - z_i)^2]^{-\frac{1}{2}} \tag{A.7}$$

$$\cong \frac{q}{4\pi\epsilon_o} \sum_{i \neq j} \int_{-\infty}^{\infty} \frac{dz}{[(x_j-x_i)^2+(y_j-y_i)^2+z^2]^{\frac{1}{2}}} \quad . \quad (A.8)$$

By Equations (A.5) through (A.8)

$$A = -\frac{4}{NR_x} \frac{q^2}{4\pi\epsilon_o m} \int_{-\infty}^{\infty} dz \sum_j \sum_{i \neq j} x_j \frac{\partial}{\partial x_j} \qquad (A.9)$$

$$\times \; [(x_j-x_i)^2+(y_j-y_i)^2+z^2]^{-\frac{1}{2}}$$

$$= -\frac{\alpha}{R_x} \int_{-\infty}^{\infty} dz \lim_{a \to 1} \frac{\partial}{\partial a} \sum_j \sum_{i \neq j} [(ax_j-x_i)^2+(y_j-y_i)^2+z^2]^{-\frac{1}{2}} \; ,$$

where

$$\alpha = \frac{4q^2}{4\pi\epsilon_o mN} \quad . \qquad (A.10)$$

We rewrite the quantity A as follows:

$$A = -\frac{\alpha}{R_x} \int_{-\infty}^{\infty} dz \lim_{a \to 1} \frac{\partial}{\partial a} \sum_j \sum_{i \neq j} R^{-1} \qquad (A.11)$$

$$\times \; [1+R^{-2}\{(ax_j-x_i)^2+(y_j-y_i)^2+z^2-R^2\}]$$

$$\qquad (A.12)$$

$$= -\frac{\alpha}{R_x} \int_{-\infty}^{\infty} dz \lim_{a \to 1} \frac{\partial}{\partial a} \sum_j \sum_{i \neq j} R^{-1} \left\{ \vphantom{\frac{(ax_j-x_i)^2}{2R^2}} \right.$$

$$\left. 1 - \frac{(ax_j-x_i)^2+(y_j-y_i)^2+z^2-R^2}{2R^2} + \ldots \right\} \; .$$

Now choose R^2 so that

$$\sum_j \sum_{i=j} [(ax_j-x_i)^2+(y_j-y_i)^2+z^2-R^2] = 0 \; , \qquad (A.13)$$

or

$$\sum_{j} \sum_{i \neq j} [a^2 x_j^2 + x_i^2 - 2ax_i x_j + y_j^2 + y_i^2 - 2y_i y_j + (z^2 - R^2)] = 0 . \quad (A.14)$$

By symmetry

$$\sum_{i} x_i = 0 , \quad \sum_{i} y_i = 0 . \quad (A.15)$$

Then by Equation (A.1) and similar definitions for R_y^2, we have

$$(a^2 + 1) \frac{N^2}{4} R_x^2 + 2 \frac{N^2}{4} R_y^2 + N^2(z^2 - R^2) = 0 . (A.16)$$

This reduces to

$$R^2 = z^2 + \frac{a^2 + 1}{4} R_x^2 + \frac{1}{2} R_y^2 . \quad (A.17)$$

Therefore

$$A \cong - \frac{\alpha}{R_x} N^2 \int_{-\infty}^{\infty} dz \lim_{a \to 1} \frac{\partial}{\partial a} R^{-1}$$

$$= \frac{\alpha}{R_x} N^2 \lim_{a \to 1} \int_{-\infty}^{\infty} dz \frac{1}{R^2} \frac{dR}{da} . \quad (A.18)$$

By Equation (A.17)

$$R \frac{dR}{da} = \frac{a}{4} R_x^2 . \quad (A.19)$$

Therefore

$$A = \frac{\alpha N^2}{R_x} \int_{-\infty}^{\infty} dz \frac{R_x^2/4}{[z^2 + \frac{1}{2}(R_x^2 + R_y^2)]^{3/2}} . \quad (A.20)$$

The desired integral is

$$\int_{-\infty}^{\infty} \frac{dz}{(z^2 + b^2)^{3/2}} = \frac{2}{b^2} , \quad (A.21)$$

so that

$$A = \frac{\alpha N^2 R_x}{4} \cdot \frac{2}{\frac{1}{2}(R_x^2 + R_y^2)} \quad .$$

By Equations (A.10) and (A.5)

$$A = \frac{4q^2 N}{4\pi\epsilon_o m} \frac{R_x}{R_x^2 + R_y^2} \tag{A.22}$$

$$= \frac{4}{NR_x} \sum_{j=1}^{N} \dot{x}_j \, \ddot{x}_j \Big|_{\text{space charge}} \quad .$$

We shall later need the quantity

$$B \equiv \sum_{j=1} \dot{x}_j \, \ddot{x}_j = - \frac{q^2}{4\pi\epsilon_o m} \sum_j \sum_{i \neq j} \dot{x}_j \frac{\partial}{\partial x_j}$$

$$[(x_j - x_i)^2 + (y_j - y_i)^2 + (z_j - z_i)^2]^{-v}.$$

$$\cong - \frac{q^2}{4\pi\epsilon_o m} \int_{-\infty}^{\infty} dz \lim_{a \to 0} \frac{\partial}{\partial a} \sum_j \sum_{i \neq j}$$

$$[(a\dot{x}_j + x_j - x_i)^2 + (y_j - y_i)^2 + (z_j - z_i)^2]$$

$$\cong - \frac{q^2}{4\pi\epsilon_o m} \int_{-\infty}^{\infty} dz \lim_{a \to 0} \frac{\partial}{\partial a} \sum_j \sum_{i \neq j} s^{-1}[1$$

$$- \frac{(a\dot{x}_j + x_j - x_i)^2 + (y_j - y_i)^2 + z^2 - s^2}{2s^2} + \ldots] \quad .\tag{A.23}$$

Let

$$(z^2 - s^2)N^2 + \sum_j \sum_{i \neq j} [(a\dot{x}_j + x_j - x_i)^2 + (y_j - y_i)^2] = 0 \tag{A.24}$$

$$(z^2 - s^2)N^2 + \sum_j \sum_{i \neq j} (a^2 \dot{x}_j^2 + 2a\dot{x}_j x_j + 2x_j^2 + 2y_j^2) = 0 \quad . \tag{A.25}$$

By Equation (A.1) through (A.3), we have

$$S^2 = z^2 + \frac{a^2 v_x^2}{4} + \frac{2a}{4} R_x \dot{R}_x + \frac{1}{2}(R_x^2 + R_y^2) \; , \tag{A.26}$$

$$2S \frac{\partial S}{\partial a} = \frac{a}{2} v_x^2 + \frac{1}{2} R_x \dot{R}_x \; , \tag{A.27}$$

$$\frac{\partial}{\partial a} \frac{1}{S} = - \frac{1}{S^2} \frac{\partial S}{\partial a} \; , \tag{A.28}$$

$$\lim_{a \to 0} \frac{\partial}{\partial a} \frac{1}{S} = - \frac{R_x \dot{R}_x / 4}{[z^2 + \frac{1}{2}(R_x^2 + R_y^2)]^{3/2}} \; . \tag{A.29}$$

Therefore Equation (A.23) becomes

$$
\begin{aligned}
B &= \frac{q^2 N^2}{4\pi\varepsilon_o m} \int_{-\infty}^{\infty} dz \; \frac{\frac{1}{4} R_x \dot{R}_x}{[z^2 + \frac{1}{2}(R_x^2 + R_y^2)]^{3/2}} \\
&= \frac{q^2 N^2}{4\pi\varepsilon_o m} \; \frac{R_x \dot{R}_x}{(R_x^2 + R_y^2)} \; ,
\end{aligned}
$$

and therefore

$$\sum_{j=1} \dot{x}_j \ddot{x}_j = \frac{q^2 N^2}{4\pi\varepsilon_o m} \; \frac{R_x \dot{R}_x}{(R_x^2 + R_y^2)} \; . \tag{A.30}$$

Emittance

Consider the quantity

$$
\begin{aligned}
E_x^2 &= R_x^2 v_x^2 - (R_x \dot{R}_x)^2 \\
&= \sum_j x_j^2 \sum_j \dot{x}_j^2 - (\sum_j x_j \dot{x}_j)^2 \; , \tag{A.31}
\end{aligned}
$$

where we have used Equations (A.1) through (A.3). We now evaluate the quantity

$$2E_x \frac{dE_x}{dt} = \sum_j x_j^2 \sum_i 2\dot{x}_i \ddot{x}_i + \sum_j x_j \dot{x}_j . 2\sum_i \dot{x}_i^2$$

$$- 2 \sum_j x_j \dot{x}_j \sum_i (x_i \ddot{x}_i + \dot{x}_i^2)$$

$$= 2[\sum_j x_j^2 \sum_i \dot{x}_i \ddot{x}_i - \sum_j \dot{x}_j x_j \sum_i x_i \ddot{x}_i] \quad , \qquad (A.32)$$

with the result

$$E_x \frac{dE_x}{dt} = \frac{R_x^3 \dot{R}_x \, q^2 N^2}{4\pi\varepsilon_o m(R_x^2 + R_y^2)} - \frac{R_x^3 \dot{R}_x \, N^2 q^2}{4\pi\varepsilon_o m(R_x^2 + R_y^2)}$$

$$= 0 \quad , \qquad (A.33)$$

where we have used Equations (A.1), (A.3), (A.22), and (A.30). We have shown that for space charge the generalized emittance defined by Equation (A.31) is constant. This is also clearly true if

$$\ddot{x}_j = \frac{q}{m} K_x X_j \quad . \qquad (A.34)$$

We can now write Equation (A.5) as

$$\frac{d^2 R_x}{dt^2} - \frac{E_x^2}{R_x^3} = \frac{4q^2 N R_x}{4\pi\varepsilon_o m(R_x^2 + R_y^2)} + \frac{q}{m} K_x R_x \quad , \qquad (A.35)$$

where we have used (A.22) and (A.34). Let

$$\frac{dz}{dt} = v_z \quad , \quad \frac{d^2 z}{dt^2} = \frac{q}{m} \frac{\partial \phi}{\partial z} \quad , \qquad (A.36)$$

so that

$$\frac{dR_x}{dt} = \frac{dR_x}{dz} v_z \quad , \qquad (A.37)$$

$$\frac{d^2 R_x}{dt^2} = v_z^2 \frac{d^2 R_x}{dz^2} + \frac{dv_z}{dt} \frac{dR_x}{dt} \quad .$$

The equation of motion becomes

$$\frac{d^2 R_x}{dt^2} + \frac{1}{2\phi} \frac{d\phi}{dz} \frac{dR_x}{dt} - \frac{E_x^2}{(2q\phi/m)R_x^3} - \frac{qN\, R_x}{2\pi\varepsilon_o \phi(R_x^2 + R_y^2)} \qquad\qquad (A.38)$$

$$- \frac{K_x R_x}{2\phi} = 0 \quad ,$$

where the total kinetic energy is

$$\frac{1}{2} mv_z^2 = q\phi \quad , \qquad\qquad (A.39)$$

and the number density is related to the current by

$$qN = \frac{I}{v_z} = \frac{I\sqrt{m}}{\sqrt{2q\phi}} \quad . \qquad\qquad (A.40)$$

In summary we have shown that the RMS beam radius obeys the differential equation

$$\frac{d^2 R_x}{dt^2} - \frac{E_x^2}{\dfrac{2q\phi}{m} R_x^3} - \sqrt{\frac{m}{2q}} \frac{I}{\phi^{3/2}} \frac{R_x}{2\pi\varepsilon_o (R_x^2 + R_y^2)} - \frac{K_x R_x}{2\phi} = 0 \quad . \quad (A.41)$$

B.   EMITTANCE ALONE

We can obtain the limitations on beam focusing due to emittance alone by considering suitably average single particle equations of motion. The x-component of a particle's displacement from the axis of the beam after the particle has passed through a lens at time t = 0 and position z = 0 is x = vt + x_o, or

$$x^2 = v^2 t^2 + 2x_o vt + x_o^2 \quad . \tag{A.42}$$

Averaging over an entire ensemble of particles at each point in the cross section of the beam, we are led to consider the average value

$$<x^2> = <v^2>t^2 + 2<x_o v> + <x_o^2> \quad . \tag{A.43}$$

These quantities are related to those used in the complete theory as follows:

$$<x^2> = R_x^2 \quad , \tag{A.44}$$

$$<v^2> = V_x^2 \quad . \tag{A.45}$$

At the focus of the beam, the value of $<x^2>$ is a minimum:

$$\frac{d}{dt} <x^2> = 0 = 2<v^2>t_m + 2<x_o v> \quad . \tag{A.46}$$

This occurs at the time

$$t_m = - \frac{<x_o v>}{<v^2>} \quad , \tag{A.47}$$

so that

$$<x_m^2> = <v^2> \left(\frac{<x_o v>}{<v^2>}\right)^2 - 2 \frac{<x_o v>^2}{<v^2>} + <x_o^2>$$

$$= <x_o^2> - \frac{<x_o v>^2}{<v^2>} \quad . \tag{A.48}$$

We shall rewrite Equation (A.48) in a more convenient form. The velocity of a particle after leaving a linear lens is

$$v = v_o - kx_o \quad , \tag{A.49}$$

where v_o is the velocity before entering the lens,

$$k = v_z/f \quad , \tag{A.50}$$

and f is the focal length of the lens. Equation (A.49) implies also

$$<x_o v> = <x_o v_o> - k <x_o^2> \quad ,$$
$$<v^2> = <v_o^2> - 2k<x_o v_o>+k^2<x_o^2> \quad . \tag{A.51}$$

Put Equation (A.51) into Equation (A.49) to obtain

$$<x_m^2> = <x_o^2> - \frac{(<x_o v_o>-k<x_o^2>)^2}{<v_o^2>-2k<x_o v_o>+k^2<x_o^2>} \quad . \tag{A.52}$$

We assume that prior to entering the lens, the beam has a velocity distribution which is constant across the cross section of the beam, i.e. is independent of the radial position of a particle in the beam. Therefore

$$<x_o v_o> = <x_o><v_o> \quad , \tag{A.53}$$

$$<v_o> = 0 = <x_o v_o> \quad . \tag{A.54}$$

With these assumptions, Equation (A.52) becomes

$$<x_m^2> = <x_o^2> - \frac{k^2<x_o^2>}{<v_o^2>+k^2<x_o^2>}$$

$$= <x_o^2> [1 - \frac{k^2<x_o^2>}{<v_o^2>+k^2<x_o^2>}] \quad ,$$

$$\langle x_m^2 \rangle = \langle x_o^2 \rangle \frac{\langle v_o^2 \rangle}{\langle v_o^2 \rangle + k^2 \langle x_o^2 \rangle} \quad . \tag{A.55}$$

We now have an expression for the minimum focal spot size of the beam in terms of the original size when entering the lens $\langle x_o^2 \rangle$, the velocity distribution prior to entering the lens $\langle v_o^2 \rangle$, and the focal length of the lens.

Similarly, we can calculate the position of the focus by Equation (A.47):

$$t_m = \frac{z_m}{v_z} = - \frac{\langle x_o v \rangle}{\langle v^2 \rangle} \quad ,$$

$$\frac{z_m}{v_z} = \frac{k \langle x_o^2 \rangle}{\langle v_o^2 \rangle + k^2 \langle x_o^2 \rangle} \quad . \tag{A.56}$$

From Equation (A.55), we see that if

$$\langle x_m^2 \rangle \ll \langle x_o^2 \rangle \quad , \tag{A.57}$$

then

$$\langle x_m^2 \rangle \cong \langle x_o^2 \rangle \langle v_o^2 \rangle / k^2 \langle x_o^2 \rangle = \langle v_o^2 \rangle f^2 / v_z^2 \quad . \tag{A.58}$$

Equations (A.50) and (A.58) now imply

$$\langle x_m^2 \rangle \cong \frac{\langle v_o^2 \rangle}{v_z^2} f^2 \quad , \tag{A.59}$$

in the limit of tight focusing. Equations (A.56) and (A.50) also imply that in the same limit,

$$z_m \cong f \quad . \tag{A.60}$$

These equations clearly show how the minimum RMS focused beam size and focal position depend on the parameters of the beam and the lens, in the limit where emittance alone is important.

SOFT X-RAY LASERS VIA ELECTRON-COLLISIONAL PUMPING*

R. A. Andrews

Naval Research Laboratory

Washington, D. C.   20375

I. INTRODUCTION

One possible technique for obtaining gain in the
soft x-ray region of the spectrum is to use electron-
collisional pumping of an appropriate ion species. This
can be done in a manner analogous to known ion lasers
which operate in the visible portion of the spectrum.
The differences being:1) more highly ionized ions are
used to obtain shorter wavelength transitions, 2) the
pumping electrons are at a higher temperature to populate the more energetic transitions, and 3) the lifetimes are shorter which implies higher pump intensity
per unit area (P/a $\propto \nu^4$). In the case of shorter wavelength transitions one can project known laser transitions isoelectronically to higher Z ions and hence
shorter wavelengths or investigate unique ionic electron
configurations that are not observed in neutral or
weakly ionized species. This technique works well for

*This work was partially supported by the Defense Ad-
vanced Research Projects Agency, ARPA Order 2694.

electronic configurations with relatively few electrons.
With many-electron systems, level crossings and other
anomalous effects with increasing Z limit the range
over which a group of levels which are a viable laser
scheme can be isoelectronically projected to higher Z.
A further problem with short wavelength lasers is that
the techniques available for generating significant
amounts of energy with very short risetimes are limited.
Discharges are limited to about 10^{-7} sec. Mode-locked
lasers, however, can be used down to the 10^{-11} sec
range with significant amounts of energy. Short wavelength requirements lead to many other problems for the
attainment of an inversion and net positive gain. Several of these are discussed in this paper along with a
particular approach to the short wavelength laser problem.

A particularly promising concept for an electron-
collisional pumped laser is the use of a picosecond
laser pulse to heat the electrons in a cold plasma
which has a large fractional population of the laser ion
species.[1,2] The initial plasma can be created by a
variety of techniques. However, if one is seriously
considering highly ionized species then a laser produced plasma offers distinct advantages since large
amounts of energy can be deposited in small volumes in
short times. This initial plasma would typically be
generated with a line focus laser to create a plasma/
laser media with a large aspect ratio and hence maximum
gain length. This plasma must be allowed to expand,
cool, and develop a maximum fractional population of a
particular laser ion species. Cooling is important since
the laser line is Doppler broadened and Boltzmann population of low lying levels may ruin a possible inversion.

Also expansion must be limited since laser heating of
the electrons will not be effective if the density is
too low. Once the initial plasma has evolved, a shorter
laser pulse propagating along the plasma axis heats the
electrons as shown schematically in Fig. 1. If the
pulse width and hence the heating time is much less
than the electron-ion equilibration time for the particular plasma conditions (temperature, density) then in
principle it is possible to obtain conditions where the
electron temperature is much higher than the ion temperature. Depending on the laser scheme and the energy
of the upper laser levels, there will be an optimum
electron temperature. Finally, once there is sufficient
inversion of the laser levels there will be superfluo-
rescent emission or traveling wave (TW) amplification
of a synchronized input pulse. Due to the general lack
of efficient reflectors for short wavelengths and the
very high power densities in short wavelength lasers,
an oscillator analogous to visible lasers does not seem
possible. Further, the short characteristic times associated with short wavelength transitions prevents
gain from being maintained for long enough times to allow
several round trip resonator transits. In fact, long
gain lengths at very short wavelengths can only be obtained by TW pumping. This is shown in Fig. 1. The
laser pulse and the gain pulse travel through the plasma
at approximately the speed of light. The width of the
gain pulse is determined by the lifetime of the laser
transition for the case of self-terminating laser action.
The heating pulse may be shorter than this.

     A discussion of this approach can be divided into
five problem areas: 1) the generation of a cold plasma
with a large fractional population of the laser ion

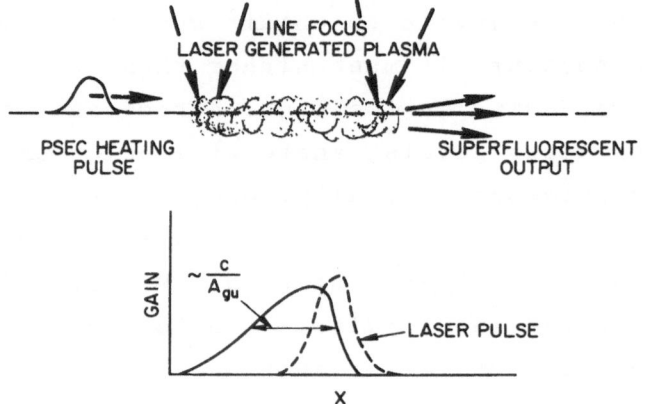

Figure 1   Schematic diagram of the electron-collisional
            laser using the picosecond pulse heating of a
            plasma approach.   The heating pulse propa-
            gates down the axis of the plasma to provide
            TW pumping.   The width of the gain pulse is
            determined by the lifetime of the upper laser
            level.

species; 2) synchronization of the electron heating
laser pulse with the plasma generation; 3) obtaining
optimum electron heating; 4) identifying potential ion
laser configurations; and 5) establishing tests and
criteria for observation of net gain. The following
sections of this paper will concentrate on areas (3),
(4), and (2) along with comments on recent experimental
work at NRL.

II. SHORT PULSE ELECTRON HEATING

Ultrashort laser pulses with peak intensity,
$I > 10^{13}$ W/cm^2 can be used very effectively to preferentially heat the electrons in a plasma. These intensities are most easily generated by modelocked lasers
generating pulses on the picosecond time scale. These
short pulses are very important for self-terminating
pulsed lasers. In fact, they limit the wavelength
where lasing can be expected since the pulse width,
$\tau \sim A^{-1}$ where A is the transition probability of the
laser transition ($A \propto \lambda^{-2}$). The short pulse width
simplifies the analysis of the electron heating since
plasma hydrodynamics and collective plasma oscillations
can be neglected.

In high intensity fields the plasma electrons os-
cillate with the frequency of the incident laser radiation. The energy of this oscillation is like "internal" energy, independent of the thermal or translational
energy. If the electron does not undergo a collision
while the intense field is present then the oscillation
energy goes back into the laser field and nothing is
changed i.e., the electron temperature is not increased
and the laser pulse is transmitted through the plasma.
However, if the oscillating electron suffers a

collision then the oscillation energy is converted to
translational or thermal energy at the expense of the
high intensity laser field. The oscillation energy is
given by

$$E_{osc} = \frac{e^2 E^2}{4m\omega^2} = \frac{2\pi e^2}{mc} \frac{I}{\omega^2} \quad ,$$

where e is the electronic charge, m the electron mass,
c the velocity of light, and ω the angular frequency of
the incident laser radiation of intensity I. At 1.06 µm

$$E_{osc} \text{ [eV]} = 1.02 \times 10^{-13} \text{ I [W/cm}^2\text{]}.$$

Hence, $E_{osc} \overset{\sim}{>} E_e$ (electron thermal energy) $\sim 3/2 \text{ kT}_e$
(electron temperature) for intensities $I > 10^{13} \text{ W/cm}^2$
and relatively cold plasmas ($T_e \sim 1$ eV). The electron's
energy is the sum of the thermal and oscillation energies.
If E_{osc} is high enough, a single collision can excite
an energetic ionic level. Alternatively, with many
collisions, the electron temperature can be increased to
higher values. This large thermal energy can then
excite more energetic ion levels in inelastic collisions.
The potentially important heating mechanisms are inverse
bremsstrahlung and stimulated Compton scattering. Potential cooling mechanisms are inverse Compton scattering, expansion, ion collisions, and bremsstrahlung.
Each of these cooling mechanisms can be shown to be
negligible on the picosecond timescale for plasma parameters of interest here.

The inverse bremsstrahlung heating rate is given by

$$\left(\frac{dE}{dt}\right)_B = E_{osc} \, \nu_{eff} \quad ,$$

where ν_{eff} is the effective electron collision rate and is given by

$$\nu_{eff} \sim 2\sqrt{3\pi/m} \ Z^3 \ e^4 \ n_e \ \ell n \ \Lambda/(E_e + E_{osc})^{3/2} \ ,$$

where Z is the net ionic charge, n_e is the electron
density, and $\ell n \ \Lambda$ is the Coulomb logarithm. The heating
rate for stimulated Compton scattering[3-5] is given by

$$\left(\frac{dE}{dt}\right)_{sc} = 3\pi\sigma_o \ \Omega \ I^2/4m\omega^2\Delta\omega \ ,$$

where $\sigma_o = 8\pi e^2/3mc^2$ is the Thompson cross section, Ω
is the solid angle of the focused laser radiation, and
$\Delta\omega$ is the width of the heating laser line. $\Delta\omega/\omega$ is
about 10^{-4} for Nd:YAG lasers and about 10^{-3} for Nd:glass
lasers. For a given laser ($\Delta\omega/\omega$) and focusing optics
(Ω) only those electrons which satisfy the condition,

$$\frac{\Delta\omega}{\omega} > 1/2 \sqrt{\frac{\Omega T_e}{mc^2}} \ ,$$

will be heated by stimulated Compton scattering. In
typical situations stimulated Compton scattering is
important for Nd:glass lasers and of minor importance
for Nd:YAG lasers. In the latter case after T_e reaches
some relatively low value inverse bremsstrahlung is the
dominant heating mechanism.   As T_e increases ν_{eff}
decreases. Hence for a given pulse width, τ, the number
of collisions $\nu_{eff} \ \tau$ decreases. For very high intensities there are insufficient collisions to obtain high
T_e. Figure 2 shows $E_e(\tau)$ for a number of ion densities
as a function of intensity, I. Note that there is a
maximum E_e for a given n_i and this occurs at an intensity where $\nu_{eff} \ \tau$ is on the order of unity. At higher
intensities there is not enough time for effective

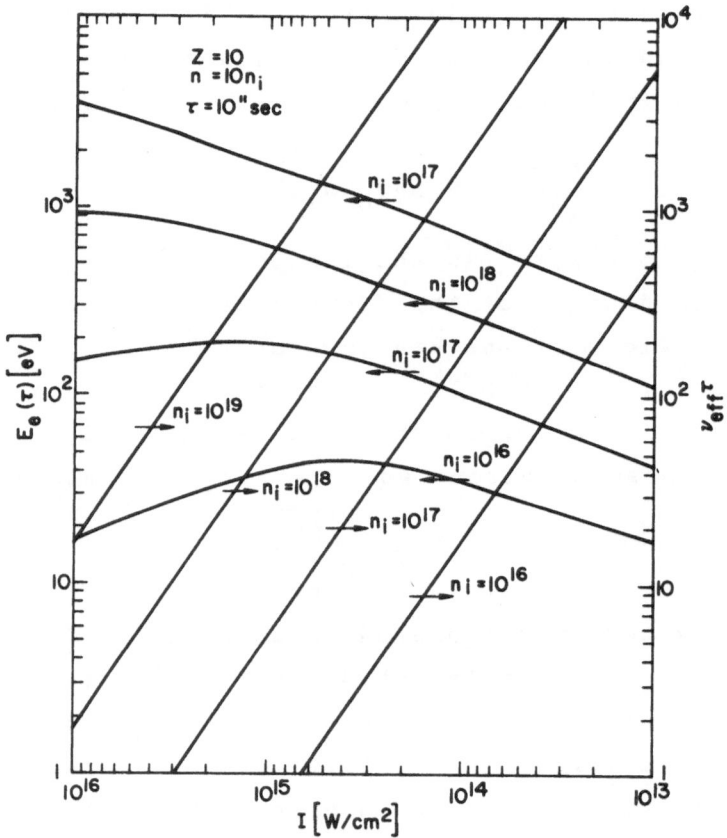

Fig. 2: Plot of final electron energy $E_e(\tau)$ after heating via inverse bremsstrahlung with an intense
pulse of duration 10^{-11} sec. for various ion
densities. $\nu_{eff}\tau$ is the number of collisions
during the pulse.

heating ($\nu_{eff} \tau < 1$). Also under typical conditions
for this problem the maximum temperature attained via
inverse bremsstrahlung heating is on the order of 1 keV.
Under conditions where stimulated Compton scattering is
important (using Nd:glass lasers) this maximum can
potentially be increased but only with extremely high
intensities. Hence, using state-of-the-art laser
technology one can produce $T_e \sim 1$ keV conditions in a
plasma without increasing T_i. This is sufficient to
pump x-ray lasers with hν in the range of 100 eV depending on the particular laser scheme.

Electron-collisional excitation probabilities for
atoms and ions can be calculated.[6] There is considerable experimental data for atoms and very little for
ions. What data exist for ions is for simple ions
with low degrees of ionization. The graph in Fig. 3
qualitatively indicates the results of theoretical
analysis of this problem. For large values of E_e/E_{exc},
cross sections (σ) can be calculated using the Born-
Coulomb approximation. Theory shows that for allowed
transitions σ falls off as $\ln E_e/E_e$ and for unallowed
transitions it falls off as E_e^{-1}. For atoms the cross
section is maximum for $E_e \sim (2-3) E_{exc}$ and depending on
the situation σ (allowed) can be comparable to σ (unallowed). In the case of ions σ has a finite value at
$E_e = E_{exc}$ due to the Coulomb field and hence the maximum cross section is at lower values of E_e/E_{exc}. Also
σ for ions scales as Z^{-4} which is an important consideration for short wavelength (higher Z) electron-collisional laser schemes. In brief, the general considerations
for collisional pumping are the following: (1) to preferentially populate an allowed transition over unallowed

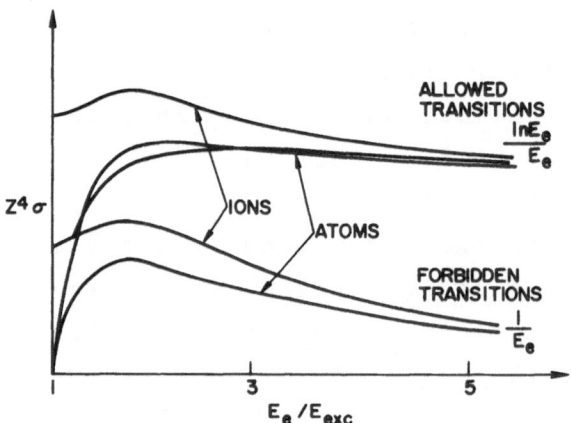

Figure 3   Characteristic curves of electron-collisional
            excitation cross sections for atoms and ions
            as a function of [electron energy (E_e)]/
            [excitation energy for a particular level
            (E_{exc})].

dipole transitions use $E_e > 3E_{exc}$, and (2) to obtain
maximum population of multipole transitions use
$E_e \sim E_{exc}$.

III. ION LASER CONFIGURATIONS

When predicting potential short wavelength laser
transitions one usually starts with known transitions
and projects along an isoelectronic sequence to higher
Z. $\Delta n = 0$ (n = principal quantum number) transitions
scale as $\lambda \propto Z^{-1}$ and $\Delta n = 1$ transitions scale as $\lambda \propto Z^{-2}$.
Hence $\Delta n = 1$ transitions are required for really short
wavelengths using reasonable states of ionization. The
laser scheme can be either "quasi-cw" where an inversion
is maintained in near equilibrium conditions or "pulsed,
self-terminating" where an inversion is developed in
transient, extreme nonequilibrium conditions. The
quasi-cw scheme has recently been discussed by Elton
both in the literature[7] and in an earlier paper in this
symposium. Since the ion populations must finally be
in equilibrium for "quasi-cw" operation (i.e. the gain
is limited only by the duration of the pumping) this
implies T_e approaches $T_i \sim 1/4 \, \chi_{Z-1} (\chi_{Z-1}$ = Ionization
potential of Z-1 ion) and an initial advantage of a
large T_e/T_i ratio disappears in a limited time $\sim (1-10)A^{-1}$.

Therefore, a true quasi-cw condition requires long
lengths of hot low density plasma to achieve reasonable
gain. In the case of the pulsed, self-terminating laser
scheme, collisional mixing is not a problem in a first
approximation, since gain only exists for time on the
order of the lifetime of the states involved. Further,
pumping times are much less than the electron-ion
equilibrium time and hence $T_i \ll T_e$. This scheme can

lead to very high gain; however, as the wavelengths become shorter the period of time over which gain can be
maintained decreases.

The self-terminating electron-collisional laser
scheme has several requirements.[8] First the laser
transition should preferably be one with $\Delta n = 1$. Second,
the upper laser level (U) should be connected to the
ground level (G) by a resonance line in order to make
use of resonance trapping, and not connected by an
elctronic-dipole transition to any lower level except
the lower laser level (L). The density of the plasma
must be sufficient to insure resonance trapping of the
spontaneous emission from the upper laser level. This
also implies that if the ground level is split, that
the splitting be small enough to maintain uniform ground
state population and therefore reabsorption of the
resonance radiation. The resonance trapping insures
efficient collisional pumping since there are effectively
no radiation losses. Third, the lower laser level should
not be connected to the ground state by an electric
dipole transition and it should be sufficiently above
the ground level so that it is not populated by the
Boltzmann distribution at T_i, the relatively cold ion
temperature. The radiative transition probability of
the laser line, A_{UL}, should be less than that of the
resonance line and the length of the pump pulse, τ_p,
should be shorter than A_{UL}^{-1}. This will insure that an
adequate inversion can be achieved for high gain. On
the other hand, A_{UL} should not be too small since this
will increase the gain length to unreasonable values.

Thus,

$$10^{-4} A_{UG} \stackrel{\sim}{<} A_{UL} \stackrel{\sim}{<} 10^{-1} A_{UG},$$

and

$$\tau_p < A_{UL}^{-1} .$$

Ions isoelectronic with potassium and ions iso-electronic with silicon represent two possible electron-collisional self-terminating laser schemes. Figure 4 illustrates the K-like isoelectronic sequence. KI and CaII have one 4s electron in the ground state. However, the ions ScIII and above have a single 3d electron in the ground state. This configuration is not found in neutral species and hence is a unique ionic configuration. The ground state is $3d(^2D)$. The first allowed excited state is $4p(^2P)$ which is also connected by an allowed transition to the lower lying $4s(^2S)$ state. These three levels make up a possible self-terminating laser which meets the criteria discussed above: the resonance line is $3d \rightarrow 4p$ and the laser line is $4p \rightarrow 4s$ as shown. This lower laser level is connected to the ground state by an electric quadrapole transition. This scheme suf-fers the disadvantage that the laser transition is $\Delta n = 0$ and hence it will not scale to very short wave-lengths. The levels in Fig. 4 are plotted through NiX using actual spectroscopic data.[9] The pump energy scales as Z^{-2} and for very moderate values of Z is much larger than the laser photon energy. It is seen that for NiX the laser transition is at a wavelength slightly less than 1000 Å while the pump energy is close to 100 eV. This isoelectronic extrapolation breaks down above NiX for another reason also. The manifold of levels corre-sponding to the $(3d)^2$ excited state starts to fall below the (4p) and (4s) excited states and the conditions

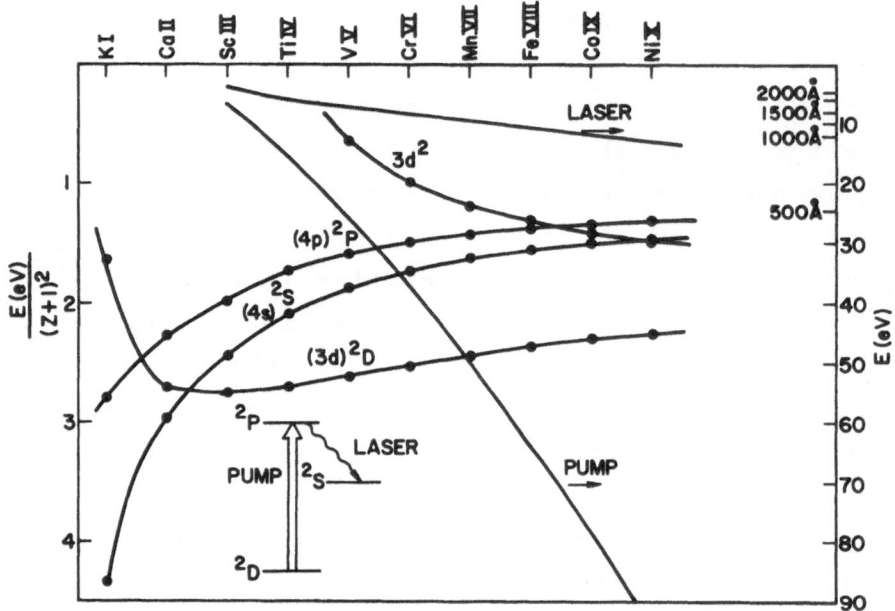

Figure 4   Energy level diagram for the K-like iso-
            electronic sequence showing the 3d, 4s, 4p
            and $(3d)^2$ levels. For a laser scheme with 3d
            ground state, 4p upper state, and 4s lower
            state the pump energy and the laser photon
            energies are plotted on the right hand ordi-
            nate.   The data for these curves was taken
            from R. Kelly and L. Palumbo, NRL Report
            7599 (1973).

discussed above are no longer satisfied.

The Si-like isoelectronic sequence has a $(3s)^2(3p)^2$ (^3P) ground state as shown in Fig. 5. For ions SIII and above the first excited states are all of the configuration $(3s)(3p)^3$. This is split into $^3S^o, ^3P^o$, and $^3D^o$ levels. Ions in this sequence have particularly strong intercombination lines connecting these excited states to the singlet ground levels $(3s)^2(3p)^2(^1S)$ and (^1D). Hence these levels fit the collisional self-terminating laser requirements discussed above. Laser transitions can be any of the following: $^3S^o \to ^1S$, $^3S^o \to ^1D$, $^3P^o \to ^1S$, $^3P^o \to ^1D$, $^3D^o \to ^1S$, $^3D^o \to ^1D$. This scheme also has $\Delta n = 0$ for the laser transition and hence the laser wavelength scales as Z^{-1}. For TiIX the $^3S \to ^1S$ transition is close to 400 $\overset{o}{A}$, and for ZnXVII it extrapolates to around 240 $\overset{o}{A}$. The ^1S state is the preferred lower laser level over the ^1D state since it is higher above the ground state and does not limit T_i to too low of a temperature. The curves in Fig. 5 were drawn from spectroscopic data.[9] Reference 9 lists several observed intercombination lines for this isoelectronic sequence. These are listed in Table I. The branching ratio shows the ratio of observed line intensities of the listed line and the corresponding line to the ground (^3P) state. This ratio approximates the ratio of the A values for these two transitions.

A general schematic of the collisional laser using the short pulse laser heating approach is shown in Fig. 6. The axial length of the gain region, L, is determined by the confocal parameter of the focused short pulse radiation. The focusing optics also determine the diameter of the gain region. For a 10^{10} Watt, 1.06 μm laser, the desired peak intensity for optimum electron

TABLE I

OBSERVED INTERCOMBINATION LINES

IN Si-LIKE IONS

ION	TRANSITION	WAVELENGTH	BRANCHING RATIO
P II	$^3D^o - \,^1D$	1772Å	1:100
	$^3P^o - \,^1D$	1473	1:170
SIII	$^3S^o - \,^1D$	789	2:3
Ci IV	$^3S^o - \,^1S$	756	1:11
	$^3S^o - \,^1D$	662	3:11
Ar V	$^3S^o - \,^1S$	651	1:70
	$^3S^o - \,^1D$	571	8:70
K VI	$^3S^o - \,^1S$	572	1:25
	$^3S^o - \,^1D$	502	1:25
Ca VII	$^3S^o - \,^1D$	448	2:45

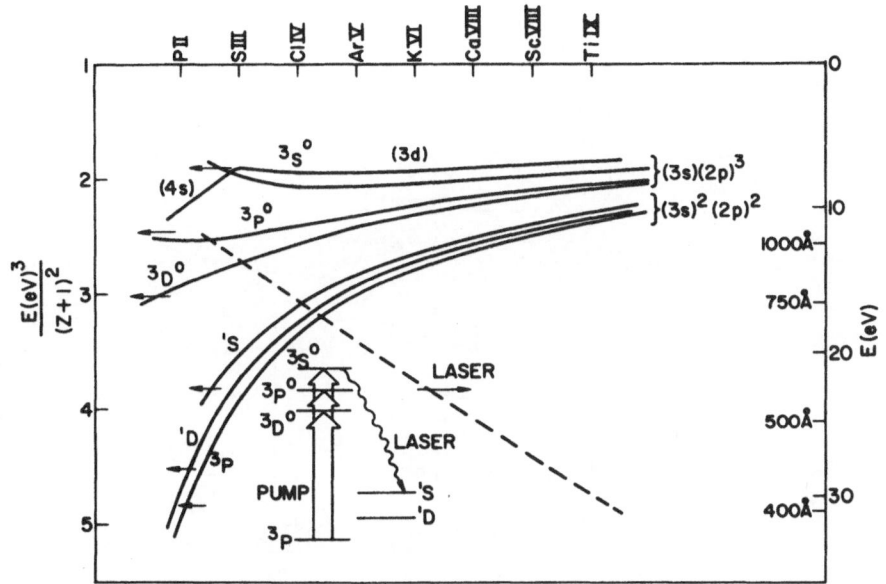

Figure 5   Energy level diagram for the Si-like isoelectronic sequence showing the $3s^2 3p^2$
levels and the $3s3p^3$ levels. For a laser
scheme using the $^3S^\circ$, state as an upper
level and the ^1S level as a lower level the
laser photon energy is plotted on the right
ordinate. The data for these curves was taken
from R. Kelly and L. Palumbo, NRL Report
7599 (1973).

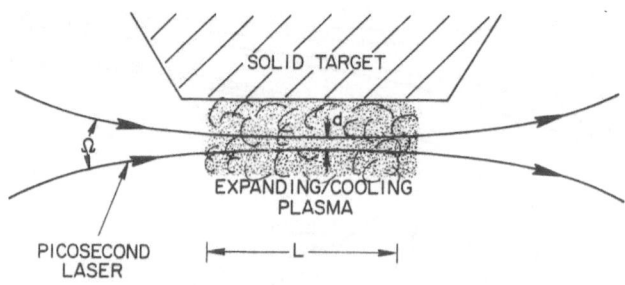

Figure 6   Schematic of the picosecond laser heated
plasma approach with laser parameters for
a 10^{10} Watt, 1.06μ laser.

heating determines the dimensions of the gain region
as shown.

As an example of this type of laser, consider ScVIII
with the $^3S \rightarrow {}^1S$ laser line at ~ 450 Å. Using a 10^{10}
Watt, 1.06μm laser focused at 10^{14} W/cm^2 and a pulse
width of 10^{-11} sec; then from Fig. 2 an electron energy
of about 150 eV can be obtained with a density of
$n_i \sim 5 \times 10^{17}$ cm^{-3}. This energy is $\sim 4.5E_{exc}$ for the
upper laser level and should insure optimum pumping.
Also, this density is sufficient to guarantee an optically thick pumping line since optical thickness, τ, is
given by

$$\tau = 5.5 \times 10^{-17} \lambda \, n_i \, d \, \sqrt{M/T_i} \quad ,$$

where M is the atomic mass number.[10] For $\tau = 10$ this
requires $n_i > 6 \times 10^{15}$ cm^{-3} for this example. This is
adequate assuming the condition $T_i \sim 1$ eV can be maintained. The electron-ion equilibration time is

$$\tau_{e,i} = 3.2 \times 10^7 \, MT_e^{3/2}/n_e \, Z \quad ,$$

and for this example $\tau_{e,i} = 6 \times 10^{-8}$ sec. Hence,
$\tau_p \ll \tau_{e,i}$ and the conditions for obtaining $T_e \gg T_i$
are satisfied. Also $\tau_p \ll A_{UL}^{-1} = 5 \times 10^{-9}$ sec [11]
which satisfies the requirement for obtaining maximum
gain. Gain can be estimated using the expression for
a doppler broadened line

$$\alpha = 1.6 \times 10^{-32} \, n_u \, \lambda^3 \, A_{UL} \, \sqrt{M/T_i} \quad .$$

Using the values given above, an estimated $n_u \sim 5 \times 10^{15}$
cm^{-3}, and a gain length $L \sim 1.5$ cm (based on a 10^{10} W,
1.06 μm laser focused to 10^{14} W/cm^2) then

$$\alpha L \sim 14.$$

If plasmas of higher density and the same temperature
can be achieved then this number should increase accordingly.

IV. SYNCHRONIZATION OF ELECTRON HEATING PULSE

This approach to a soft x-ray laser requires two
laser pulses, one to produce the plasma and the second
to preferentially heat the plasma electrons. Energy is
a primary consideration in the first pulse which must
ionize neutral target atoms to produce the desired stage
of ionization. Hence, typically this pulse can be a
nanosecond pulse assuming that extremely high values of
Z are not required. The heating pulse must be a picosecond pulse to achieve the desired heating. These two
laser pulses must be synchronized with a jitter time
short compared to d/v where d is the diameter of the
gain medium and v is a typical plasma velocity. Hence,
the jitter must be $< 10^{-9}$ sec. This problem has recently
been solved at NRL. Figure 7 shows a schematic diagram
of the NRL dual laser system. The picosecond laser consists of a modelocked Nd:YAG oscillator and four Nd:YAG
amplifiers: one 1/4", two 3/8", and one 1/2". The
performance of this laser is indicated in Table II. The
laser produces 30 psec pulses at an intensity $> 10^{10}$
Watts. The second laser consists of a Q-switched
Nd:YAG oscillator followed by three Nd:glass amplifiers.
It produces 3 nsec pulses at an intensity of 10^9 Watts.
Its characteristics are indicated in Table III. The
synchronization technique is shown schematically in
Fig. 8. The sequence of events is as follows: (1) the
flashlamps for both oscillators are fired, (2) the modelocked pulse train begins to appear, (3) an early pulse
in the modelocked train triggers the pockels cell of the

TABLE II

MEASURED PARAMETERS OF PICOSECOND LASER SYSTEM

STAGE	PULSE ENERGY (mJ)	PULSE DURATION (PSEC)	POWER (W)	BEAM PROFILE	INTENSITY (W/CM^2)	AMPLIFIER GAIN
OSCILLATOR	.20	25	8×10^6	1mm (Circular Gaussian)	10^9	
AMPLIFIER 1st Stage	4.8	25	1.9×10^8	2mm	6×10^9	30 (30)
2nd Stage	13.2	25	5×10^8	$.6 \times .3$ CM (ELLIPTICAL AIRY DISC)	3.5×10^9	10.6 (12)
3rd Stage	114.0	30	4×10^9	$.5 \times .9$ CM	1.1×10^{10}	9.4 (12)
4th Stage	210	30	7×10^9	$.5 \times 1.2$ CM	1.25×10^{10}	4.1 (7)
TARGET	190	30	6.3×10^9	$30 \mu \times 60 \mu$	5×10^{14}	

NOTE: The parameters of the picosecond laser system are tabulated at various stages. Large signal gain under actual operation conditions is given in the last column with small signal gain in parenthesis. Total gain is reduced from the product of individual gains by power loss at the apertures and reflections.

TABLE III
PICOSECOND LASER SYSTEM PARAMETERS
Q-SWITCHED LASER (SYCHRONIZED)

STAGE	PULSE ENERGY (J)	PULSE DURATION (ns)	POWER (W)	BEAM PROFILE	INTENSITY (W/cm^2)
DESIGN GOALS					
OSCILLATOR OUTPUT	0.05	50	1×10^6 (PEAK)	2mm (CIRCULAR GAUSSIAN)	
OUTPUT FROM GATED SHUTTER	0.003	3	1×10^6	2mm	
AMPLIFIER OUTPUT (3 STAGE)	3.0	3	1×10^9	2 cm	
TARGET	3.0	3	1×10^9	50μ × 50μ (CIRCULAR FOCUS)	5×10^{13}
				50μ × 2mm (LINE FOCUS)	1.25×10^{12}
SYCHRONIZATION ACCURACY		< 1 ns			
MEASURED OUTPUT					
OSCILLATOR	.02	30	$.67 \times 10^6$	2 mm	

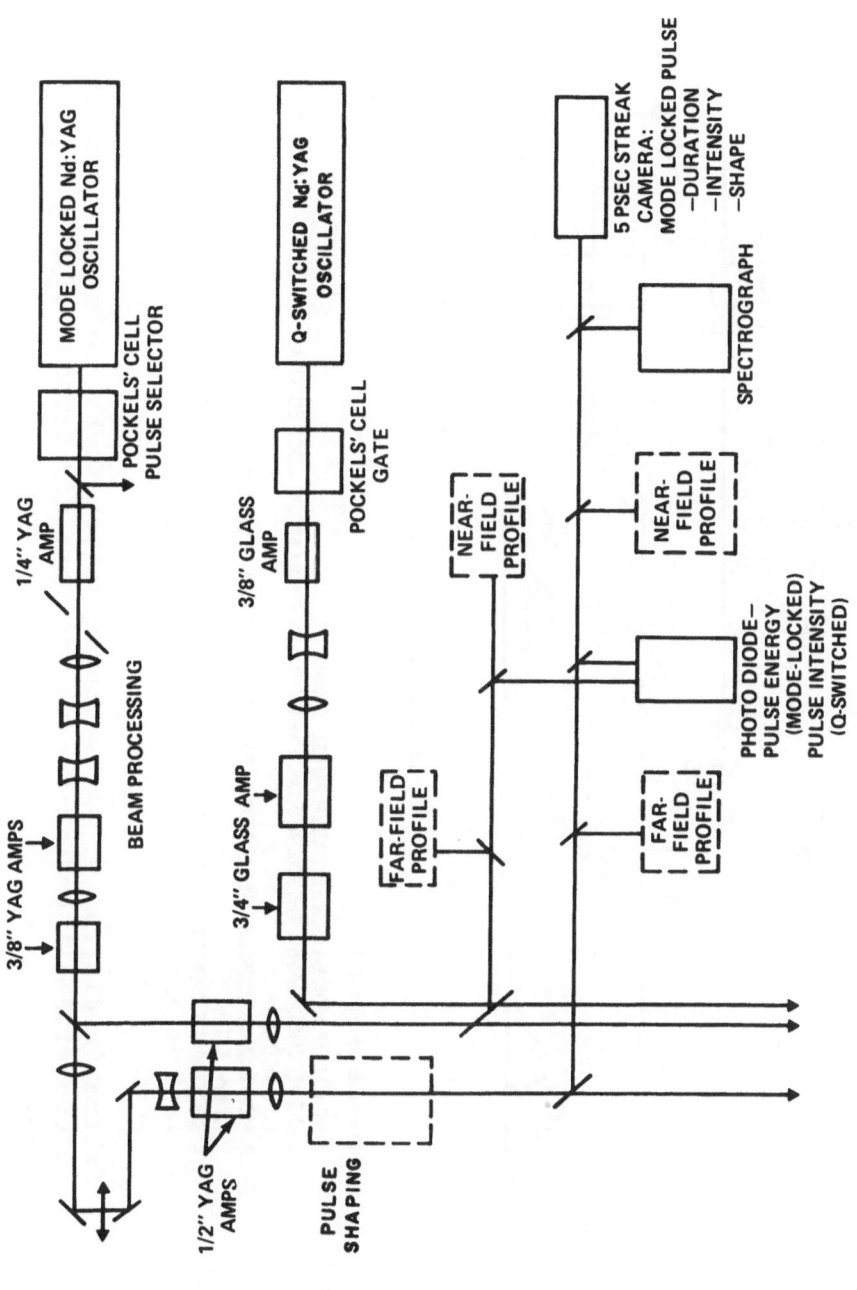

Fig. 7: Schematic diagram of the NRL dual synchronized laser system which generates a 30 psec, 10^{10} W pulse synchronized with jitter < 1 nsec to a 3 nsec
10^9 W pulse.

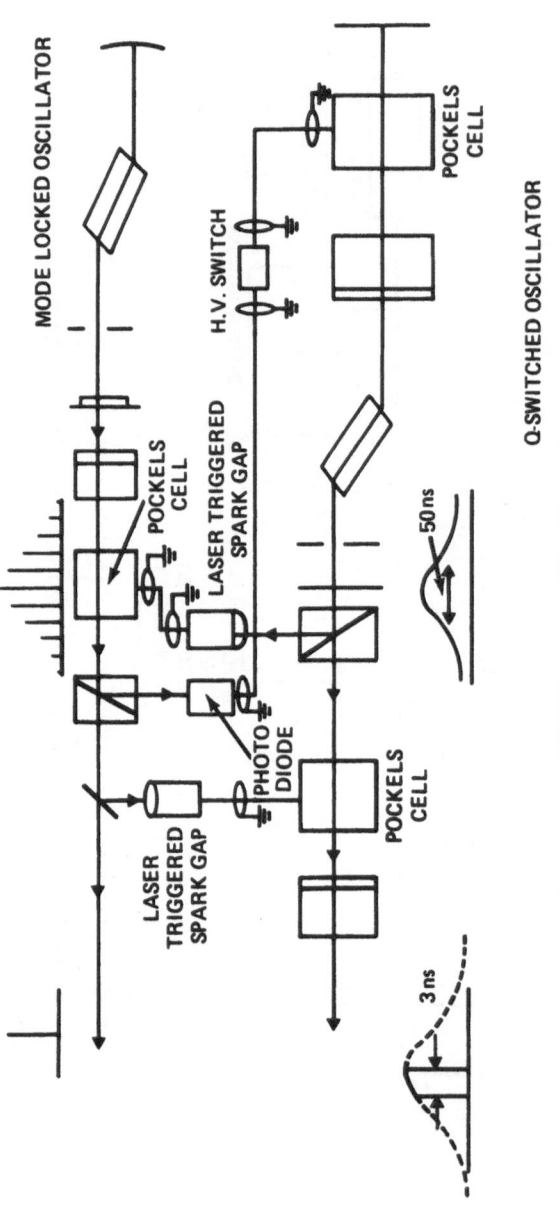

SEQUENCE OF EVENTS:

1. START OF MODE LOCKED LASER PULSE TRAIN
2. Q-SWITCHED LASER TRIGGERED ON EARLY PULSE FROM MODE LOCKED TRAIN
   ($\Delta t \sim$ 10-20ns)
3. MODE LOCKED PULSE SELECTED AFTER START OF Q-SWITCHED PULSE
   ($\Delta t \sim$ 6-7ns)
4. SECTION OF Q-SWITCHED PULSE GATED OUT BY SELECTED MODE LOCKED PULSE
   ($\Delta t <$ 1ns)

Fig. 8: Schematic diagram of the synchronization technique used for obtaining two pulses with jitter < 1 nsec. See text for details of the sequence of events.

Q-switched oscillator (at this point jitter $\sim$ 10-20 nsec), (4) the Q-switched pulse appears and triggers the pulse selecting pockels cell in the modelocked train (at this point jitter = interpulse spacing in modelocked train $\sim$ 6-7 nsec) and, (5) the selected modelocked pulse triggers the gate which passes a 3 nsec portion of the Q-switched pulse. At this final point the jitter has been measured to be < 1 nsec. This short jitter time is achieved by overdriving the final laser triggered spark gap with the single selected modelocked pulse. Figure 9 shows the two laser pulses at various stages of the sychronization process. Of course, once these pulses are generated and amplified the delay between them can be varied by changing their respective optical path lengths to the target.

V. EXPERIMENTAL STUDIES

Experiments are underway at NRL to investigate the possibility of demonstrating gain in the soft x-ray region using collisional pumping in a picosecond laser pulse heated plasma. The early experiments have been aimed at (1) demonstrating the attainability of $T_e >> T_i$ conditions and (2) investigating the effect of laser prepulses on the x-ray yield from a laser produced plasma.

The 30 psec, 10^{10} Watt laser has been used to generate laser produced plasmas. These plasmas have been diagnosed with soft x-ray spectrometers for a variety of medium Z targets. A typical x-ray spectrum for aluminum is shown in Fig. 10 which shows He-like and H-like aluminum lines. This series of experiments has been modeled using the NRL "Hot Spot" code which has been described elsewhere.[12] The result of this

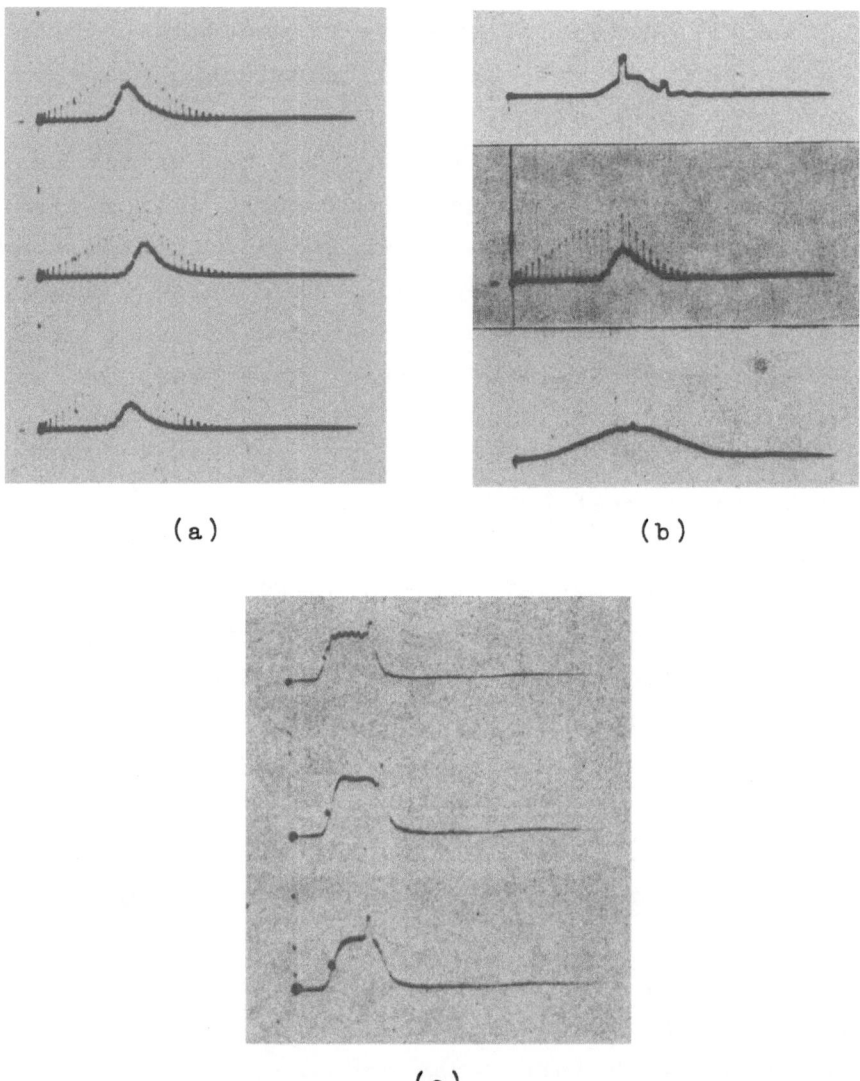

(a)   (b)

(c)

Fig. 9:   Laser pulses at various stages of synchroniza-
          tion. (a) Q-switched pulse with mode locked
          train, (b) selected single mode locked pulse
          with Q-switched pulse, (c) selected mode locked
          pulse with shuttered Q-switched pulse.

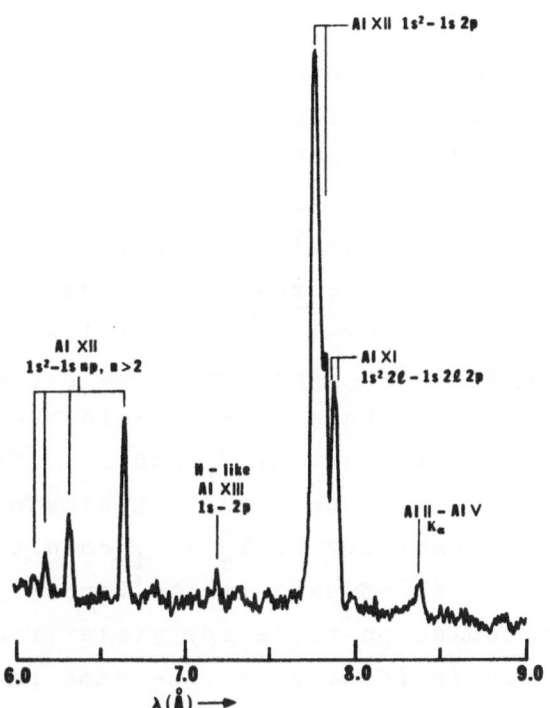

Figure 10   A typical x-ray spectrum of a laser produced plasma generated using a 30 psec, 0.1
Joule pulse.   This spectrum shows H-like
and He-like aluminum lines.

modeling is shown in Fig. 11 where the integrated intensities of the aluminum lines shown previously are compared with theory. The experimental data has been
normalized to the AlXII ($1s^2$ - 1s5p) line. It is seen
that the agreement is quite good. The low experimental
value for the ($1S^2$ - 1s2p) line is due to film saturation and the low experimental value for the (1s - 2p)
line is due to an optically thick plasma. This same
code has been used to predict the time history of the
evolution of the various Al ions. These results are
shown in Fig. 12. Projections of this type along with
experimental data will be required to determine the
optimum time for pumping a particular laser ion species
in the development of a laser produced plasma. Based on
the experimental data this code has also been used to
trace the time development of T_i and T_e. These results
are shown in Fig. 13. The results indicate quite
clearly the attainability of $T_e > T_i$ conditions.

A second series of experiments investigated the
effect of a prepulse on the x-ray yield from the plasma.
The prepulse was incident along the same path as the
main pulse. It preceeded the main pulse by $\sim$ 6 nsec
and was 1/3 the intensity of the main pulse. Both
pulses were 30 psec in duration. The results of these
experiments are shown in Fig. 14. Both with and without
a prepulse there seems to be two temperatures, one
$\sim$ 1 kev and one < 0.5 kev. With the prepulse the total
x-ray flux is increased by almost an order of magnitude.
These results evidence how effectively picosecond pulses
can couple into a preexisting plasma.

VI. SUMMARY

Electron collisional pumping of ions in a plasma

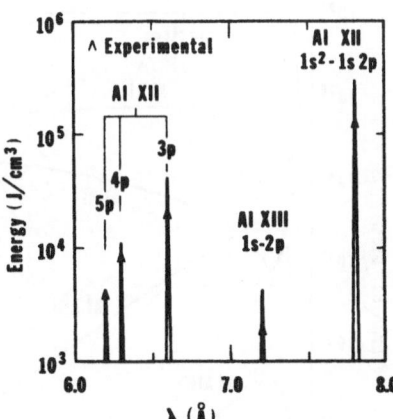

Fig. 11:   Comparison of theoretical model with inte-
           grated experimental data for Aℓ. The data
           was normalized to the He-like 5p line. The
           2p line was saturated on the film which ac-
           counts for the low value for this line. The
           plasma was optically thick for the H-like
           1s-2p line accounting for the low experi-
           mental value of this line.

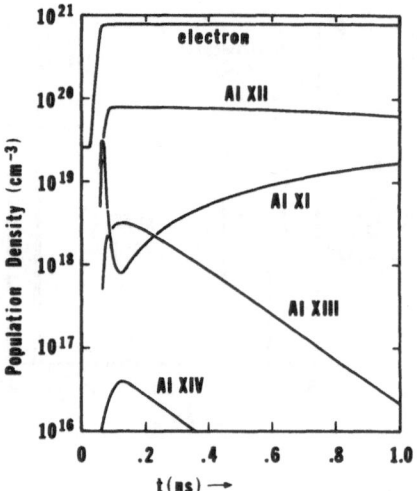

Fig. 12:   Theoretical predictions made with the "Hot
           Spot" code of the time history of the various
           Al ion populations.

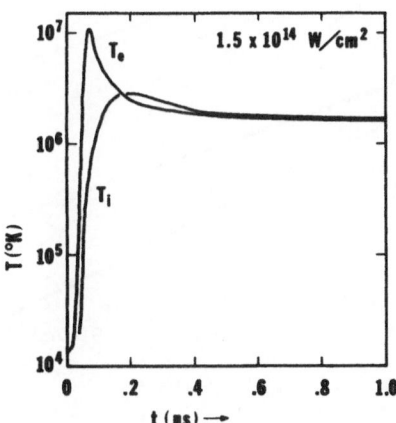

Fig. 13: Theoretical predictions made with the "Hot
Spot" code of the time history of T_i and T_e.

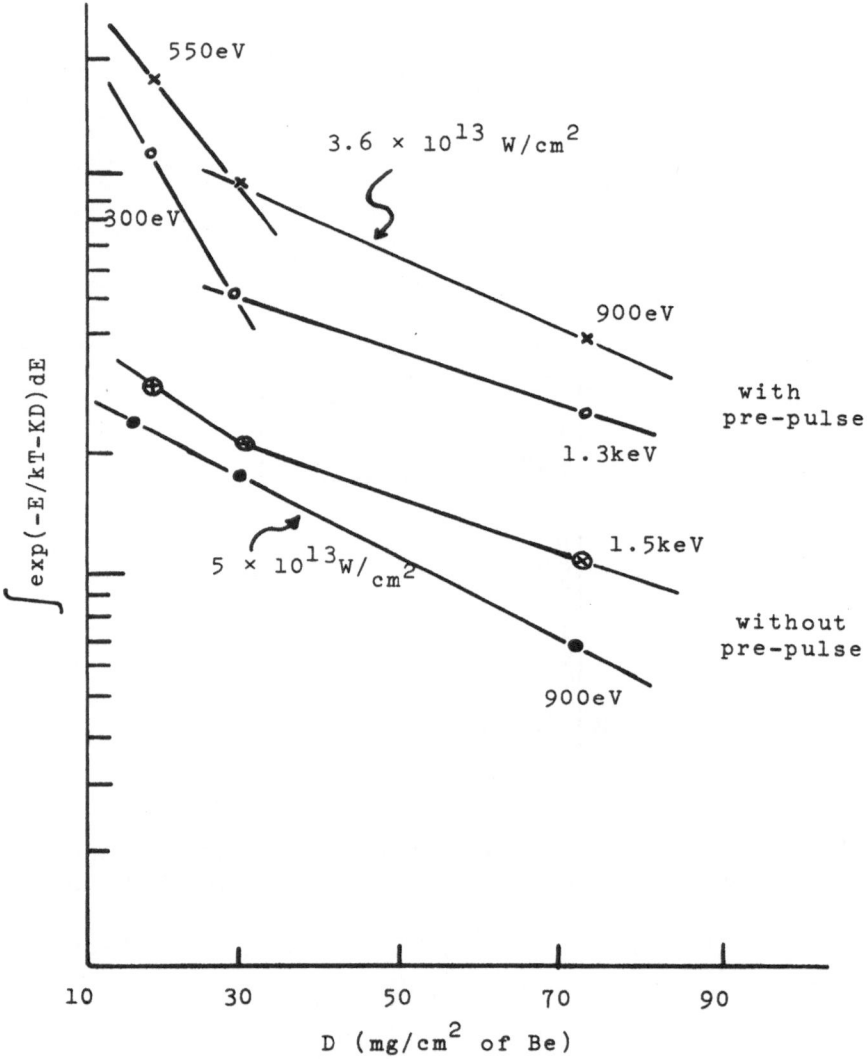

Fig. 14:   Experimental data showing the effect of a pre-
           pulse on the total x-ray output of the laser
           produced plasma.   Both pulses are 30 psec in
           duration; the prepulse is 1/3 of the main
           pulse which follows the prepulse by ∿ 6 nsec.

heated by picosecond laser pulses is potentially a means of generating gain in the soft x-ray region. The approach uses isoelectronic projections of known laser transitions to higher Z and shorter wavelengths or in some cases electronic configurations unique to ionic species such as K-like ions with a (3d) ground state. The advantages of this approach are that there are no photionization losses in the laser medium; no Auger effects; and for the case of self-terminating lasers, collisional effects are not limiting in a first approximation. In the self-terminating case very high gains can be achieved.

The disadvantages are that the electron heating is probably limited to $\sim$ 1 KeV and hence the laser transitions to $\sim$ 100 eV. Isoelectronic projections are difficult over large ranges of Z due to level crossing and anomalous effects. This approach tends to be inefficient at longer wavelengths since the densities are low and the laser absorption low. Finally, it is very difficult to obtain high densities of a single "cold" ion species which is the optimum media for this approach.

ACKNOWLEDGEMENT

This paper describes work done as part of the NRL/ARPA x-ray laser program which is a combined effort of many individuals besides the author. These include Drs. R. Elton, J. Reintjes, and T. Lee and Mr. R. Eckardt who were involved with the work described in this paper. The NRL "Hot Spot" code was developed by Drs. J. David and K. Whitney of the NRL Plasma Physics Division.

REFERENCES

1.   I. N. Knyazev and V. S. Letokhov, Optics Comm. $\underline{3}$,
     332 (1971).

2.   T. P. Donaldson, R. J. Hutcheon, M. H. Key, and
     C. Lewis, VII Intn'l Quantum Electronics Conf.
     Montreal, paper G.9, (1972).

3.   P. I. Peyraud, J. de Phys. $\underline{29}$, 88 (1968).

4.   P. I. Peyraud, J. de Phys. $\underline{29}$, 306 (1968).

5.   P. I. Peyraud, J. de Phys. $\underline{29}$, 872 (1968).

6.   J. B. Hasted, "Physics of Atomic Collisions"
     (American Elsevier, New York, 1972).

7.   R. C. Elton, NRL Memo Report No. 2799, (1974);
     Appl. Optics $\underline{14}$, 97 (1975).

8.   W. T. Walter, N. Solimene, M. Piltch, and G.
     Gould, IEEE J. Quantum Electronics $\underline{QE-2}$, 474 (1966).

9.   R. L. Kelly and L. J. Palumbo, NRL Report No.
     7599 (1973).

10. R. C. Elton in "Methods of Experimental Physics -
     Plasma Physics", Ch 4, Vol 9A, eds. H. R. Griem
     and R. H. Lovberg, (Academic Press, New York,
     1970).

11. W. L. Wiese, M. W. Smith, and B. M. Miles, Nat.
     Bur. Stand. Report NSRDS-NBS 22 (1969).

12. K. G. Whitney, J. Davis, J. Appl. Phys., $\underline{45}$, 5294
     (1974).

X-RAY AND γ-RAY LASERS

George Chapline and Lowell Wood

(Presented by George Chapline)

Lawrence Livermore Laboratory

University of California, Livermore, California

The x-ray laser talks heard so far have concentrated on the problem of producing lasing action in the region 1000 A° > λ > 10 A°. Since this region is the one for which experimental demonstrations of lasing are most likely in the near future it is, of course, proper that most work on x-ray lasers concentrate on this regime. In this talk, however, I will look a little further off in the future and discuss the prospects for producing coherent radiation at wavelengths λ < 10A°. Since coherent radiation is most easily produced by producing a population inversion between discrete levels of a radiator, it is most natural to think of generating short wavelength coherent radiation by means of an x-ray laser (XRL) using innershell atomic transitions or a γ-ray laser (γRL) using transitions between discrete levels of nuclei.

Lasing action at very short wavelengths will be hard to achieve because of three difficulties not present

at longer wavelengths. First, matter has a high opacity
at short wavelengths. The nearest analogue to a transparent optical medium such as glass is a hot, fully
stripped plasma where the opacity is due only to
Compton scattering. Second, the stimulated emission
cross-section, being upper-bounded by $\lambda^2/8\pi$, will in
general be smaller at shorter wavelengths. This implies
that rather high inversion densities will be required
to produce lasing action. In the case of x-ray lasers
this in turn means that enormous pumping powers will be
required. In the case of γ-ray lasers using long lived
isomers high pumping powers may not be required, however
even under the most optimistic assumptions, inversion
densities must be close to solid densities. Third,
a laser operating at very short wavelengths must do so
without the benefit of mirrors. This is due both to
the fact that all materials have a low reflectivity for
for $\lambda < 10$ A$^\circ$ and also to the fact that an XRL or γRL
would operate at such a high flux level that any mirrors
would be destroyed.

If a laser is to operate without mirrors then the
gain down the length of the laser must be large enough
to allow a traveling wave to grow by stimulated emission.
In practice, the gain will probably have to exceed
100db; i.e., the small signal gain coefficient α times
the length ℓ of the laser must satisfy:

$$\alpha\ell \gtrsim 25 \quad . \tag{1}$$

For the case of an XRL using an allowed transition,
Eq. (1) translates into the following threshold condition on the net inversion density $N^* = N_2 - (g_2/g_1)N_1$:

$$N* > \frac{\Delta\nu(eV)}{\ell(cm)} \, 10^{18} \ cm^{-3} \qquad\qquad (2)$$

where $\Delta\nu(eV)$ is the width of the transition measured
in electron volts.

The exact inversion density required for an x-ray
laser will be determined by the line width $\Delta\nu$. In
cold matter this will be equal to the Auger width which
for transitions of interest will be on the order of 1
eV. Thus the net inversion density in cold matter must
exceed $\sim 10^{18} \ cm^{-3}$. In hot matter the width will be
determined by the Stark effect at most densities of
interest (at low densities the Doppler width will be
important). In a fully stripped high Z plasma this
width will be given by

$$(\Delta\nu)_{Stark} \approx 30\eta^{2/3} \ eV \qquad\qquad (3)$$

where η is the density in units of $5 \cdot 10^{22} \ cm^{-3}$. The
main conclusion to be drawn from Eqs. (3) and (2) is
that inversion densities much smaller than $10^{18} \ cm^{-3}$
are possible only if $N*/N$ is close to unity; i.e., one
has a very efficient scheme for producing population
inversions. The charge exchange scheme being pursued
by the group at the University of Arizona may actually
be efficient enough to allow threshold inversion
densities low enough to be attainable in low Z ion beams,
and thus be a means for generating very soft coherent
x-rays. As one goes to higher Z, however, the attainment of the necessary inversion densities with ion beams
will become extremely difficult. For most kinds of
pumping schemes proposed to date it is difficult to
achieve fractional inversions much larger than 10^{-3};
thus threshold inversion densities for XRL's will

typically exceed 10^{19} cm^{-3}.

   A schematic picture of an XRL is shown in Fig. 1.
The pumping source -- which might be a focussed laser
beam or the x-rays produced by focussing a laser on a
nearby high Z target -- is swept along the lasing medium
at the speed of light, in order to minimize the pumping
energy required. It can be shown that in order to pump
an XRL operating at wavelengths $\leq$ 10 A° the pumping
laser must have a power > 10^{12} W and the pumping laser
beam must be focused to a spot less than 30 μm across.
Such lasers are not now available but soon will be as
a result of laser fusion research.

   What about γ-ray lasers? For a given energy the
lifetime of a γ-ray transition will be much longer than
the corresponding allowed x-ray transition. This means
that, just as for forbidden x-ray transitions, the
threshold inversion will be higher than that given in
Eq. (2) by a factor $(\tau_{spon}/\tau_{allowed})$ where τ_{spon} is the
spontaneous radiative lifetime of the transition and
$\tau_{allowed} \sim (\hbar\omega)^{-2}$ is the lifetime of a corresponding
allowed x-ray transition ($\sim 10^{-13}$ sec. at $\hbar\omega$ = 1 keV).
As before, the actual threshold inversion densities will
be determined by the achievable line widths Δν, which
for γ-ray transitions are fortunately not so strongly
affected by their atomic environments. Indeed, in most
circumstances γ-ray line widths are equal to the Doppler
width. For a "vigorously" pumped γ-ray laser medium,
we would have $\Delta\nu = (\Delta\nu)_{Doppler}$ where

$$\left(\frac{\Delta\nu}{\nu}\right)_{Doppler} \approx 10^{-3} \left[\frac{\theta(keV)}{A}\right]^{\frac{1}{2}} \tag{4}$$

where θ(keV) is the temperature of the emitting nuclei
in kilovolts and A is the mass number of the nuclei.

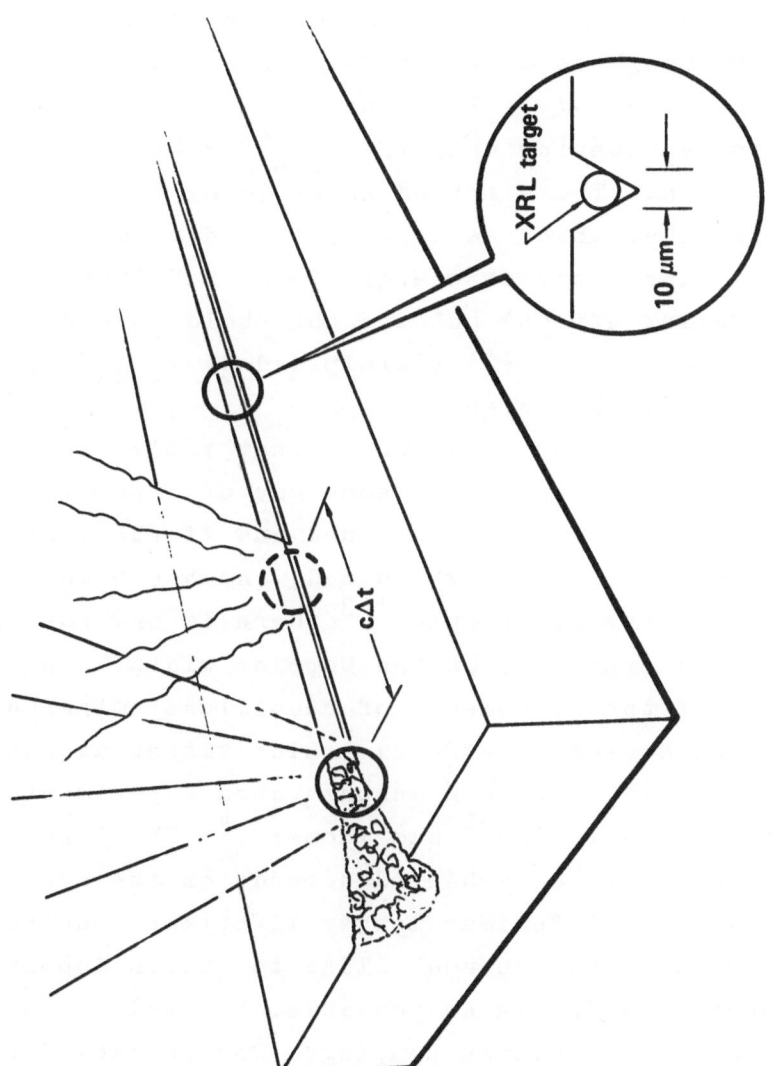

Figure 1 X-Ray Laser.

Unfortunately because γ-ray transitions are extremely
weak relative to x-ray transitions the gains implied by
Eq. (4) are going to be small even at low temperatures.
In fact, at temperatures that one can realistically expect in a vigorously pumped γ-ray medium the gain would
not be high enough to overcome Compton scattering. For
example, assuming $\Delta\nu/\nu = 10^{-4}$ one will be able to overcome Compton scattering only if $\tau_{spon}^{-1} \gtrsim Z(10Z)E1(M1)$
Weisskopf units. There are no known γ-ray transitions
with $\hbar\omega < 200$ keV which satisfy this condition. There
are strong transitions at energies ∿ 15 MeV (the
Goldhaber-Teller effect) but the threshold inversion
density implied by Eq. (2) (multiplied by $\tau_{spon}/\tau_{allowed}$)
would be unattainably high.

The conclusion to be drawn is that γRL's appear
feasible at this time only if some way of improving on
the Doppler line width can be found and if the γ-ray
laser medium can be "gently" pumped. As has been
pointed out by several people [1-3] there is one possible
way of greatly improving on the Doppler width: namely,
to make use of the phenomenon of recoilless emission in
crystals (the Mossbauer effect). Line widths as small
as 10^5 Hz have in fact been demonstrated experimentally
using short-lived $\tau \sim 10^{-6}$ sec isomers.[4] The line
width that needs to be achieved depends on the spontaneous radiative lifetime; longer lifetimes require
smaller widths. Thus we would like to use an isomer
with as short a lifetime as possible, but still long
enough to allow for gentle pumping. One possible method
of gentle pumping would be to excite long lived isomeric
states in, for example, a nuclear reactor and then con-
dense the excited isomeric nuclei into a crystal (see
Fig. 2). If one assumes that such a process would

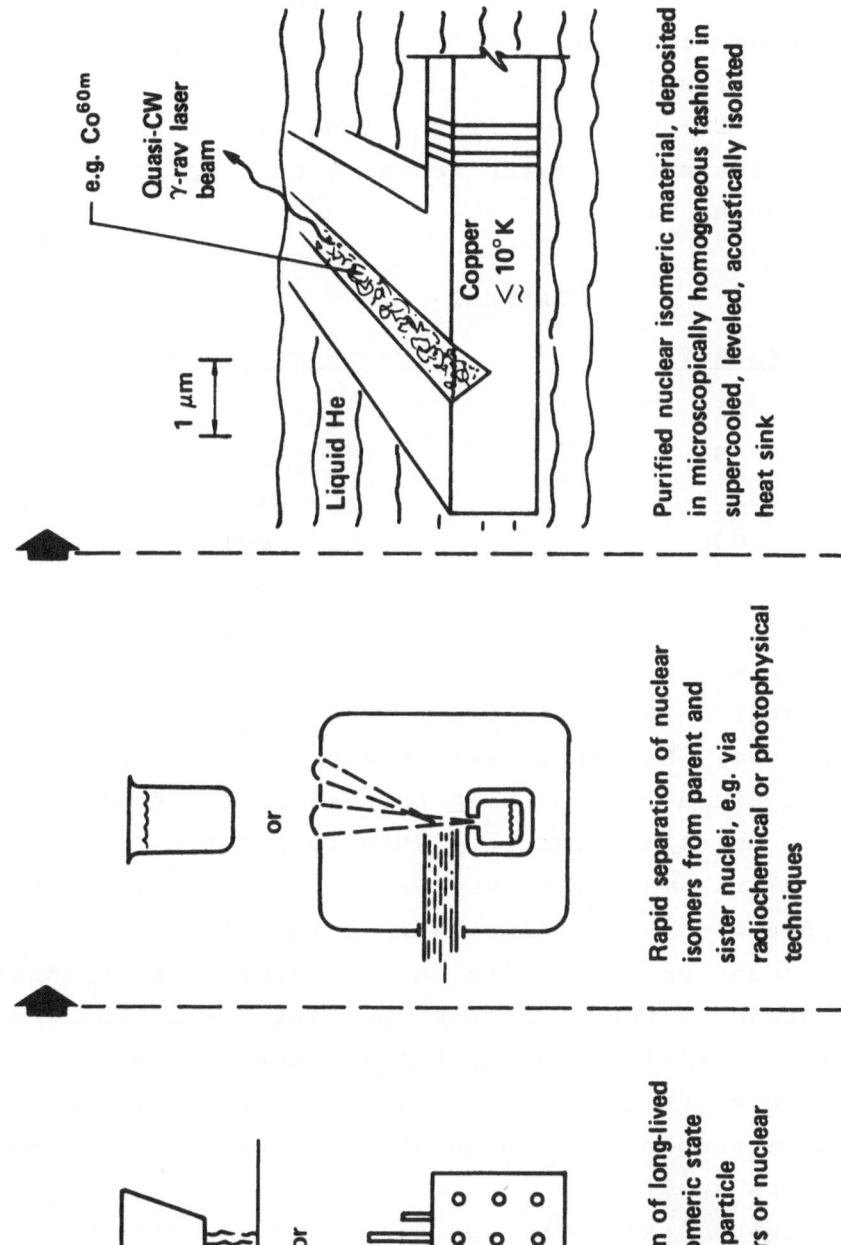

Preparation of long-lived nuclear isomeric state species in particle accelerators or nuclear reactors

Rapid separation of nuclear isomers from parent and sister nuclei, e.g. via radiochemical or photophysical techniques

Purified nuclear isomeric material, deposited in microscopically homogeneous fashion in supercooled, leveled, acoustically isolated heat sink

Figure 2 Gently Pumped γ-Ray Lasers.

consume at least 10 min. then one would be interested
in spontaneous radiative lifetimes on the order of 10^4
sec (the difference in times being due to internal
conversion). The maximum allowed line widths for such
a lifetime, assuming a laser medium 1 cm long, are listed
in the following table:

TABLE I

$\hbar\omega$(keV)	Maximum $\Delta\nu$(Hz)
10	10.
50	.3
100	.1
1000	.001

Whether such small line widths are attainable is not
known. They are probably not attainable in bulk samples
due to crystal imperfections and impurities. It has
been suggested,[5] however, that it might be possible to
attain line widths approaching natural line widths in
small, pure crystal whiskers. When techniques for
measuring ultra-small line widths become available, it
will be possible to experimentally pursue this approach.
The requirement on line width can, of course, be relaxed
if some method of rapid "gentle" pumping can be devised.
The Russian physicists Goldanskii and Letokhov have
made some interesting suggestions[6,7] on how to achieve
rapid gentle pumping, but none of these schemes has been
shown to work in detail.
        A computer search through all γ-ray transitions
whose energies and lifetimes are known has recently
been carried out at the Livermore Laboratory to identify
those transitions which might be used in a gently pumped

γRL. A total of 22 candidates were found, some of the
more promising of which are shown in Table II. The
quantity $\zeta \equiv (\lambda^2/8\pi) [(1+\alpha) \sigma_a]^{-1}$, where α is the internal conversion coefficient and σ_a is the sum of the
photoelectric absorption and Compton scattering cross-
sections, is a "figure of merit" for candidate nuclei.
It represents the multiple of the minimum attainable
line width, viz. $(1+\alpha)/\tau_{spon}$, for which the gain is
positive. Operation of a γRL at the maximum allowed
line width listed in Table I would typically require a
figure of merit of few hundred. Obviously, only for
values of ζ much larger than 1 is there a good chance
of being able to achieve the threshold condition for
laser action. Although the practical problems that
must be overcome in order to operate a gently pumped
γRL are formidable, the spectacular properties of such
a laser justifies a serious effort to overcome the
obstacles. Because of time-bandwidth limitations, a
gently pumped γRL would emit a pulse on the order of 1
sec duration. A gently pumped γRL would therefore be
a quasi-CW source of fantastically monochromatic radia-
tion with $1A° \gtrsim \lambda \gtrsim .1A°$. Such a laser would be ideal
for molecular holography.

ACKNOWLEDGEMENT*

The authors wish to acknowledge helpful discussions
with their colleagues at LLL concerning x-ray and γ-ray
lasers. We are particularly grateful to J. Katz,
J. Marling, J. Nuckolls, C. Rhoades, P. Sommerville, and
E. Teller.

---

*This work was performed under the auspices of the
 United States Atomic Energy Commission.

TABLE II

Isomer	k	λ	σ_{Max}	σ_{abs}	α	ζ	$\tau_{1/2}$	Class	Production
Co^{60m}	.059	.21	$1.8 \cdot 10^5$	120	41	37	10.7min	M3	Co^{59}, 100%; 18b
Se^{79m}	.096		$5 \cdot 10^4$	84	7	112	3.9min	E3	Se^{78}, 24%; 0.4b
Se^{81m}	.103	.12	$5.8 \cdot 10^4$	70	9	92	57 min	E3	Se^{80}, 49%; 0.1b
Br^{77m}	.108	.15	$5.2 \cdot 10^4$	70	6	124	4.2min	E3	Se^{77}(p,n)Br^{77}
Tc^{99m}	.143		$3 \cdot 10^4$	70	30	14	350 min	M4	Mo^{99}, β$^-$ decay

TABLE II. Nuclear isomers considered most promising for production of gently pumped YRL action. The gamma energy k in Mev, the corresponding wavelength λ in Å, the approximate maximum stimulated emission cross-section possible σ_{Max} in barns, the photoelectric absorption plus Compton scattering cross sections at the transition energy σ_{abs} in barns per atom, the total internal conversion coefficient α, the figure-of-merit of the isomer $\zeta \equiv \sigma_{Max}/(1+\alpha)\sigma_{abs}$, the excited state half-life $\tau_{1/2}$, the multipolarity of the radiative decay, and the most promising means of production of the isomer, including the fractional natural abundance and thermal neutron capture cross section (in barns) to the isomeric state of its precursor. Co^{60m}, Se^{79m} and Tc^{99m} are considered especially interesting as they are formed substantially population-inverted.

REFERENCES

1. Vali, V. and Vali, W., Proc. IEEE <u>51</u>, 182 (1963).

2. Baldwin, G. C., Neissel, J. P., and Tonks, L.,
    Proc. IEEE <u>51</u>, 1247 (1963).

3. Chirkov, B. V., Soviet Phys. JETP <u>17</u>, 1355 (1963).

4. Perlow, G. in <u>Perspective in Mossbauer Spectroscopy</u>,
    (Plenum Press, New York, 1973).

5. Wood. L. and Chapline, G., Nature <u>252</u>, 447 (1974).

6. Goldanskii, V., and Kagan, Y., Sov. Phys. JETP <u>37</u>,
    49 (1973).

7. Letokhov, V. S., JETP <u>64</u>, 1555 (1973).

Some of the participants of the Orbis Scientiae II in attendance during the High Energy Physics Session

SCATTERED RADIATION AND NEUTRON PRODUCTION IN LASER-PLASMA INTERACTION*

P. L. Mascheroni

Fusion Research Center

University of Texas at Austin, Austin, Texas

INTRODUCTION

The study of the interaction of a strong wave with a plasma is of interest to fusion physics. In laser-pellet-driven-fusion,[1] one heats a microsphere of properly tamped D-T fuel in less than a microsecond by electromagnetic radiation of the order of 10^{17} W/cm^2(Nd). The particle distribution function at the absorbing plasma layer (the energy deposition layer) provides the initial distribution function for the hydro-regime that describes the pellet, and hence is an input in its design (the pellet design and pulse shape must keep a spherical, symmetric hydro-regime). We can find many other examples of this,[2] like the heating of magnetically confined plasma by intense radio frequency fields,[3] current driven turbulence, etc. In current driven turbulence, for example in a Tokamak,[4] the heated particle distribution function provides the initial data for the study of the relaxation of the plasma to a Tokamak

*This work was supported in part by the U. S. Atomic Energy Comission Contract No. AT-(40-1)-4478.

contained one.

Nonlinear processes provide an important vehicle
for transfer of energy to the particles.[5] In laser
fusion at the deposition layer inverse Bremsstrahlung
is much smaller than the nonlinear heating for intensity
less than 10^{13}W/cm^2. In a turbulent heated Tokamak,
the turbulent resistivity is much larger than the classical if the drift velocity is larger than that of
sound.[4] Nonlinear processes have a long history[6] since
they have been observed in solids, liquids and gases.
In an inertia contained plasma (or laser created plasma)
the wide range for values of the density, temperature
and expansion velocity provide a quite complex area of
research for these processes, usually called parametric
instabilities,[6] since the initial radiation supported
by the plasma, the pump, is quite strong. These parametric processes are considered to occur in an unmag-
netized plasma, which may not be the case due to generation of magnetic fields[7] from particle streaming and
resonance absorption.[8] These parametric processes can
produce anomalous reflection or stimulated scattering
(Raman or Brillouin,[9] for example) or absorption through,
for example, decay instabilities: electron-electron and
electron-ion (ion acoustic instatiblity). We also have
stimulated Compton, filamentation, oscillating two stream,
etc. effects, which can be classified as absorptive or
reflective parametric instatiblities. The resonance absorption of a wave is a linear wave conversion due to in-
homogeneities.[8,10] In this case we also have reflection
of radiation. The absorptive processes are heating
mechanisms which produce fusion reaction, i.e., neutron
production for the reaction of interest.

The parametric processes are approximations. Due

to the different mechanisms that can produce harmonic
generation, frequency shifts, etc., we would rather take
a general approach. Diagrams seem to be useful to keep
a record of the processes considered.[11]

BASIC FRAMEWORK

Suppose that one can write

$$\psi = \psi_s + g\psi_t$$

where g is very small and ψ could be the fields $\vec{E}$ or $\vec{B}$
or a particle distribution function like $f_e(\vec{r}, \vec{v}, t)$.
The linearized version of the Maxwell Equation is[12]

$$\overleftrightarrow{D}(k,\omega),\ \vec{E}(\vec{k},\omega) = 0$$

which contains the physics of small waves. The
zeroes of the linear dielectric function give the dispersion relation for the eigenmodes of the plasma. Here
$\overleftrightarrow{D}$ is a function of ψ_s, and

$$\omega^2 \overleftrightarrow{D}(k,\omega) = c^2 (\vec{k}\vec{k}-\overleftrightarrow{I}k^2)+\overleftrightarrow{\varepsilon}(\vec{k},\omega)\omega^2 \tag{1}$$

where $\overleftrightarrow{\varepsilon}$ is the linear dielectric permittivity

$$\overleftrightarrow{\varepsilon} = 1+\overleftrightarrow{\chi}_e+\overleftrightarrow{\chi}_i ,$$

and the χ's are the susceptibilities to be calculated
from the linearized solution of the Vlasov equation.

The generalization in the case where one needs to
keep nonlinear terms due to large wave phenomena is[13]

$$\omega^2 \overleftrightarrow{D}(\vec{k},\omega)\cdot \vec{E}(\vec{k},\omega) =- \frac{4\pi i}{c^2}\ \omega J(\vec{k},\omega) \tag{2}$$

where $\vec{J}(k,\omega)$ is the nonlinear current density. In an isotropic plasma the susceptibilities can be diagonalized. Associated with this we have a longitudinal component

$$\chi_\sigma^\ell \ (\vec{k},\omega) \ = \ (4\pi e^2{}_\sigma/m_\sigma k^2) \int \frac{\vec{k}\cdot(\partial/\partial\vec{v})f_{s\sigma}d^3\vec{v}}{\omega-\vec{k}\cdot\vec{v}}$$

and a transverse one $\chi_\sigma{}^t$ (which we do not write). If f_s is Maxwellian, the χ's can be expressed in terms of the plasma dispersion function Z.

The nonlinear current can be obtained from a kinetic or fluid approach. This arises from terms like $\vec{v} \times \vec{B}$ in the Vlasov or the fluid equation, or from $\vec{v} \cdot \nabla\vec{v}$, $\nabla\cdot(n\vec{v})$ in the fluid description. For the time being let k be a symbol which indicates $\{\vec{k},\omega\}$. In the general case we have many pump frequencies: let $\mu_{k_o} \ll 1$, where $\mu_{k_o} = 2e\vec{k}_o \cdot \vec{E}_k/m\omega_o^2$. The nonlinear current in the Fourier transform is

$$4\pi\vec{J}_k = \sum^{(k)}_{k_1} (1+\chi_{i_{k_1}})\omega_1 \vec{k}_1\cdot\frac{\vec{E}_{k_1}}{\omega_1} \vec{V}_{k_o} +\chi_{e_k} \omega\vec{k}(\frac{\vec{E}_{k_1}}{\omega_1} \cdot\vec{V}_{k_o}), \qquad (3)$$

where the summation is over all the values of k_1 and k_o such that $k = k_1+k_o$, $\vec{V}_{k_o} = e\vec{E}_{k_o}/m\omega_o$, and the χ's indicate the electron or ion susceptibility to be calculated by using the expression in terms of the plasma dispersion function. The Maxwell wave equation for the electric (or magnetic) field with the expression for the nonlinear current given above but with just one pump, k_o, correctly describes the parametric instabilities as well as the wave conversion of interest here.

As an example we now consider stimulated scattering in an homogeneous medium from ion modes which are eigenmodes in the presence of the pump. Using

equations (1) and (2) and (3) after truncation and separation of the fields into longitudinal and transverse
components, we have

$$\omega^2 \epsilon \vec{A} = -i\chi_e \omega \vec{k}(\vec{A}_{-1} \cdot \vec{V}_o),$$

$$[-c^2 k^2_{-1} + \omega^2_{-1} \epsilon_{-1}]\vec{A}_{-1} = -i(1+\chi_i)\omega(\vec{k} \cdot \vec{A})\vec{V}_{-o}, \qquad (4)$$

where $\epsilon = 1+\chi_e+\chi_i$, $\vec{A} = \vec{E}_k/\omega$, and the two-mode coupling
is valid if

$$|\vec{k}| \simeq 2|\vec{k}_o|\sin\theta/2; \cos\theta = (\hat{k}_o \cdot \hat{k}_1) = \phi.$$

In a weakly inhomogeneous medium in which the inhomo-
geneities are a function of one coordinate, say x, it
is still possible to keep the classification of the
fields in longitudinal and transverse, as we will describe later. Equations (4) represent a 3-wave parametric coupling or interaction when the pump depletion
is ignored. A representation of this interaction is[11]

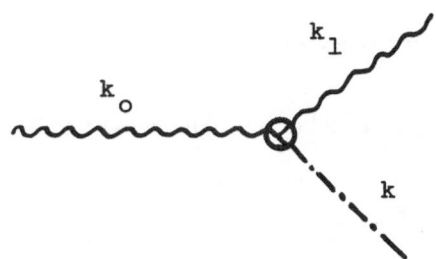

The wiggly line describes the photon field, the point-
broken line. _._._ a phonon. (A broken line - - -,
a plasmon.)

Consider now the region where $\omega_0 \sim \omega_p$, i.e., where anomalous absorption takes place. For $k\lambda_D \ll 1$ the dielectric permitivity is

$$\varepsilon = \varepsilon_0 - 3v_e^2 \frac{\vec{k}\vec{k}}{\omega^2}$$

$$\varepsilon_0 = \overleftrightarrow{I} (1 - \frac{\omega_p^2(x)}{\omega^2}) .$$

Neglecting the nonlinear current, from Maxwell's equations we have

$$\omega^2 \overleftrightarrow{\varepsilon} \cdot \vec{E}_\omega = c^2 \nabla \times \nabla \times \vec{E}_\omega$$

$$\nabla^2 \vec{B}_\omega + \nabla \ell n \ \varepsilon_0 \times \nabla \times \vec{B}_\omega + (\omega^2/c^2) \ \varepsilon \vec{B}_\omega = 0. \qquad (4a)$$

Then if $\vec{k} \rightarrow -i\vec{\nabla}$ and

$$\vec{B}_\omega = \hat{e}_z B(x) \exp(i\chi y), \vec{E}_\omega = \hat{e}_x E(x) \exp(i\chi y) + \hat{e}_y E_y(x) \exp(i\chi y);$$

$$\beta^2 = 3v_e^2/c^2, \ \chi = k \sin\theta.$$

One obtains Piliya's equations[10]

$$H'' - (\varepsilon'_0/\varepsilon_0)H' + (k^2\varepsilon_0 - \chi^2)H = -k\chi\varepsilon_0 E - \chi^2 H,$$

$$H = B + \chi/k\beta^2 E, \qquad (5)$$

$$\beta^2 E'' + k^2\varepsilon_0 E = -k\chi H.$$

Assume a linear profile $\omega_p^2(x) = \omega_p^2(0)(1+\frac{x}{L})$, and define $p = \frac{x}{L}(kL)^{2/3}$, $q = \chi^2(kL)^{2/3}/k^2$. Piliya's solution is

in terms of the Airy function $H(p) \simeq -\sqrt{p}\, A_i(q+p), -p\gg1$
and $H(p) \sim \sqrt{-p}\{c_+A_+(q+p) + c_-A_-(q+p)\}$, $-p\gg1$ where

$$A_\pm = A_i(x)\pm B_i(x), \quad \S = p\beta^{-2/3},$$

thus:

$$E = -(\chi/k)(kL)^{2/3}\beta^{-2/3}H(p)\{\int_0^\infty e^{\S t-t^3/3}dt-i\pi[A_i(\S)-iB_i(\S)]\}$$

$$(6)$$

The reflection is $|c_-/c_+|^2$, and the absorption of the
radiation is $A = 1 -|\frac{c-}{c+}|^2$. This, as calculated by various
investigators, is about 50% with $kL\sim100,$[10] $10°\leq\theta\leq20°$,
and is independent of β. The light is thus reflected and
generates a nearly backward longitudinal wave of the
same frequency. Since the absorption is independent of
β one can study the case $\beta\ll1$ and extrapolate most of
the results to larger β. For example, it is easy to
observe the singular behavior of the longitudinal field
near $x = 0$ when $\beta\ll1$. As long as $\beta \neq 0$ Eq. (6) shows
a coupling of the "longitudinal" component E and the
perpendicular component B due to the resonance. A
process of this type will be indicated by a diagram

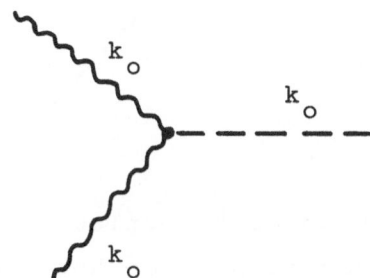

which indicates that the incident radiation is reflected
and generates a longitudinal wave with the same frequency. By changing the boundary conditions we can describe the reflection of a plasma wave and generation of

a photon wave.

In the case that the radiation is over the threshold for parametric effects one has to consider $j \neq 0$.
Then we have

$$(1-\beta^2)\nabla\nabla\cdot\vec{E}-\nabla^2\vec{E} = k^2\epsilon_o\vec{E}+\frac{4\pi i\omega}{c^2}\vec{j}$$

which leads to

$$\beta^2 E'' + k_o^2\epsilon_o E = -k\chi B - \frac{4\pi i\omega}{c^2} j_x$$

and a corresponding equation for B. Through the nonlinear current one has coupling with other modes, for
example, ion modes. We can represent this type of interaction by a diagram:[14]

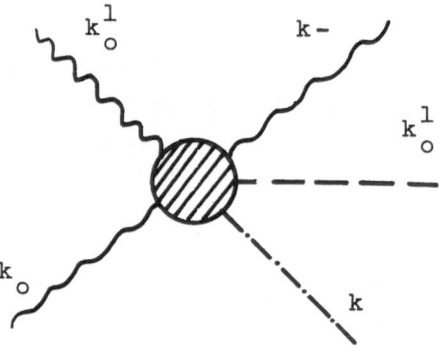

which implies a system of 4 coupled second order equations. If the plasma is not too hot and the pump has a
moderate power, linear wave conversion is dominant and
the excited electrostatic mode can be a pump for a
parametric decay. This is a separable process which we
schematize as [11]

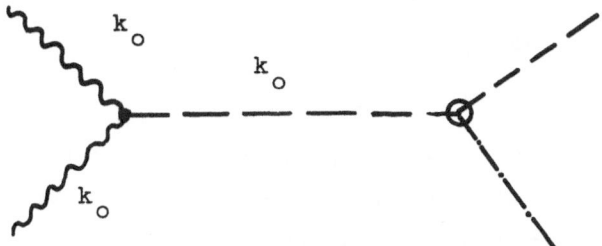

and treat each vertex as independent of the other.

Since the vertex is a function of the susceptibility, (or local plasma density) which if the electrostatic field is intense is strongly modified,[15,16] one can expect another source of nonlinear effects through the requirement of self-consistency for the density profile. The cavities in the density profile can trap waves and a new source of regenerative effects may be obtained.[16] For the description of caviton formations due to resonance absorption (or the oscillating two stream), one can use the electrostatic field near the resonance coupled with itself,[17] i.e.,

$$\omega^2 \epsilon E = -\frac{\omega_p^2 \delta n_e}{n_o} E .$$

Since the strongest component in f_s is that due to the ponderomotive force[18] we have

$$\delta n_e = -\frac{\chi_e}{4\pi} \frac{k^2}{m\omega_o^2} |E|^2; \quad \chi_e = \frac{1}{k^2 \lambda_D^2} .$$

Then

$$2i\omega_o \frac{\partial E}{\partial t} + 3V_e^2 \nabla\nabla \cdot E + (\omega_o^2 - \omega_{pe}^2)E = \frac{\omega_p^2}{4\pi N T_e}|E|^2 E$$

the solution of which describes the solutions. In the
presence of a pump E_o,

$$E = (E_o + \tilde{E})e^{i\omega_o t}$$

one obtains a generalized equation,[19] which describes the
"spikons"; (the addition of a damping term is straight-
forward). This entity is characteristic of strong
electron plasma turbulence.[20]

A glance at Eq. (2) tells us that the general
structure of the equation for mode coupling can formally
be written as

$$\omega^2 \vec{\varepsilon}^{\ell} \cdot E^{\ell} = -i\omega \sum \chi^{(k)\vec{\ell}}(k; k_1, k_2) \cdot \vec{E}(k_1) \cdot \vec{E}(k_2)$$

$$-(k^2 c^2 - \omega^2 \vec{\varepsilon}^t) \cdot \vec{E}^t = -i\omega \sum \chi^{\vec{t}}(k; k_1, k_2) \cdot \vec{E}(k_1) \cdot \vec{E}(k_2)$$

where the χ's are quantities which can be obtained from
expression of the nonlinear current. Thus an eigenmode
of a plasma of eigenfrequency ω will couple with a big
wave ω_o (the pump, a resonant wave, etc.) to produce new
eigenfrequencies $\omega + j\omega_o$, $j = 1,2,\ldots$etc. This leads
to harmonic generation. There are many ways of obtaining
a given harmonic and graphical representations are
straightforward to draw. If density modifications are such
that ion waves are invoked in the wave transformation process it seems natural to include this wave in a diagram,
hence we have for example

SCATTERED RADIATION AND NEUTRON PRODUCTION

The study of the scattered radiation, "back-scattered", "sidescattered", etc. depending on the angle has received considerable attention from experimentalists.[21] We will consider here in some detail the experimental results reported by Bobin (since they were the first to come to our attention) in which correlations with the neutron production and hard x-rays were found.[21b]

Bobin, et al., observed the light emitted in a backward direction as well as 45° from this direction, from the interaction of an Nd-glass laser and a solid target (H_2 or D_2). Intense lines were found at $2\omega_o$ and $(3/2)\omega_o$. These lines when observed at 45° were found to have a broadened spectrum toward lower frequency. Although no spectral study has yet been done for the backward spectrum, we will assume that it is the same as that for the 45° direction. The polarization of the wave in the backward direction is, also, the same as for the incident. Furthermore, they have measured the intensity of these lines as a function of the position of the focal spot on the target. The neutron production is then correlated with the $2\omega_o$ line and the hard x-rays with that of the $(3/2\omega_o)$.

In this section we propose a mechanism to explain the backscattered radiation and neutron production,[22] with some numerical estimates, and in the next, some alternatives.

The graph we consider here is

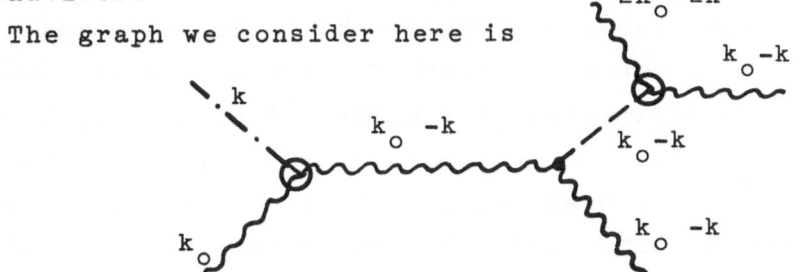

Giving more details than can be inferred from reading
the figure, we consider the following scheme. A parametric nearly forward Brillouin decay produces ra-
diation at $\omega_o-\omega_a$, where ω_o is the pump frequency and
ω_a the ion wave. Due to the density gradient, the
linear wave conversion produces the backscattered light
at $\omega_o-\omega_a$ and by a non-linear effect the harmonic at
$2(\omega_o-\omega_a)$ is then generated. The strong pump drives the
low frequency modes ω_a producing a broad spectrum of
them. The spatial domain where these waves are produced
goes up to the critical density where through the
electron-ion decay the pump also drives a broad spectrum
of ion waves. The direct turbulent heating of the ions
then produces the neutron yield. The spatial structure
of the focal region gives a length in which the laser can
be focused and still allow radiation over the threshold
for parametric effects to reach the critical density region where neutrons and the second harmonic are generated.
This explains the correlations. The direct turbulent
heating of the electrons will not explain the clean
correlations in Bobin's experiments. Numerical estimates
seem to support this picture.

The plasma density profile, expansion velocity, and
temperature are, presumably, a function of the position
of the focal spot and the spatial-temporal structure of
the laser pulse. In our model we neglect the dependence
of the position of the focal spot on the characteristic
of the plasma produced. Consequently, in this approximation we have the same plasma for the different shots
in Bobin's experiments. For numerical estimates we use
the spatial structure of the laser pulse at the focus
by Sigal,[21] which uses the same kind of lens and a laser
of similar characteristics to that used by Bobin. When

one considers the dependence on the shape of the laser
pulse of the processes considered here, there is room for
explanation of disagreement between different experimental
results. For example, if the length of the pulse is too
short there may not be time for temporal instabilities
to grow, say near saturation (which since the plasma is
collisionless must be from nonlinear effects), and
convective effects may be dominant.[23] Also the spatial
length of the pulse, about 200μ, is large enough that
(in some approximation) we may apply Rosenbluth's criteria
for the boundary conditions below.[24,25]

We then start with Eq. (4). To transform back to
the $(\vec{x},t)$ space, one replaces $\vec{k}$ and ω by $-i\nabla$ and $i\frac{\partial}{\partial t}$
respectively, so that χ is represented by a different
operator:

$$-\frac{\partial^2}{\partial t^2}\epsilon(i\frac{\partial}{\partial t},-i\nabla)A(x,t)=-i\frac{\partial}{\partial t}[\chi_e(i\frac{\partial}{\partial t},-i\nabla)]\nabla[\vec{A}_{-1}(x,t)\cdot\vec{V}_o]$$

with a similar equation for $\vec{A}_{-1}$. There are two cases of
interest. One is stimulated Brillouin scattering
(SBS) which corresponds to a regime in which the
eigenfrequency $\tilde{\omega}$ for the ion mode is in the range
$\omega_{pi}^2<<\tilde{\omega}^2\gtrsim k^2c_s^2$, $\tilde{\omega}$ = Realω, $k\lambda_D<<1$, $c_s^2 = T_e/M$. If $\tilde{\omega}\sim kc_s$,
the light scatters a natural ion mode. If $\tilde{\omega}>kc_s$ the
scattering is due to a pump driven mode. Hence

$$(-\frac{\partial^2}{\partial t^2}+c_s^2\nabla^2)\vec{A} = -i\frac{\partial}{\partial t}\nabla(\vec{A}_{-1}\cdot\vec{V}_o) \qquad (7)$$

$$(c^2\nabla^2-\omega_p^2-\frac{\partial^2}{\partial t^2})A_{-1} = i(1+\chi_i)\vec{V}_{-o}\frac{\partial}{\partial t}\nabla\cdot\vec{A} \qquad (8)$$

The other is stimulated Compton scattering (SCS):
$\omega_1-\omega_0 \simeq kV_i$; V_i being the ion thermal velocity. In this

case since ε is nonresonant equation (7) has to be re-
placed by

$$\frac{\partial^2}{\partial t^2}\vec{A} = i\frac{\partial}{\partial t}\frac{\chi e}{\varepsilon}\nabla(\vec{A}_{-1}\cdot\vec{V}_o) \qquad (9)$$

with the same equation (8) for the electromagnetic field.
With the proper modifications both cases can be treated
simultaneously. Hence we consider a plasma which is
weakly inhomogeneous where the density, temperature,
and expansion velocity are functions of a coordinate x.
Following Rosenbluth,[24] one writes $\vec{A}(r,t) = \vec{a}(x,t)$ exp
$i[\vec{k}(o)\cdot\vec{x}-\omega t+\int_o^x\Delta k(x)dx]$, with a similar expression for
$\vec{A}_{-1}$, and substitute this in the above equations. Here
ω is the real part of the eigenfrequencies $\vec{k}(x) = \vec{k}(o)$
$+\Delta k(x)$, etc. and $\omega_o = \omega_1+\omega$. If the frequency shift due
to the pump is small then $\omega\sim c_s k$. For pump driven modes
then $\omega_o-\omega_1 >> kc_s$ and k(x) is set to zero because there
is no first order dispersion relation for the unperturbed
mode. The equation for the longitudinal field with
damping is

$$[\Omega^2+2i\Gamma\omega+2ic_s^2 k\phi\frac{d}{dx}+c_s^2\frac{d^2}{dx^2}+2i\omega\frac{d}{dx} - \frac{d^2}{dt^2}]a =$$

$$= -i\ ka_{-1}V_o\alpha\exp\ iK'x^2/2$$

where

$$K(x) = k_o(x)-k_1(x)-k(x) = K'x, \Omega^2 = \omega^2-c_s^2(x)k^2, \alpha=(\hat{V}_o\cdot\hat{A}_{-1}).$$

We assume $\frac{da}{dx} < ka, \frac{d^2a}{dx^2} < k\phi\ \frac{da}{dx}$. For near forward
Brillouin scattering one needs to check that k is not
so small that the approximation breaks down. For the
electromagnetic wave we have

$$(\frac{\partial}{\partial t}+\Gamma_{-1}+V_{-1}\frac{\partial}{\partial x})a_{-1}(x,t)=(1+\chi_i)V_{-o}\frac{\omega}{2\omega_{-1}}k\alpha\exp(-iK'x^2/2) \quad (9a)$$

When one neglects the second derivatives in (2), we obtain

$$(\frac{\partial}{\partial t}+\tilde{\Gamma}+V\frac{\partial}{\partial x})a(x,t) = -(ka_{-1}V_o\alpha/2\omega)\exp[iK'x^2/2] \quad (9b)$$

$\tilde{\Gamma} = \Gamma-i\Omega^2/2\omega$ and the V's are the group velocity in the x direction. Equations (9) are Rosenbluth's equations, if $\Omega = 0$ which is the case of non-driven modes.

Here, we consider the case of light scattered in the near forward direction with damping included. Hence $V_{-1}V>0$ and we consider an initial value problem where at t = 0 a fluctuation at the origin is given. We perform a Laplace transformation with P the transform variable, eliminate a and write

$$A = (P+\Gamma_{-1})/V_{-1}-(P+\tilde{\Gamma})/V, \lambda = \gamma_o^2/(V_{-1}VK'),$$

$$\gamma_o^2 = (1+\chi_i)V_o^2K^2\alpha^2/4\omega\omega_{-1} .$$

We have $[d^2/dx'^2+\frac{1}{4}x'^2+(\frac{1}{2}-\lambda)]\bar{a}_1(x) = c\delta(x)$, where c is a constant. The solutions for x>0 well behaved at ∞ are the parabolic cylindric functions $D_{i\lambda}(x'e^{i\pi/4})$, and $D_{-i\lambda-1}(x'e^{-i\pi/4})$. A linear combination of them is null at the origin. If this is multiplied by the step function, the solution is null for x<0 for the decay waves. For strong pumping $\gamma_o>2(V_{-1}V)^{\frac{1}{2}}(\Gamma/V+\Gamma_{-1}/V_{-1})$; with $\lambda\gg1$ the peak amplification is $\lambda[\pi-2/\sqrt{\lambda})(\Gamma_{-1}/V_{-1}+\Gamma/V)]$.

Consequently, even if the damping is strong in the regime defined above, we need to consider only the amplification factor λ. For purposes of comparison, we shall consider only λ and ignore damping from now on.

We consider first the case $\Omega^2 = 0$, i.e., SBS from natural ion modes. Liu, Rosenbluth and White have found[24] that (a) for near-forward scattering $\lambda \simeq (V_o/V_e)^2 (\omega_p/\omega_o)^2 k_o L_T \alpha^2$; (b) for near-side scattering $\lambda \simeq (V_o/V_e)^2 k_o L_N \alpha^2$; and (c) for near backscattering $\lambda \simeq (V_o/V_e)^2 (\omega_p/\omega_o)^2 k_o L_W \alpha^2$; where L_T, L_N and L_W are the temperature, density, and expansion velocity scale length $L_N^{-1} = \frac{d}{dx} \ln N$, etc., ($V_e$ is the electron thermal velocity). To obtain the above expressions one writes $\omega = -\vec{k} \cdot \vec{w} + k c_s / (1 + k_2^2 \lambda_D^2)^{1/2}$ for the ion wave, where $\vec{W}$ is the expansion velocity, and observes that $k \sim 2k_o \sin \theta/2$. We will use the scaling $L_T > L_N$, $L_T > L_W$, depending in what region of the plasma density profile we are considering the process since the electrical thermal conductivity may behave in an anomalous way[26] (due to particle streaming, self-induced magnetic fields, etc.). With the scaling above forward scattering is dominant in the region defined by $\omega_p^2(x) > \omega_o^2 L_N/L_T, L_T > L_W$.

Consider now the case in which $\Omega^2 \sim \omega^2, \omega^2 >> k^2 c_s^2$ i.e., the case of ion-driven modes. As was stated before the mismatch $K(x)$ does not depend on the ion wave, and the growth rate at the beginning is dominated by side scattering $\lambda \sim (\frac{V_o}{V_e})^2 k_o L_N \alpha^2$, $\omega < \omega_{pi}$.

When the pump strongly drives eigenmodes ω such that $\omega_{pi} \gtrsim \omega >> k c_s$ the convective term can be ignored and then, we have

$$\frac{\partial^2}{\partial t^2} \vec{A} = i \nabla \frac{\partial}{\partial t} (\vec{A}_{-1} \cdot \vec{V}_o) ,$$

which we replace in Eq. (8) to get

$$(\mathrm{d}^2\nabla^2-\omega_p^2-\frac{\partial^2}{\partial t^2})A_{-1} = -k^2(1+\chi_i)V_o^2\alpha^2A_{-1}.$$

With the WKB approximation for the electromagnetic wave, we get

$$\frac{\partial A_{-1}}{\partial u} = -ik^2(1+\chi_i)V_o^2\alpha^2A_{-1}/2\omega_{-1} \quad , \quad u = t+x/V.$$

Thus, the instability growth rate is
$$\gamma = \sqrt{\pi}\ (T_e/T_i)(\omega^2_{pi}V_o^2/c_s^2\omega_o)[(\omega/\sqrt{2}kV_i)\exp(-\omega^2/2k^2V_1^2)]$$
with $k\lambda_D<<1$. This is not modified by gradients and, necessarily has a maximum in the backward direction.

Following similar steps we obtain that SCS gives $\dfrac{\partial}{\partial u}A_{-1} = \dfrac{-i(1+\chi_i)V_o^2\alpha^2k^2}{2\omega_{-1}\varepsilon}A_{-1}$ which gives a growth rate $-\dfrac{V_o^2k^2}{2\omega_{-1}}I_m\dfrac{(1+\chi_i)\chi}{\varepsilon}$ and which is again not modified by gradients. Two limits are of interest. For $k\lambda_D<<1$, we have $\gamma \simeq (1/2)(T_e/T_i)(V_o/c_s)^2(\omega^2_{pi}/\omega_o)\alpha^2$ which is independent of the angle and of the same order as the result for SBS. For $k\lambda_D>>1$ we get $\gamma\simeq\sqrt{\frac{\pi}{2}}e^{-1/2}(V_o/c_s)^2(k\lambda_D)^{-4}\times$ $(T_e/T_i)\alpha^2\omega^2_{pi}/2\omega_o$ which shows a maximum near forward scattering. It can then be inferred from the results presented above that: (1) The electronic thermal conductivity, which determines the value of L_T, makes forward Brillouin scattering dominant over side or backscattering (as well as over self-focusing, not discussed here). On the other hand, computer simulations for homogeneous plasmas have shown that the unstable ion wave grows and breaks and allows the radiation to penetrate deeper into the plasma and so reach the critical density region. (2) That SCS[9] is not affected by gradients; for $k\lambda_D<<1$ it is rather isotropic and for

$k\lambda_D$>>1 it shows a maximum toward the forward direction.
(3) At strong pumping driven SBS allows the formation of a broad spectrum of ion waves which goes
up to ω_{pi}. These points support the picture already
outlined.

We shall consider next the forward oblique Brillouin
wave $\omega_o-\omega_a$, where ω_a is a frequency of the ion wave,
incident at an angle θ with respect to the density
gradient. From the previous section Eq. (6) we know
that absorption of radiation is about 50% with
$kL\sim100$, $10°\leq \theta \leq 20°$, and independent of β. Computer
simulation has indicated that the angular range is much
wider when one uses a self-cossistent density profile. The light is thus reflected and generates a
nearly backward longitudinal wave of the same frequency.
The reflected transverse and longitudinal waves are the
sources for the non-linear current at the second harmonic
which must be emitted at the reflected angle.

From Eq. (6) and the comments we have made concerning
β in the previous section, it follows that the strongest
component for the term in the current is that which involves the largest derivative over $E(x)$. Hence, we use
Eq. (4a) with a source term

$$(4\pi/c)\nabla x\vec{j} \simeq -(i/c)(\partial^2 E(x)/\ x^2)V_k\hat{e}_z, \quad V_k = eE_y/m\omega \ .$$

Consequently,

$$B''_2-(\epsilon'_2/\epsilon_2)B'_2+k^2_2B_2 = (i/c)E''V_k \equiv F(x) \ ,$$

$$\epsilon_2 = \epsilon_o(2\omega_o,x); \quad k_2^2 = (2\omega_o/c)^2(\epsilon_2-\sin^2\theta).$$

The WKB solutions of the homogeneous equation normalized

to unity when $x \to \infty$ (vacuum) are

$$B_2^{\mp}(x) = (2\omega_0\epsilon_2\phi/ck_2)^{\frac{1}{2}} \exp \pm i\int k_2 dx.$$

Then the solution is

$$B_2(x) = \int_{-\infty}^{x} \frac{F(x')B_2^{+}(x')}{\epsilon_2(x')} dx' B_2^{-}(x) + \int_{x}^{\infty} \frac{F(x')B_2^{-}(x')}{\epsilon_2(x')} dx' B_2^{+}(x).$$

Using Eqs. (5) one obtains for the ratio of the second
harmonic flux to the incident forward-scattered Brillouin
wave

$$\frac{s_2(-\infty)}{s_1} = \frac{e}{2m\omega_0^2\phi} {}^2 \left| \frac{1}{H(-\infty)} \int_{-\infty}^{\infty} dx' \frac{B_2^{+}(x')}{\epsilon_2} E_y E''(x') \right|^2 . \quad (10)$$

Vinogradov and Postovalov and Erokhin and Moiseev[27]
calculated this expression. To make a numerical estimate assume that the intensity of the laser light is
about 10^{15}W/cm^2, and $T_e \sim 1 KeV$. If $L_N \sim 100$ μm, and $L_T \sim 4L_N$,
the amplification factor is about 40 and the forward
Brillouin intensity is about 10^{12}W/cm^2. From the above
expression the coefficient of transformation into the
second harmonic for this value of the intensity is about
1/1000, at an angle of about 10°. Since 50% of the
Brillouin wave is transformed back into reflected radiation, the observed ratio of the second harmonic to the
reflected fundamental is about 1/500. We are ignoring
the contribution to the backscattered light at $\omega_0 - \omega_a$ of
the direct backscattered Brillouin radiation since we are
assuming $L_W \ll L_T$. If the main contribution for this wave
comes from a region in the density profile where $L_W \sim L_T$,
the backscattered Brillouin radiation has about twice
the intensity of the reflected light and has an asymmetric
broadening of the same type as the forward scattered
Brillouin Radiation. (However the angular depend-

ence for this wave and for that reflected from the critical density region are different, allowing for experimental differentiation.) The observed ratio in Bobin's experimental result is of the order of 10^{-2}, while that
of Lee, et al. is about 10^{-3}. The shift and asymmetric
broadening toward lower frequency (red) is due to the
excitation of ion waves, the spectrum of which increases
with increased intensity, and to SCS. (We are ignoring
the Doppler shift toward the blue due to the expansion
of the plasma.) A glance at our equations shows that
the scattered radiation should have the same polarization as the incident one, $\alpha \sim 1$, however the presence
of turbulence can modify this behavior.

The correlation of the neutron yield with the $2\omega_o$
radiation is due mainly to the spatial structure of
the focal region which has a radius of about 150 µm.
Ion waves which travel forward and are able to reach
the critical-density region may also contribute to the
neutron output. To explain the correlations we consider
direct heating of the ions[28] by ion waves as a result
of parametric instabilities: SBS, SCS, etc. The
electron ion decay in the presence of a strong pump has
also a broad spectrum of ion waves as a result of the
same kind of mechanism, i.e., pump driven modes. Assume
that $\eta^2 = E_o^2/4\pi \ NkT<1$; we have for the turbulent
heating

$$D(V) \sim \sqrt{\pi} \ (e/M)^2 \sum_{k\omega} \frac{|E_{k\omega}|^2}{\Delta\omega} \ \exp[-(\omega-kV)^2/\Delta\omega^2]$$

where $\Delta\omega = 2[1/3k^2 D(V)]^{1/3}$. We assume that $T_e \gtrsim T_i$, and
$\Delta\omega \sim \omega_{pi}$. The ion distribution function evolves as
$\partial f_i/\partial t = (\partial/\partial V)[D(V)(\partial/\partial V)f_i]$.

Then $\partial Ti/\partial t = \sum_k^{\omega_{pi}} e^2|E_{k\omega}|^2/M\Delta\omega \approx e^2 E_i/M\omega_{pi}$. Since parametric instabilities are the drive for the turbulence, we can roughly estimate E_i^2 by using the Manley-Rowe relations: $\Delta E_i^2/\Delta E_e^2 \simeq (m/M)^{\frac{1}{2}}$. Here $\Delta E_{e,i}^2$ is the energy stored in electron (ion) plasma waves and $(m/M)^{\frac{1}{2}}$ is the ratio of the frequencies. Consequently, if I is the pump intensity, we write $I \sim (c/8\pi)\Delta E_e^2 \sim (c/8\pi)E_e^2$. If the instability region has a length ℓ and the plasma moves through this region at the expansion velocity W, one then estimates an increase in the ion energy of order $\Delta T_i \sim (m/M)^{-\frac{1}{2}} \omega_{pi}(I/Nc)\ell/W$. Assume $\ell \sim 10\mu m$, $W \sim c_s \sim 3\times10^7 cm$ sec^{-1}. Then for $I = 10^{14} W/cm^2$ we have $\Delta T_i \sim 1KeV$. The total number of neutrons per unit volume and time is $1/2N^2 [\sigma V_r]_{Av}$, where V_r is the relative apeed and σ the ion cross section, and the brackets indicate the average over the distribution function that we are taking as Maxwellian. Consequently, the total number of neutrons is $1/2N^2 [\sigma V_r]_{Av}(\ell/W)\pi r^2 c_s \tau$, where r is the radius of the focal spot, c_s the speed of sound at the focal spot, and τ the duration of the laser pulse. In Bobin's experiments, the duration was 3×10^{-9} sec. Hence for the d(dn) reaction channel $[\sigma V_r]_{Av} \sim 8\times10^{-23} cm^3 sec^{-1}$ at $T_i \sim 1KeV$. If $r \sim 100\mu m$ and $c_s \sim W$, we get a neutron yield of 10^4. This is the order of magnitude observed.

ALTERNATIVE SCHEMES AND THE $3/2 \omega_o$ COMPONENT (OR LINE)

In the previous section we assumed that the ion acoustic parametric instability was the dominant process. Alternatively the resonance can be dominant and provide the pump for the parametric Brillouin scattering and decay. We have for example

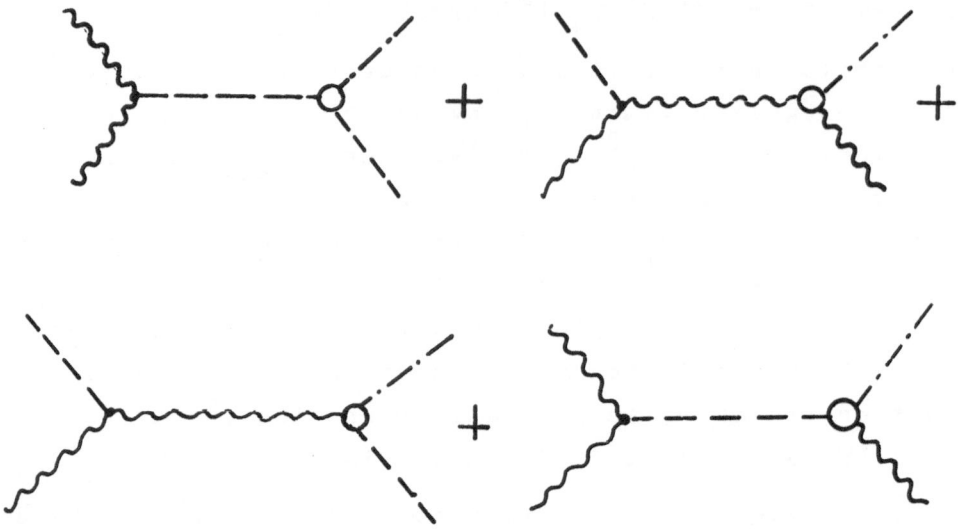

The parametric vertex in the last graph can be seen as
the inverse of parametric decay. We have seen
that linear wave convection produces density modifications. If the scale δ of this modification can be
considered a modulation of the plasma, one should
seek an eigenmode with a wavelength $\lambda \simeq \delta$, since
this will be the most unstable.[14] A quick estimate
can be made if we recall the comments about β and take
$\beta \ll 1$. The electrostatic field near the resonance is

$$E \sim E_o (2\pi k_o L)^{-\frac{1}{2}} L / (x + i\delta)$$

$$\delta = (\lambda_D^2 L)^{1/3}$$

where we take the optimal angles for resonance absorption. The ion acoustic wave near the resonance has a wave length $k_a = \frac{2\pi}{\delta}$. Hence, the ion acoustic frequency is fixed $\omega_a = c_s k_a$. Next use the basic equations that describe the Brillouin scattering but with E as the pump. Make WKB approximations for the electromagnetic wave. Since the growth rate could be of the order of ω_a or larger, keep up to the second order derivative in time and first order in x in the equation for the ion mode. The growth rate for this instability is

$$\gamma \simeq (\omega_{pi}/\omega_{pe})(v/c)(2\pi c \omega_o / \delta)^{\frac{1}{2}}$$

$$v = eE/m\omega_o.$$

Thus the growth rate for a strong pump (we were neglecting damping and convective effects which produce a threshold) is proportional to E_o, a result which agrees with computer simulations. We have also that the scattered wave has the same polarization and shift toward the red as the incident.

We will now discuss a way of explaining the $3/2\omega_o$ line.[11] Near $(1/4)n_{cr}$ we have Raman scattering and two-plasmon decay instabilities. In this case since $\omega_o > \omega_p(x)$ it is safer to assume that the parametric effect is dominant. We then have

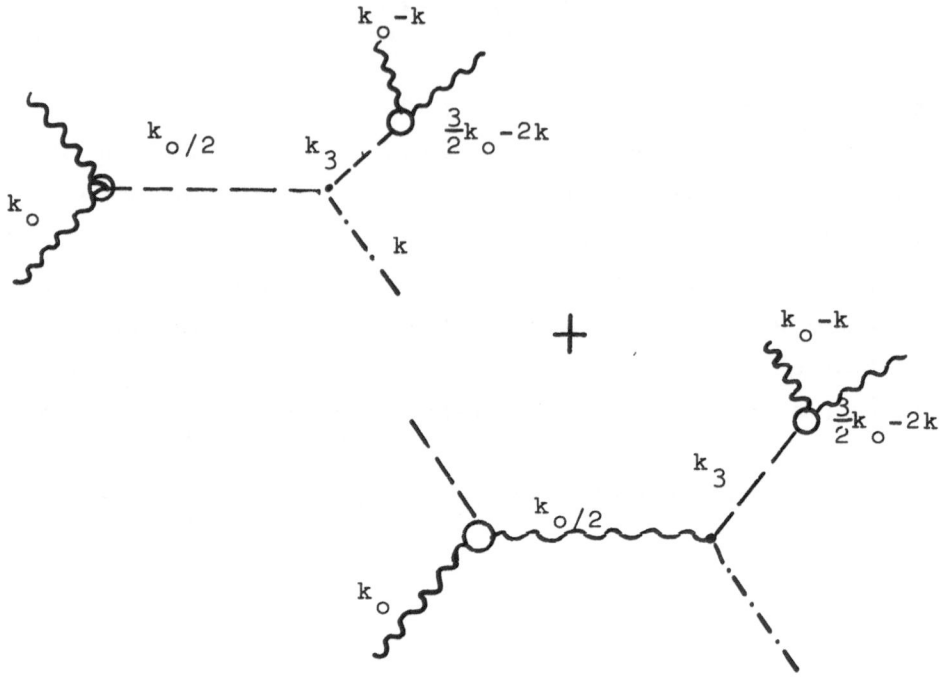

The decay waves are near their critical surface, and
linear wave absorption, as well as reflection, will take
place.

        Consider any of the diagrams. This indicates that
one of the daughter waves (after decay or scattering)
moves in a plasma oblique to a density gradient. Linear
wave conversion takes place, with the associated density
modification, and ion wave excitation. Consequently,
k_3 represents a resonant wave of frequency about $\frac{1}{2}\omega_o - \omega_a$,
which can interact nonlinearly with any strong wave.
The candidate is the backscattered wave at the fundamental frequency ω_o. We can use Eq. (10) with the
proper modifications. The backscattered intensity for
the $3/2\omega_o$ line is then of the same order as that for
the second harmonic and with the same characteristics.

REFERENCES

The literature is quite wide. With apologies to many this list is rather selective of recent papers.

1. a) Many references of interest are in "Laser Interaction and Related Plasma Phenomena", Vol. III, eds. H. Schwarz and H. Hora (Plenum, New York, 1974).

b) R. Kidder, Proc. of International School of Physics "Enrico Fermi", Course XLVIII (Academic, New York, 1971); Nucl. Fusion $\underline{14}$, 53 (1974); and ref. (1a).

c) J. Nuckolls, J. Wood, L. Thiessen, G. Zimmerman, "Nature" $\underline{329}$, 139 (1972) and ref. (1a).

d) K. Bruekner, S. Jorna, Rev. Mod. Phys. $\underline{46}$, 325 (1974) and ref. (1a).

e) J. S. Clarke, H. Fisher, R. Mason, Phys. Rev. Lett. $\underline{30}$, 89 (1973); R. Mason, W. Gula, G. Fraley, R. Malone, R. Morse, IAEA-CN-33/F5-5.

2. "Status and Objectives of Tokomak Systems for Fusion Research," U.S. AEC, WASH-1295 (1974).

3. a) J. Brownell, H. Dreiser, R. Ellis, J. Ingraham, IAEA-CN-33/H4-2(1974).

b) R. Chang, M. Porkolab, Phys. Rev. Lett. $\underline{32}$, 1227 (1974);

c) H. Hendel, J. Flick, Phys. Rev. Lett. $\underline{31}$, 199 (1973).

4. R. Bengston, K. Gentle, J. Jancarik, S. Meadly, P. Nielsen, & P. Phillips (to be published in Phys. Fluids).

5. D. F. Dubois, ref. (1a).
W. L. Kruer, K. G. Estabrook, J. J. Thompson, ref. (1a) and Phys. Rev. Lett. $\underline{31}$, 918 (1973).
J. Dawson, A. Lin, UCLA-PPG 191 (1974).

6. V. N. Tsytovich, <u>Non-Linear Effects in Plasma</u>, (Plenum, N. Y. 1970) and <u>The Theory of Plasma Turbulence</u>, (Pergammon, N. Y., 1972).

7.   J. A. Stamper, et al, ref. (la).
     J. A. Stamper and D. Tidman, Phys. of Fluids <u>16</u>,
     2024 (1973).

8.   a) W. Kruer, E. Valeo, K. Estabrook, J. Thomson,
     B. Langdon, B. Lainski, IAEA-CN-33/F5-3, (1974).
     b) D. Forslund, J. Kindel, K. Lee, E. Linsman, and
     R. Morse, Fourth Annual Anomalous Absorption Conf.,
     Livermore, 1974, paper C.2.

9.   a) D. W. Forslund, J. Kindel, E. Lindman, LASL pre-
     print LA-UR-73-500 (to be published) and Phys. Rev.
     Lett. <u>30</u>, 739 (1973).
     b) W. L. Kruer, K. G. Estabrook and K. H. Sinz,
     Nucl. Fusion <u>13</u>, 952 (1973).

10. A. D. Piliya, Sov. Phys. Tech. Phys. <u>11</u>, 609, (1966)
     and Sov. Phys. USPEKHI <u>14</u>, 413, (1972).
     D. Forslund, J. Kindel, K. Lee, E. Lindman, R. Morse,
     LASL-LA-UR-74-1628.
     D. Kelly, A. Banos, UCLA, PPG-170.
     R. White, F. F. Chen, Plasma Physics <u>16</u>, 565 (1974).

11. P. L. Mascheroni, Fourth Annual Anomalous Absorption
     Conference, Livermore, 1974, paper A2. (A dia-
     grammatic representation for the coupling of dif-
     ferent processes was given there.)

12. See for example N. Krall, A. Trivelpiece, <u>Principles
     of Plasma Physics</u>, (McGraw-Hill, N. Y., 1973).

13. D. Dubois, ref. (la); V. N. Tsytovich, ref. (6);
     P. L. Mascheroni, ref. (la).
     J. Drake, P. Kaw, Y. Lee, G. Schmidt, C. Liu, M.
     Rosenbluth, UCLA-PPG (1973).

14. D. Forslund, J. Kindel, K. Lee, E. Lindman, LASL
     report LA-UR-74-1628, to be published. They con-
     sider in addition the self-consistent equation for
     the pump, as well as the anti-stoke component for

the radiation.

15. a) See ref (8a); also E. Valeo, W. Kruer, Phys.
    Rev. Lett. 33, 750 (1974).
    b) D. Forslund, J. Kindel, K. Lee, E. Lindman,
    LASL-LA-UR 1636.

16. R. Stenzel, A. Wong, H. Kim, Phys. Rev. Lett. 32,
    654 (1974).

17. See for example A. Kaufman, LBL-3098; G. Schmidt,
    S.I.T.-285, preprint.

18. This is the beating term in Eq. (3).

19. G. Morales, Y. Lee, R. White, Phys. Rev. Lett. 32,
    457 (1974).

20. K. Nishikawa, Y. Lee, C. Liu, UCLA-PPG 168.
    A. S. Kingsep, L. Rudakov, R. Sudan, Phys. Rev.
    Lett. 31, 1482 (1973).

21. a) J. W. Shearer, et al, Phys. Rev. A, 6, 764 (1972).
    See the paper by R. P. Godwin, K. Eidman and R. Sigel,
    M. Lubin, J. Stamper, et al, in ref. (1a).
    b) J. L. Bobin, M. DeCroisette, B. Meyer, Y. Vitel,
    Phys. Rev. Lett. 30, 594 (1973).
    J. L. Bobin in ref. (1a).
    c) P. Lee, D. Giovanelli, R. Godwin, G. McCall,
    Appl. Phys. Lett. 24, 406 (1974).
    B. H. Ripin, J. McMahon, E. Lean, W. Manheimer, J.
    Stamper, Phys. Rev. Lett. 33, 634 (1974).
    L. M. Goldman, J. Soures, M. Lubin, Phys. Rev. Lett.
    31, 1184 (1973).

22. P. L. Mascheroni, Phys. Rev. Lett. 34, #3 (1975).
    Bull. Am. Phys. Soc. Ser. II, Vol. 19, #9, abstracts
    2H1, 2H2, 3G5; 1974 CTR Theory Meeting, pg. 58.

23. B. Fried, R. Gould, G. Schmidt, UCLA-PPG 146.

24. a) M. N. Rosenbluth, Phys. Rev. Lett. 29, 565,
    (1972).

b) R. White, C. Liu, M. Rosenbluth, Phys. Rev. Lett. 31, 520, (1973);

M. N. Rosenbluth, R. White, C. Liu, Phys. Rev. Lett. 31, 1190 (1973);

R. White, et al, Nucl. Fusion 14, 45 (1974).

25. D. F. DuBois, D. Forslund, E. Williams, Phys. Rev. Lett. 33, 1013 (1974).

26. A. Ehler, D. Giovanelli, R. Godwin, G. McCall, R. Morse, S. Rockwood, LASL report LA-5611-MS.

27. A Vinogradov, V. Postovalov, Sov. Phys. JETP 36 (1973); E. Erokin, S. Moiseev, Sov. Phys. Tech. Phys. 15, 885 (1970).

28. S. Bodner, G. Chapline, J. DeGroot, Plasma Phys. 21, 15 (1973); H. Hendel and J. Flick, Phys. Rev. Lett. 31, 199 (1973).

THE THEORY OF POWERFUL LASER LIGHT PROPAGATION IN A MATERIAL MEDIUM*

V.N. Lugovoi and A.M. Prokhorov

Academy of Sciences of the USSR

P.N. Lebedev Physical Institute, Moscow, USSR

I. INTRODUCTION

Recently propagation of intense light beams in a nonlinear medium has attracted great attention. The interest in this problem is due, first of all, to the fact that specific features of beam propagation in a medium greatly affect practically all the phenomena of nonlinear optics which are now being widely studied, e.g. stimulated Raman scattering, stimulated Mandelstam-Brillouin scattering, stimulated scattering in a wing of a Rayleigh line, optical breakdown in gases and dielectrics, broadening of a laser pulse spectrum, etc. The correct interpretation of these phenomena in many

---

*This paper was prepared on the request of the Center for Theoretical Studies, University of Miami, Coral Gables, Florida, and complements the series of Review Articles sponsored by the Army Research Office in Durham, North Carolina.

This paper was not presented during the Orbis Scientiae 1975.

cases entirely depends on the beam propagation picture.

An insight into this picture even without reference
to the above phenomena is also particularly important.
It suffices to compare it with linear optics, whose
development would be impossible without knowing the
basic specific features of light propagation in linear
media. Similarly, the understanding of the light propagation picture in a nonlinear medium is extremely
important for the development of nonlinear optics. Of
special concern are light beams generated by lasers in
a pulse regime, in which medium nonlinearity is largely
due to the practically inertialess Kerr effect (Lamb,
1971)*. Therefore, most works on light propagation in
nonlinear media which followed the first one by Chiao,
Garmire and Townes (1964), considered the Kerr nonlinearity where the refractive index of a medium is a
function of light intensity.

Chiao et al (1964) introduced a concept of beam
critical power and it was then shown (Kelley, 1965) that
propagation of a supercritical-power light beam in a
medium with the Kerr nonlinearity begins in the following
way: the intensity on the beam axis increases infinite-
ly (within the parabolic equation used), as it approaches
some point on the axis ("collapse" point). However, the

---

*Here it should be mentioned that in general the Kerr
 effect determines the type of medium nonlinearity in
 nonresonant interaction of light with substances. In
 the special case of resonant interaction, which is observed, for example, in gases at a light frequency
 close to one of those of molecular (atomic) transitions,
 the nonlinearity is of another type. This present review is not concerned with light pulse propagation in
 the latter class of media. A review of relevant works
 is given in Lamb (1971).

beam propagation picture behind the "collapse" point was
not considered. The generally accepted point of view at
that time was that behind the "collapse" point the beam
was self-trapped, i.e. propagated as a waveguide (Chiao
et al, 1964). The beam intensity profile in a waveguide
propagation through the Kerr medium was calculated by
Chiao et al (1964). Thin light filaments observed experimentally in liquids, glasses and later in gases were
considered to be waveguides. Note that the possibility
of an electromagnetic beamed self-trapped into a waveguide was pointed out as early as 1958 by Volkov, who
was the first to calculate the beam intensity profile
at self-trapping in a plasma. Later such a possibility
was also mentioned by Askaryan (1962) and Talanov (1964)
(the intensity profile in the latter was just that of
Volkov (1958)).

However, many experimental results could not be explained in terms of beam being self-trapped into a waveguide behind the "collapse" point. First of all, one
could not account for the very short "lifetime" of filaments which typically was 10^{-10} sec for giant laser
pulses. Furthermore, filaments in such substances as,
for example, carbon bisulphide, toluene, nitrobenzene,
etc., were longer than 5-10 cm. It was still not clear
why such filaments are possible under the conditions of
intense stimulated Raman scattering occuring in these
media. In fact, according to some experimental results,
at the end of the cell up to 90% of the energy was
pumped from the filament into a Stokes scattering
component. The calculations predicted such pumping over
a path of only 0.1 cm, which was in disagreement with
lengths of 5 to 10 cm. No explanation was found for
generation of the anti-Stokes components of stimulated

Raman scattering. According to the waveguide concept,
the apex angles of cones of anti-Stokes component radiation should be determined by conditions of the Cerenkov
type (see Szöke (1964) and Lugovoi and Sobelman (1970)).
In fact, the radiation of these components observed was
not of the Cerenkov type but intermediate between the
latter and the so-called volume-type. Loy and Shen
(1971) discovered that the ultrashort pulses in backward stimulated Raman scattering occur inside the cell
containing the investigated substance. This pulse
generation cannot be explained in terms of the waveguide
picture of beam propagation. The picture fails to in-
terpret the discrete character of optical damage in
transparent media, some specific features of laser pulse
spectrum broadening on traversing a substance, and other
phenomena.

     Dyshko, Lugovoi and Prokhorov (1967), based on the
numerical solution of the problem, suggested a new
(multifoci) picture of light beam propagation in media
with the Kerr nonlinearity. Then Lugovoi and Prokhorov
(1968) stated that the thin light filaments being experimentally observed were a moving foci trajectory
rather than a waveguide. This new concept of the important features of light beam propagation in nonlinear
media avoids all contradictions and provides an explanation to many experimental results.

     The multifoci, stationary in the time structure of
a light beam, is a finite row of single foci on its axis
which is formed by successively focusing various annular
zones of the beam. Here the "collapse" point itself is
the center of the first focus rather than the beginning
of a waveguide filament. The detailed investigation by
Dyshko et al (1971) of the effect of various nonlinear

absorptions (Lugovoi and Prokhorov, 1968) in a medium
(i.e. of the imaginary part of the refractive index) on
the beam propagation picture showed that multifoci
structure always arises independently of the specific
type of this absorption. The authors also investigated
the effect of deviations (possible under real conditions)
from the square-law field dependence of the real part of
the refractive index associated with the so-called Kerr
nonlinearity "saturation" (see Akhmanov et al, 1966,
1967; Brewer et al, 1968; Gustafson et al, 1968;
Gustafson and Townes, 1972), or with nonlinear absorption in a medium. Here, based on the numerical calculations it has been established that the light beam multifoci structure is preserved not only qualitatively but
also quantitatively: foci parameters differ very slightly. Thus, for media with Kerr nonlinearity, the multifoci picture of light beam propagation has proved to be
universal, i.e. it should be observed under various
physical conditions.

In reality the incident beam is not stationary. Its
power varies in time within a laser pulse envelope.
Since the foci positions along the beam axis depend on
the initial power and the power itself does vary with
time, the foci move along the beam axis. For giant
pulses under typical conditions the foci move at a speed
of 10^9 cm/sec. As a result of this movement, if one continuously observes the beam propagation one finds that
one side of the film will show thin filaments --traces
of moving foci. A theory was developed according to
which the cross-sections of these filaments depend on
the foci cross-sections, and their "lifetime" is simply
the time it takes the foci to pass the output medium
plane. A typical lifetime of filaments was calculated

to be about 10^{-10} sec (Lugovoi and Prokhorov, 1968),
which is in full accord with the experimental values.

Thus, a multifoci picture of this propagation
provides a consistent theory of high-power laser propagation in Kerr nonlinearity media and leads to the
concept of moving foci. This theory, which is directly
based on the Maxwell equations, is presented below.

II. METHODS OF INVESTIGATION AND SOME RESULTS
ON POWERFUL LASER PULSE PROPAGATION IN A MATERIAL MEDIUM

1. Initial Equations

Light wave propagation in a nonlinear medium is
described by the set of Maxwell equations

$$\text{curl } \vec{E} = -\frac{1}{c}\frac{\partial \vec{H}}{\partial t} , \quad \text{curl } \vec{H} = \frac{1}{c}\frac{\partial \vec{D}}{\partial t} , \text{ div } \vec{D} = 0 \qquad (1)$$

and by the material equation representing a nonlinear
relation between the displacement vector $\vec{D}$ and the electric field vector $\vec{E}$, which can be written as

$$\vec{D} = \varepsilon\vec{E} , \qquad (2)$$

where ε is a nonlinear function of $\vec{E}$. Of greatest interest are high-power linearly polarized beams generated
by lasers in a pulse regime (pulse width is about 10^{-8} sec
or less). As a rule, such intervals are too short for
the strictions-- or thermal and other related mechanisms
of nonlinearity at a normal beam diameter > 0.1~0.3cm--
to show up because of a relatively long time of redistribution of the substance density due to striction
forces or nonuniform heating (Zeldovich and Raizer,
1966; Shen, 1966, 1967; Raizer 1967).

In this* case the orientational effect (Chiao et al, 1964; Zeldovich and Raizer, 1966; Shen, 1966, 1967;) or the (Raizer, 1967; Lallemand and Bloembergen, 1965; Bloembergen and Lallemand 1966) Kerr effect (Brewer and Lee, 1968; Duguay et al, 1970) contribute most to medium nonlinearity.

The characteristic setting time of the electronic Kerr effect does not exceed 10^{-15} sec. Thus, this nonlinear mechanism can show up even at picosecond pulses with a pulse width down to 10^{-12} sec. For typical giant pulses with durations of 10^{-8} sec, it is the orientational Kerr effect, whose characteristic setting time is 10^{-10} - 10^{-12} sec, that often dominates in medium nonlinearity. If the characteristic setting time τ_k of the Kerr effect is much less than those of light intensity I of a given polarization, the refractive index at any point $\vec{r}$ of a medium is a function of I:

$$n = n(I) = n_o + n_2' I + \ldots, \quad (n_2' > 0) . \qquad (3)$$

For dielectric permeability it gives

$$\varepsilon = \varepsilon(I) = \varepsilon_o + \varepsilon_2' I + \ldots, \quad (\varepsilon_o = n_o^2, \ \varepsilon_2' = 2n_o n_2') . \qquad (4)$$

Under conditions (9) to be discussed below, the type of polarization determined by an incident light pulse is almost entirely preserved while this pulse propagates in

---

*Fabelinski and Starmor (1967) and Polloni et al (1969) also suggest possible rapid mechanisms of the Kerr effect associated with molecule librations. Hellwarth (1966) considers the fast mechanism consisting in "microscopic dusting" of molecules.

a medium. So, for simplicity, eqs. (3) and (4) can be considered to be independent of the polarization type.

By excluding the magnetic field $\vec{H}$ and electrical induction $\vec{D}$ from equation (1) and by taking into account (2) and (4) one has for the electric field $\vec{E}$:

$$\text{grad}\,[\frac{\vec{E}\,\text{grad}\varepsilon}{\varepsilon}]\,-\,\Delta\vec{E}\,+\,\frac{1}{c^2}\,\frac{\partial^2}{\partial t^2}\,(\varepsilon\vec{E})\,=\,0\ ,\ \varepsilon\,=\,\varepsilon(<E^2>)\ .$$

$$(5)$$

Here $<E^2>$ is the value of $E^2(<E^2>\propto I)$ averaged over a period. By choosing a coordinate system whose Z axis is along the main direction of the light beam propagation and by assuming

$$\vec{E}\,=\,\frac{1}{2}\,\vec{E}\,(\tau,t)e^{ikz-i\omega t}\,+\,\text{c.c.}\ ,\qquad\qquad(6)$$

where $k=\frac{\omega}{c}\sqrt{\varepsilon_o}$; ω is the central frequency of the field oscillations in a beam and considering that $<E^2>\propto|\vec{E}|^2$, one finds the following exact equation for the new unknown function $\vec{E}$:

$$\frac{ik}{\varepsilon}\,(\vec{E}\cdot\text{grad}\,\varepsilon)\,+\,\text{grad}\,[\frac{\vec{E}\cdot\text{grad}\,\varepsilon}{\varepsilon}]\,-\,[\Delta\vec{E}\,+\,2ik\,\frac{\partial\vec{E}}{\partial z}$$

$$+\,(\frac{\omega}{c})^2(\delta\varepsilon)\vec{E}\,+\,\frac{\varepsilon}{c^2}\,(2i\omega\,\frac{\partial\vec{E}}{\partial t}\,-\,\frac{\partial^2\vec{E}}{\partial t^2})\,+\,\frac{2}{c^2}\,\frac{\partial\varepsilon}{\partial t}\,(i\omega\vec{E}\,-\,\frac{\partial\vec{E}}{\partial t})$$

$$-\,\frac{1}{c^2}\,\frac{\partial^2\varepsilon}{\partial t^2}\,E]\,=\,0\ ,$$

where

$$\varepsilon\,=\,\varepsilon(|\vec{E}|^2)\,=\,\varepsilon_o\,+\,\frac{1}{2}\,\varepsilon_2|\vec{E}|^2\,+\,\ldots,\ \delta\varepsilon\,=\,\varepsilon-\varepsilon_o\,=$$

$$\frac{1}{2}\,\varepsilon_2\,|E|^2\,+\,\ldots,\ (\varepsilon_2>0)\qquad\qquad\qquad(8)$$

and the vector $\vec{k}$ is equal in magnitude to k and directed along the z axis.

In what follows it is assumed that for all $\vec{r}$ and t
the following (and in practice the most interesting)
conditions are fulfilled:

$$|\delta\epsilon| << \epsilon_o \quad , \tag{9a}$$

$$\lambda << \Lambda_\perp << \Lambda_{||} \quad , \tag{9b}$$

$$\frac{\Lambda_\perp}{\lambda} T << \tau \quad , \tag{9c}$$

$$\tau_k << \tau \quad , \tag{9d}$$

where $\Lambda_\perp$ and $\Lambda_{||}$ are the characteristic scales of the
variations in amplitude of E along the z axis and perpendicular to it, respectively; τ sets the scale of the
E variation as a function of t; $\lambda = \frac{2\pi}{k}$ is the wave length
of light; $T = \frac{2\pi}{\omega}$ is the oscillation period of the
light, τ_k is the characteristic setting time of the Kerr
effect in a medium. Condition (9d) provides an expression for dielectric permeability in the form given
by (8). Under conditions (9a), (9b), (9c), some of the
terms in (7) are negligible. First, consider the projection of this vector equation on the xy-plane. In this
case the first term is zero. If one considers that
$\frac{\partial\epsilon}{\partial r_\perp} = \frac{\partial\epsilon}{\partial|E|^2}\frac{\partial|E|^2}{\partial r_\perp} \sim \frac{\delta\epsilon}{\Lambda_\perp}$ and similarly $\frac{\partial\epsilon}{\partial t} \sim \frac{\delta\epsilon}{\tau})$ an estimate
of the other terms gives:

$$\left|\text{grad}_\perp\left[\frac{E \cdot \text{grad}\epsilon}{\epsilon}\right]\right| \sim \frac{\delta\epsilon}{\epsilon_o}\frac{E}{\Lambda_\perp^2} \quad , \quad \Delta_\perp E_\perp \sim \frac{E_\perp}{\Lambda_\perp^2} \quad , \quad \frac{\partial^2 E'}{\partial z^2} \sim \frac{E_\perp}{\Lambda_{||}^2} \quad ,$$

$$2ik \, \frac{\partial E_\perp}{\partial z} \sim i \, \frac{4\pi E_\perp}{\lambda \Lambda_\|} \quad , \qquad\qquad \left(\frac{\omega}{c}\right)^2 (\delta\epsilon)E_\perp \sim \frac{4\pi^2}{\lambda^2}(\delta\epsilon)E_\perp \quad ,$$

$$\frac{\epsilon}{c^2} \, 2i\omega \, \frac{\partial E_\perp}{\partial t} \sim 4\pi i \, \frac{\epsilon_0 T}{\lambda^2 \tau} \, E_\perp \quad , \qquad \frac{\epsilon}{c^2} \, \frac{\partial 2 E_\perp}{\partial t^2} \sim \frac{\epsilon_0 T^2}{\lambda^2 \tau^2} \, E_\perp \quad ;$$

$$\frac{2}{c^2} \, \frac{\partial \epsilon}{\partial t} \, i\omega E_\perp \sim 4\pi i \, \frac{(\delta\epsilon)T}{\lambda^2 \tau} \, E_\perp \quad , \qquad \frac{2}{c^2} \, \frac{\partial \epsilon}{\partial t} \, \frac{\partial E_\perp}{\partial t} \sim 2 \, \frac{(\delta\epsilon)T^2}{\lambda^2 \tau^2} \, E_\perp \quad ,$$

$$\frac{1}{c^2} \, \frac{\partial^2 \epsilon}{\partial t^2} \, E_\perp \sim \frac{(\delta\epsilon)T^2}{\lambda^2 \tau^2} \, E_\perp \quad ,$$

(here $E_\perp = (E_x, E_y)$ $\Delta_\perp = \frac{\partial^2}{x^2} + \frac{\partial^2}{\partial y^2}$). So to an accuracy of this order the equation for $E_\perp$ is as follows:

$$\Delta_\perp E_\perp + 2ik \left(\frac{\partial E_\perp}{\partial z} + \frac{1}{v} \, \frac{\partial E_\perp}{\partial t}\right) + k^2 \, \frac{(\delta\epsilon)}{\epsilon_0} \, E_\perp = 0, \quad v = \frac{c}{n_0} \quad . \quad (10)$$

Here $\delta\epsilon = \delta\epsilon(|E_\perp|^2 + |E_z|^2)$. The equation obtained by projecting vector equation (7) on the z axis direction completes a set of equations for the three values E_x, E_y, E_z. However, it is more convenient to derive the third equation directly from $\mathrm{div}(\epsilon \vec{E} e^{ikz-i\omega t}) = 0$ (see (1), (2), (6)), from which

$$E_z = \frac{i}{k\epsilon} \, \mathrm{div}(\epsilon \vec{E}) \quad . \tag{11}$$

From eq. (11) it is evident that under conditions (9a), (9b) at all $\vec{r}$ and t the following inequality is valid:

$$|E_z| << |E_\perp| \quad . \tag{12}$$

Here to an accuracy of higher order one has

$$E_z = \frac{i}{k} \left(\frac{\partial E_x}{\partial x} + \frac{\partial E_y}{\partial y} \right) \quad .$$

Taking into account (12), one can assume that in eq. (10) $\delta\epsilon = \delta\epsilon(|E_\perp|^2)$. Thus, light beam propagation in a nonlinear medium under conditions (9) is described by a parabolic equation [eq. (10)] for the transverse component E of the electric field in this beam*. Here the longitudinal component E_z can be found in the first approximation by means of eq. (13) provided that the solution of eq. (10) is known.

From eq. (10) for $E_\perp$, it immediately follows that if in the initial plane z = 0 the field at any t has some (constant) polarization, the latter is also pre- served at z > 0, i.e. throughout beam propagation in a nonlinear medium. In particular, the propagation of linearly polarized light beams generated by pulse lasers is determined by an equation for one field component, e.g. for component E_x E (the x-axis is assumed to be directed along the electric field):

$$\Delta E + 2ik \left(\frac{\partial E}{\partial z} + \frac{1}{v} \frac{\partial E}{\partial t} \right) + k^2 \frac{\delta\epsilon(|E|^2)}{\epsilon_o} E = 0 \quad . \qquad (14)$$

Function $\delta\epsilon(|E|^2)$ in this equation has the following simple form for media with a pronounced Kerr effect

$$\delta\epsilon(|E|^2) = \frac{1}{2} \epsilon_2 |E|^2 \quad . \qquad (15)$$

This expression is the first term of the expansion of

---

*Earlier the parabolic equation was extensively used for describing wave processes in linear (homogeneous and inhomogeneous) media. In a weakly inhomogeneous medium it is just (10), where $\delta\epsilon = \delta\epsilon(\vec{r},t)$.

$\delta\varepsilon$ in powers of $|E|^2$. For refractive index n it gives

$$n = n_o (1 + \frac{1}{2} n_2 |E|^2), \quad n_2 = \frac{\varepsilon_2}{2\varepsilon_o} \quad . \qquad (16)$$

Medium nonlinearity of the form of (15) or (16) is usually referred to as Kerr nonlinearity. The $|E|^2$ dependence of $n(|E|^2)$ for large values $|E|^2$ at $|E|^2 \gtrsim \frac{1}{n_2}$ (i.e. when expression (16) breaks down) was considered in detail in work [15]* and in Warburger and Dawes (1968) and Dawes and Warburger (1969) it was approximated by the function

$$n = n_o(1 + \frac{1}{2} \frac{n_2 |E|^2}{1 + |E|^2 / |E_s|^2}) \qquad (17)$$

or other functions (Zakharov et al, 1971) not differing in principle from eq. (17). According to eq. (17), at $|E|^2 >> |E_s|^2$ (or ε)n no longer depends on $|E|^2$. In this case one speaks of the Kerr nonlinearity saturation. Since, however, $|E_s|^2 \sim \frac{1}{n_2}$ (which, in its turn, is due to the nonresonant character of the Kerr effect), because of (9a) eq. (14) holds only at $|E|^2 << |E_s|^2$ with practically no saturation of nonlinearity (i.e. expressions (15), (16) are valid). Therefore, to describe light beam propagation in a medium one should use the Maxwell equations directly and avoid eq. (14). In Section 4 the same saturation is shown to be preserved also in the stationary case when $\frac{\partial E}{\partial t} \equiv 0$**.

At very low $|E|^2$ values expressions (15), (16) can

---

*Gustafson and Townes also consider refractive index of a medium vs. light intensity in a substantially non-stationary case where (9d) breaks down.

**See next page.

be considered real. At the same time, Lugovoi and
Prokhorov (1968) noted that it was necessary to take
into account not only the real part, $\delta\epsilon$, but also an
imaginary one, ($\delta\epsilon''$), associated with nonlinear absorption in a medium when describing light beam propagation in nonlinear media at relatively high values of
$|E|^2$ (though $|E|^2 << |E_s|^2$). Generally, the value of $\delta\epsilon''$
is a nonlinear function of E. If the main mechanism of
nonlinear absorption is a multiphoton one whose
characteristic setting time is under 10^{-15} sec, the value
of $\delta\epsilon''$ (as well as the real part $\delta\epsilon'$) under condition
(9c) can be simply written as a function of $|E|^2$. If
the major contribution to nonlinear absorption is made
by energy conversion from the beam into stimulated
scattering components, the dependence of $\delta\epsilon''$ on E cannot
be reduced to a function of $|E|^2$ (see Section 5b). In
what follows one can see that in the most interesting
cases, the value of $\delta\epsilon''$, being very important for beam
propagation, remains for all $\vec{r}$ and t much less than $\delta\epsilon'$.
The latter point enables us to ignore the corrections to
the real part of $\delta\epsilon$ associated with nonlinear absorption
in a medium (see Section 6). Thus, $\delta\epsilon$ in equation (14)
can be written as:

$$\delta\epsilon = \frac{1}{2} \epsilon_2 |E|^2 + i\delta\epsilon'' \quad . \tag{18}$$

In some cases of practical interest the following
condition is valid

$$\frac{\Lambda_{\parallel}}{\tau} << v \quad , \tag{19}$$

**This fact was not taken into account in some works
(e.g. [30-32]) considering light beam propagation in
terms of the parabolic equation at a considerable
saturation of the Kerr nonlinearity of a medium.

where the term $\frac{1}{v}\frac{\partial E}{\partial t}$ of equation (14) is small compared to the term $\frac{\partial E}{\partial z}$. By neglecting the first of these terms one derives the equation

$$\Delta_\perp E + 2ik\frac{\partial E}{\partial z} + k^2\frac{\delta\varepsilon}{\varepsilon_o}E = 0 \quad . \tag{20}$$

The time dependence remains only in the boundary condition

$$E\Big|_{z=0} = \phi(r_\perp,t) \ , \tag{21}$$

where $\phi(r,t)$ is a given function determined by the electric field of a light pulse incident on the boundary z = 0 of a nonlinear medium. Thus, the solution of the initial nonstationary problem under condition (19) is reduced to that of a stationary problem, eqs. (20) and (21), where the time t is only a parameter in the boundary condition. That is why we first consider the stationary problem. The conditions under which inequality (19) holds will be called quasi-stationary.

2. Analytic Investigation

Of special practical concern is the case of an axially symmetric beam. A deviation $r_\perp$ from the beam axis will be designated as r. Then, for simplicity, we assume

$$\delta\varepsilon'' = \varepsilon_o m(|E|^2), \ (m(0) = 0) \ , \tag{22}$$

and consider that the beam incident on the nonlinear medium boundary has a flat phase front and a Gaussian intensity distribution. Then equation (20) has the form

$$\frac{\partial^2 E}{\partial r^2} + \frac{1}{r}\frac{\partial E}{\partial r} + 2ik\frac{\partial E}{\partial z} + k^2[n_2|E|^2 + im(|E|^2)]E = 0 \ , \tag{23}$$

with the boundary condition

$$E\big|_{z=0} = E_o e^{-\frac{r^2}{2\bar{a}^2_o}} \quad . \qquad (24)$$

Here the field E_o is a parameter and the value $\bar{a}_o$ is the initial beam radius.

For an analytic investigation we take the field E in the form

$$E = e^A \quad . \qquad (25)$$

From (23) one derives the following equation for A:

$$\frac{\partial^2 A}{\partial r^2} + \frac{1}{r}\frac{\partial A}{\partial r} + \left(\frac{\partial A}{\partial r}\right)^2 + 2ik\frac{\partial A}{\partial z} + k^2[n_2 e^{2A^r} + im(e^{2A^r})] = 0, (A^r = ReA).$$
$$(26)$$

We assume that A is an analytic function of x, y, or in other words, that the expansion of A in a Taylor series contains only integer non-negative powers of $q = x^2 + y^2$,

$$A(q,z) = A_o(z) + qA_1(z) + q^2 A_2(z) + \ldots \quad .(27)$$

By substituting (27) into (26) one derives a system of equations for $A_n(z)$

$$2ik\frac{\partial A_{n-1}}{\partial z} + 4[n^2 A_n + \sum_{k=0}^{n} k(n-k)A_k A_{n-k}] + (k^2 n_2 e^{2A^r_o}) \qquad (28)$$

$$\cdot(L_{n-1} + iM_{n-1}) = 0, (n = 1,2\ldots) ,$$

where the real coefficients $L_k(z), M_k(z)$ are determined by the equalities $\exp[2(qA_1^r + q^2 A_2^r + \ldots)] = 1 + qL_1 + q^2 L_2 + \ldots, n_2^{-1}\exp(-2A_o^r)m\exp(2A_o^r + 2qA_1^r + 2q^2 A_2^r)]$
$= M_o + qM_1 + q^2 M_2 + \ldots \quad .$
$$(29)$$

By introducing the symbols

$$a_o = e^{2A^r_o}, b_o = \frac{1}{k} \frac{dA^i_o}{dz} \; , \; k^{-2m}A_m = a_m + ib_m, \; A^i = \text{Im}A, \; (m=1,2..),$$

(30)

from equation (28) one finds

$$2b_o = n_2 a_o + 4a_1 \quad .$$

(31)

Relation (31) determines a correction to the longitudinal wave number on the beam axis. For the other coefficients a_k, b_k eq. (28) reduces to an infinite system of equations. To make the main properties of this system clear we assume $m(|E|^2) \equiv 0$, i.e. ignore absorption in a medium. In this case this system has the form

$$a'_o + 4b_1 a_o = 0$$

$$a'_1 + 4b_1 a_1 = -8b_2$$

$$b'_1 + 2b_1^2 = n_2 a_o a_1 + 2a_1^2 + 8a_2$$

$$a'_2 + 8b_1 a_2 = -8a_1 b_2 - 18b_3$$

$$b'_2 + 8b_1 b_2 = n_2 a_o (a_1^2 + a_2) + 8a_1 a_2 + 18a_3$$

$$a'_3 + 12b_1 a_3 = -4(3a_1 b_3 + 4a_2 b_2) - 32b_4$$

$$b'_3 + 12b_1 b_3 = n_2 a_o (\tfrac{2}{3} a_1^2 + 2a_1 a_2 + a_3) + 8(a_2^2 - b_2^2) + 12a_1 a_3 + 32a_4$$

$$a'_4 + 16b_2 a_4 = -8(2a_1 b_4 + 3a_2 b_3 + 3a_3 b_2) - 50 b_5$$

(32)

. . . .

Here the prime means a derivative with respect to $u = kz$.

First we consider the case without medium non-linearity, i.e. $n_2 = 0$. Here it is evident that system (32) has a class of solutions such that $a_2 \equiv b_2 \equiv a_3 \equiv b_3 \ldots \equiv a$ and the rest of the coefficients (a_0, a_1, b_1) satisfy the equations

$$a_0' + 4b_1 a_0 = 0 \quad,$$

$$a_1' + 4b_1 a_1 = 0 \quad, \tag{33}$$

$$b_1' + 2b_1^2 = 8a_1^2 \quad.$$

According to (25), (27), (33), this class of solutions corresponds to Gaussian beams and the fact that it exists means that a Gaussian beam propagating in a linear homogeneous medium preserves its shape. It is the width of the intensity distribution over a cross section (coefficient a_1), the intensity on the axis (coefficient a_0) and the phase front curvature (coefficient b_1) that are changed. A general solution of (33) gives the following expression for amplitude E:

$$E = \frac{E_0}{1 - \frac{z}{R} + i\frac{z}{\ell_d}} \cdot \exp\left[\frac{\frac{1}{2a_0^{-2}} + \frac{ik}{2R}}{1 - \frac{z}{R} + i\frac{z}{\ell_d}} z^2\right] \quad, \tag{34}$$

where

$$\ell_d = ka_0^{-2} \quad.$$

In the particular case of $\frac{1}{R} = 0$, the solution eq. (34) satisfies boundary condition (24). Note that nonzero values of $\frac{1}{R}$ (as seen from eq. (34)) correspond to a more general boundary condition

$$E|_{Z=0} = E_o \exp\left[-\frac{1}{2}\left(\frac{1}{\bar{a}_o^2} + i\frac{k}{R}\right)r^2\right] \quad , \tag{36}$$

which in turn corresponds to a focused beam with a point $z = R$ of geometrical convergence of the rays.

   In the case of a nonlinear medium when $n_2 \neq 0$, the situation changes considerably: this class of solutions (Gaussian beams) does not exist. Even in the case of a Gaussian initial distribution its form will change in the course of beam propagation in a nonlinear medium, i.e. the nonzero coefficients $a_2, b_2, a_3, b_3 \ldots$ will appear. We expand the coefficients a_k in a Taylor series in u: $a_k = \Sigma a_k^{(n)} u^n$. Proceeding from eqs. (28) to (32) it is clear that in order to determine the coefficient $a_o^{(n)}$ at $n_2 \neq 0$ one needs the first $2n-1$ equation of system (32), and for coefficient $a_1^{(n)}$, $(2n+1)$ of these equations. As is shown by Lugovoi (1967) this fact strongly influences the character of the solution provided that

$$E_o \gtrsim E_{cr} \quad , \tag{37}$$

where

$$E_{cr} = \frac{1}{\sqrt{n_2(k\bar{a}_o)^2}} \quad . \tag{38}$$

Condition (37) can be also written as

$$P \gtrsim P_{cr} \tag{39}$$

where $P = \frac{cn_o}{8\pi} \int |E|^2 dr_{\perp}$ is the beam power and

$$P_{cr} = \frac{cn_o}{2n_2 k^2} \quad . \tag{40}$$

Thus, the condition obtained by Chiao et al (1964),

where the nonlinear medium effect is important, is just
condition (39) describing essentially the intensity redistribution across a beam. In the process of its
propagation this redistribution is accompanied by a
change in the initial beam shape. The latter fact was
not taken into account by Talanov (1965) (see also the
review by Akhmanov et al, [1967]), so the author failed
to obtain a correct picture of beam propagation in the
medium which he considered. The same objection also
applies to some recent work by Kerr (1971), where beam
propagation in a nonlinear medium is assumed to be
Gaussian.

The correct analytical solution of the problem was
obtained by Lugovoi (1967), which however holds only
under conditions

$$|N-1| \ll 1, \quad z \ll ka_o^{-2}, \quad (N = E_o/E_{cr}) \; ,$$

i.e. over some range of initial fields E_o and near the
medium boundary. This solution at $r = 0$ has the form

$$\frac{|E|^2}{E_o^2} = 1 + \frac{N^2 - 1}{(k\bar{a}_o^{-2})^2} z^2 - \frac{11}{3(ka_o^{-2})^4} z^4 \; . \qquad (41)$$

If $E_o^2 \gg E_{cr}^2$, then from (32) one has the expansion

$$\frac{|E|^2}{E_o^2} = 1 + \frac{n_2 E_o^2}{\bar{a}_o^2} z^2 + \dots \; , \qquad (42)$$

which determines the characteristic length ℓ_x of the
initial change over z of axial field $|E|$:

$$\ell_x = \frac{\bar{a}_o}{\sqrt{2n E_o^2}} \; . \qquad (43)$$

Expression (43) was first obtained by Kelley (1965).

Due to their limited validity, the analytical results, as will be shown later, do not display some of the essential features of beam propagation in a nonlinear medium. Therefore it is the numerical solution of this problem that is given below.

3. Numerical Solution of Parabolic Equation with No Absorption in a Medium

Consider the numerical solution of eq. (23) with boundary condition (24). First, following Kelley (1965), we assume that $m(|E|^2) \equiv 0$, i.e. absorption in the medium will be ignored. By introducing, for convenience, the quantities

$$X = \frac{E}{E_o} \; , \quad N = \frac{E_o}{E_{cr}} \; , \quad z_1 = \frac{z}{\ell_x} \; , \quad r_1 = \frac{r}{\bar{a}_o} \; , \tag{44}$$

where E_{cr} and ℓ_x are determined by expressions (38) and (43), one arrives at the following equation for x.

$$\frac{\partial^2 X}{\partial r_1^2} + \frac{1}{r_1} \frac{\partial X}{\partial r_1} + 2iN \frac{\partial X}{\partial z_1} + N^2 |X|^2 X = 0 \; , \tag{45}$$

with boundary condition

$$X|_{z=0} = e^{-\frac{1}{2} r_1^2} \; . \tag{46}$$

According to Dyshko et al (1967) at $|N-1| \ll 1$, $z_1 \ll 1$, the numerical solution of eqs. (45)-(46) supports the analytical expression (41). Then at $N > N_1$ (where $N_1 \approx 2$), according to Chiao et al (1964), Dyshko et al (1967), Goldberg et al (1967), the solution acquires an important singularity: the intensity $|X|^2$ on the beam axis (i.e. at $r_1 = 0$) dependence on z_1 infinitely increases as z_1

approaches some point $z_1{}^*$. The expression for $z_1{}^*$, obtained in Kelley (1965) by approximation of the numerical
results has the form

$$z_1{}^* \approx \frac{0.7N}{N-N_1} \qquad\qquad (47)$$

or, using eqs. (44) and (38),

$$z^* \simeq 0.4 \frac{k\bar{a}_o^2}{\sqrt{(\frac{P}{P_{cr}} - 1)}} \qquad . \qquad (48)$$

Here $z^*=z_1^* \ell_x$, P is the incident beam power, the value
of P_{cr} is close to the value given by (40).

Note that, strictly speaking, the numerical calculations do not settle the question of an existing
singularity in the mathematical sense at $z_1 = z_1^*$, since
in such calculations, when approaching z_1^*, one obtains
an increasing but always finite (within the limit of
available calculation time and computer memory) value
$|X|^2$. In spite of this, according to Kelley (1965) and
Dyshko et al (1967), the rate of increase in z near r=0
and at z values close to z^* proves to be so large that
condition $\Lambda_{\parallel} \gg \lambda$ (see (9)), necessary for parabolic
equation (45) to be valid, breaks down under real con-
ditions. From this point of view the question of an
existing mathematical peculiarity in the solution is of
a formal character and will not be discussed any further.
But in the following section we shall consider the factors which, under real conditions can limit an increase
in $|X|^2$ as $z_1 \to z_1^*$.

III. STATIONARY-IN-TIME MULTIFOCI STRUCTURE OF LIGHT BEAMS IN A NONLINEAR MEDIUM

4. Formulation of the Problem with Allowance for Absorption

In the work by Dyshko et al (1967), on the basis of the numerical solution of eq. (45) with boundary condition (46), it was established that the power "flowing into" the singularity present at $z \to z^*$ is only a fraction of the initial beam power which is close to the critical value P_{cr}. Under these circumstances the scale $\Lambda_\perp$ of the spatial change of field E in a cross-section proves to be minimum near the beam axis and is

$$\Lambda_\perp = \sqrt{\frac{32 P_{cr} \ell n_2}{C n_o |E|^2}} \quad , \tag{49}$$

or the diameter of the near-axis beam intensity distribution over this cross-section. From eqs. (49) and (40) one derives at once

$$\Lambda_\perp = \frac{\lambda}{2 \sqrt{n_2} |E|^2} \quad . \tag{50}$$

Thus, with condition

$$\Lambda_\perp \gg \lambda \tag{51}$$

fulfilled, the first of conditions (9) for the validity*

---

*If on the other hand, $n_2 |E|^2 \gtrsim 1$ one obtains from (50) $\Lambda_\perp \lesssim \frac{\lambda}{2}$. Therefore, to describe light beam propagation in media with allowance only for saturation of the Kerr nonlinearity (when $n_2 |E|^2 \gtrsim 1$), one should use directly, even in a stationary case, the Maxwell equations, without the parabolic equation.

of the parabolic equation is fulfilled also.

Later it will be also shown (see Section 6) that inequality (51) ensures that condition $\Lambda_{||} >> \Lambda_{\perp}$ is satisfied. Therefore, in the vicinity of the point z*, the parabolic equation is applicable insofar as inequality (51) is valid*).

If condition (51) breaks down, i.e. at $\Lambda_{\perp} \sim \lambda$, the parabolic equation does not hold any longer. On assuming that as $z \rightarrow z*$ the values $\Lambda_{\perp} \sim \lambda$ are reached, the intensity I on the beam axis is found from eqs. (40) and (49) to be $I \sim 3 \times 10^{12} w/cm^2$ to $3 \times 10^{14} \, w/cm^2$ for typical media, say, carbon bisulphide, toluene, nitrobenzene, some glasses, etc. (i.e. media with $n_2 \sim 10^{-11}$ to 10^{-13} CGSE). At such (and in the most cases much lower) intensities, considerable nonlinear light absorption can take place in a substance (e.g. multiphoton absorption; absorption due to energy conversion into stimulated scattering components; absorption due to substance breakdown; etc.). Such nonlinear absorption in the beam regions with intensities from along 10^{10} to $10^{11} w/cm^2$ has been registered in many experiments (see, for example, Shen and Shaman 1965, 1967; Brewer and Lifsits, 1966; Chiao et al, 1966; Brewer and Townes, 1967)). This means that nonlinear absorption can be the main factor limiting the energy density as $z \rightarrow z*$. Here the minimum value $\Lambda_{\perp}$, is much longer than the wavelength λ, which is also in accord with the experimental results. The relation $\Lambda_{\perp} >> \lambda$ may also be obtained from eq. (50) if one takes

---

*Conditions (9c), (9d), and (19) for the validity of equation (20) which are associated with a beam non-stationary-in-time will be considered in what follows (see Section 7).

into account that the measured value $n_2|E|^2$ is of order 10^{-3} (see Bloembergen, 1971).

Under the above conditions describing the light beam propagation, the parabolic equation may be applied but with $\delta\varepsilon$ written in complex form (18). The imaginary part $\delta\varepsilon''$ in (18) is determined by the dominant type of nonlinear absorption in the medium. In what follows three kinds of this type of absorption will be considered in detail: a) three-photon absorption; b) absorption due to energy inversion into the first Stokes component of the stimulated Raman scattering (SRS); c) two-photon absorption.

5. Numerical Calculation of the Parabolic Equation
in a Medium with Kerr Nonlinearity and
Nonlinear Absorption
a) three-photon absorption

In the case of three-photon absorption the imaginary part of the dielectric permeability medium is known to be

$$\delta\varepsilon'' = \varepsilon_0 m_4 |E|^2 \quad , \tag{52}$$

where m_4 is a real coefficient. By introducing the notation (44) and also taking

$$\mu_4 = \frac{m_4 E_o^2}{n_2} N^2 \tag{53}$$

and considering eqs. (22), (23), (24) one has the following equation for the variable X:

$$\frac{\partial^2 X}{\partial r_1^2} + \frac{1}{r_1} \frac{\partial X}{\partial r_1} + 2iN \frac{\partial X}{\partial z_1} + (N^2|X|^2 + i\mu_4|X|^4)X = 0 \quad , \tag{54}$$

with boundary condition (46). At $\mu_4 = 0$, eq. (54) reduces
to eq. (45). Parameter N, according to (44) and (38),
is associated with incident the beam power P_o through
the relation:

$$N = N_1 \sqrt{\frac{P_o}{P_{cr}^{(1)}}} \quad , \quad (55)$$

where

$$P_{cr}^{(1)} = \frac{cN_1^2 n_o}{8n_2 k^2} \approx P_{cr} \quad . \quad (56)$$

The results of the numerical calculation of the
problem are available in Dyshko et al (1971) for values
of the parameter N ranging from 2 to 10, which corresponds to an incident beam power from about $P_{cr}^{(1)}$ to
27 $P_{cr}^{(1)}$ and for parameter μ_4 in the range of
$10^{-3} < \mu_4 < 0.2$. Note at once that at $\mu_4 = 0$ the energy
density at $z_1 \rightarrow z_1^*$ increases only to finite values as
was expected. The numerical calculations also show that
the process of light beam propagation in the medium is
qualitatively reduced to formation of a beam multifoci
structure. The multifoci structure is a finite row of
single foci on the axis which result from successive
focusing of various annular regions of the beam. Here
the point $z_1 = z_1^*$ itself determines the center of the
first focus of a multifoci structure.

The longitudinal section of the beam propagating
in the medium is given graphically in Fig. 1, from
which one can see that there are a number of foci on
the beam axis. It is also seen that only a fraction of
the initial beam power flows through the first focus.
The energy "inflowing" into the focus is partially

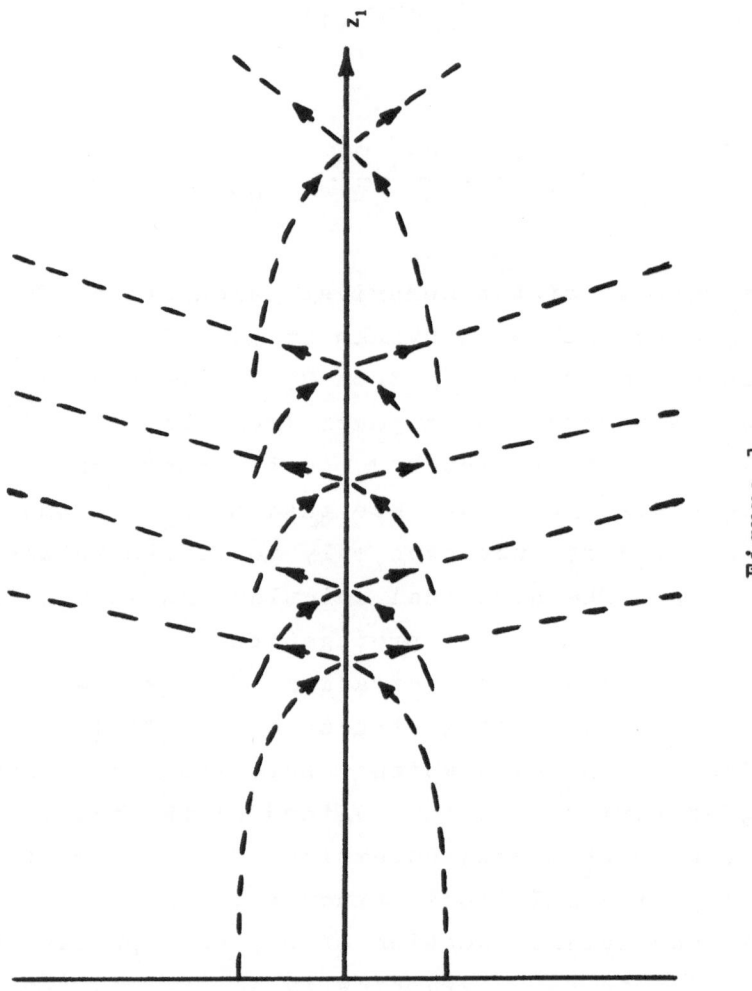

Figure 1

absorbed due to nonlinear absorption in the medium and leaves the focus in the form of a rapidly diverging (from the region of the incident beam) annular wave. That part of the beam which has passed the first focus forms a second focus some distance away and in the same manner. This focus receives power close to the critical value $P_{cr}^{(1)}$. Similarly part of the energy is absorbed at this focus due to nonlinear absorption in the medium and the rest of it flows out in the form of a diverging annular wave. Third, fourth foci, etc., are formed in the same way. Thus the beam passing through each focus is somewhat depleted, and foci are formed until the entire power of the initial beam has been spent. When it has, foci are no longer formed. This means that the total number of foci is finite and depends on the initial beam power. The m focus is formed at power $P = P_{cr}^{(m)}$ or, in other words, at $N = N_m$. The quantities $P_{cr}^{(m)}$ and N_m, following eqs. (55) and (56), are connected through the relation

$$P_{cr}^{(m)} = \frac{cN_m^2 n_o}{8n_2 k^2} \quad . \tag{57}$$

The numerical calculations show that at sufficiently low values of coefficient μ_4, $P_{cr}^{(m)}$ can be evaluated from:

$$P_{cr}^{(m)} \approx mP_{cr}^{(1)} \quad . \tag{58}$$

For a better understanding of the character of this solution, we look at Fig. 2 the upper curve of which gives the ratio of beam power P for $z_1 > 0$ to power P_o of the incident beam vs. z at $N = 6$, $\mu_4 = 0.05$. The three lower plots of this Figure give $|X|^2$ as a function of z_1 on the beam axis (i.e. at $r_1 = 0$)

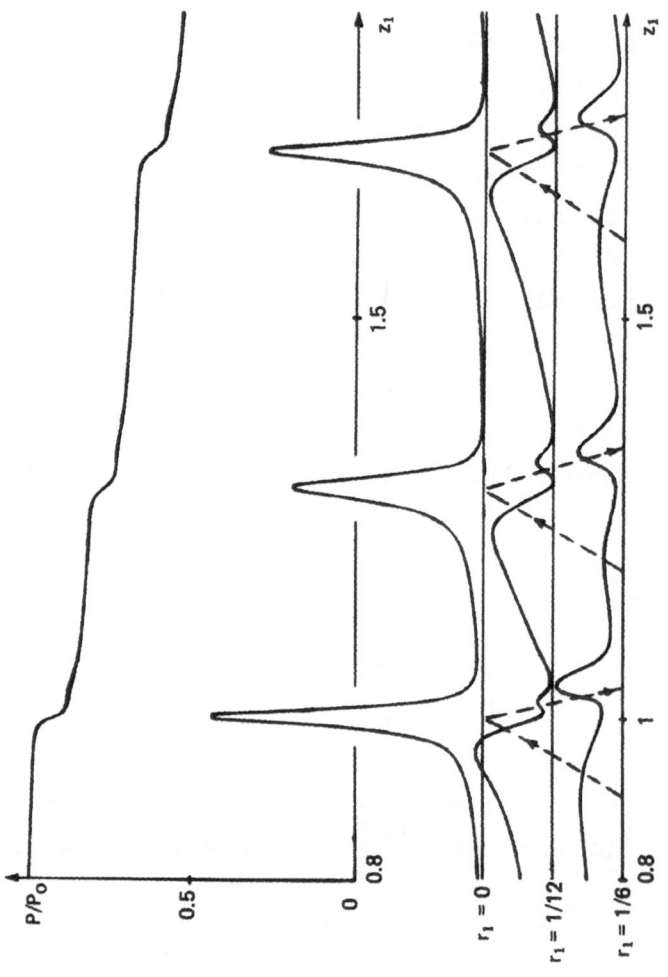

Figure 2

and on two cylinders off the axis ($r_1 = \frac{1}{12}$ and $r_1 = \frac{1}{6}$)
with z_1 lying between 0.8 and 1.9 (this corresponds to
the range of the first three foci). For convenience the
plots are drawn to different scales (larger z_1 correspond to a larger scale). The upper plot gives relative
beam power P/P_o vs. z_1 over the same range of values z_1.
The dependence of $|X|^2$ on z_1 at $r_1 = 0$ has three sharp
peaks, corresponding to three focal regions on the beam
axis. The curve of $|X|^2$ vs. z_1 at $r_1 = \frac{1}{12}$ and $r_1 = \frac{1}{6}$
describe the process of foci formation with annular
waves leaving them (the latter process is represented
by the dashed lines just as in Fig. 1).

As regards the character of solution for greater
distance from the beam axis ($r_1 \gtrsim 1$), the annular waves
diverging from different foci (which, in general, inter-
fere with the wave passing out of the foci situated before plane z_1) form a complex annular structure characterized by a number of maxima and minima of the variable
$|X|^2$ which depend on r_1. The above annular structure is
observed after the first focus (i.e. at $z_1 > z_{\phi 1}$). Its
bounding surface is well approximated by the surface of
a cone with an apex dependent on the angle at which the
wave leaves the first focus.

Now consider the curve of relative beam power P/P_o
vs. z_1 given in Fig. 2. One can see that power P after
the beam passes through each focus is decreased by a
value of the order of $P_{cr}^{(1)}$ (in our case N = 6,
according to (56) we have $P_{cr}^{(1)}/P_o \approx 0.1$). This power
decrease seems to result from absorption of an appreciable part of the electromagnetic energy received by
the focal region. Therefore, it is clear that energy
flow $P_{\phi m}$ through the "central" plane of this region
(i.e. plane $z_1 = z_{\phi m}$ where $z_{\phi m}$ is the point in the m-th

focal region where $|X|^2$ is maximum should be equal only
to a fraction of $P_{cr}^{(1)}$. According to Dyshko et al
(1971), over the whole range of values N and M_4 and for
all numbers of focal regions, $P_{\phi m}$ has practically the
constant value $P_{\phi m} \approx \frac{2}{3} P_{cr}^{(1)}$ (with absolute accuracy
$0.025 P_{cr}^{(1)}$).

The dependence of $|X_{\phi m}|^2$ on the parameter μ_4 in the
range of values $|X_{\phi m}|^2$ considered is approximately an
inverse proportion ($|X_{\phi m}|^2 \propto \frac{1}{\mu_4}$). More exact information
about the character of this dependence is given in Fig.
3 by the curve of $|X_{\phi 1}|^2$ vs. $\frac{1}{\mu_4}$ at N = 6.

The total number of foci in the beam and their
positions on the z_1 axis depend on N and μ_4. However,
at sufficiently low values of μ_4 (which corresponds to
a well-defined multifoci structure where values of
$|X_{\phi m}|^2$ are rather high) foci positions $z_{\phi m}$ are almost independent of μ_4 and are close to the foci positions
described by Dyshko et al (1971). Foci positions on
the z_1 axis depend strongly only on parameter N. Fig.
4 illustrates a family of curves which determine $z_{\phi m}$ as
a function of N. Here the value of parameter μ_4 for each
N was chosen so that value $|X_{\phi 1}|^2$ was about 170, i.e.
rather large (the corresponding ratios μ_4/N^2 vs. N are
also shown in Fig. 4). For comparison, a similar family
of curves obtained in Dyshko et al (1967) is presented
in Fig. 5. These families of curves have the same
character and very similar quantitative features (dis-
crepancies are mainly observed for foci of great multi-
plicity and which can be attributed to the fact that
for given values of μ_4 the power absorbed between
adjacent focal regions (though not great)
is comparable with that absorbed in the focal regions
themselves (see Fig. 2). At lower μ_4, the power

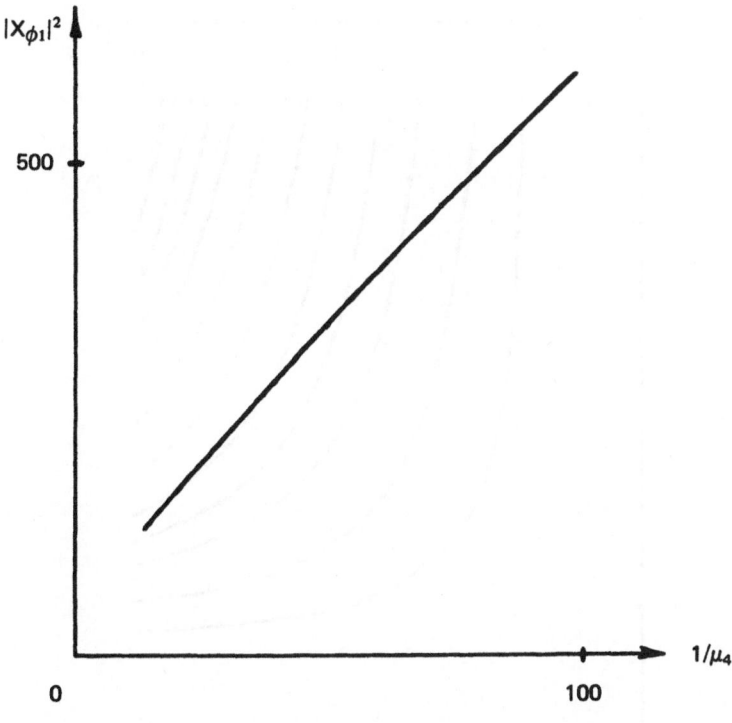

Figure 3

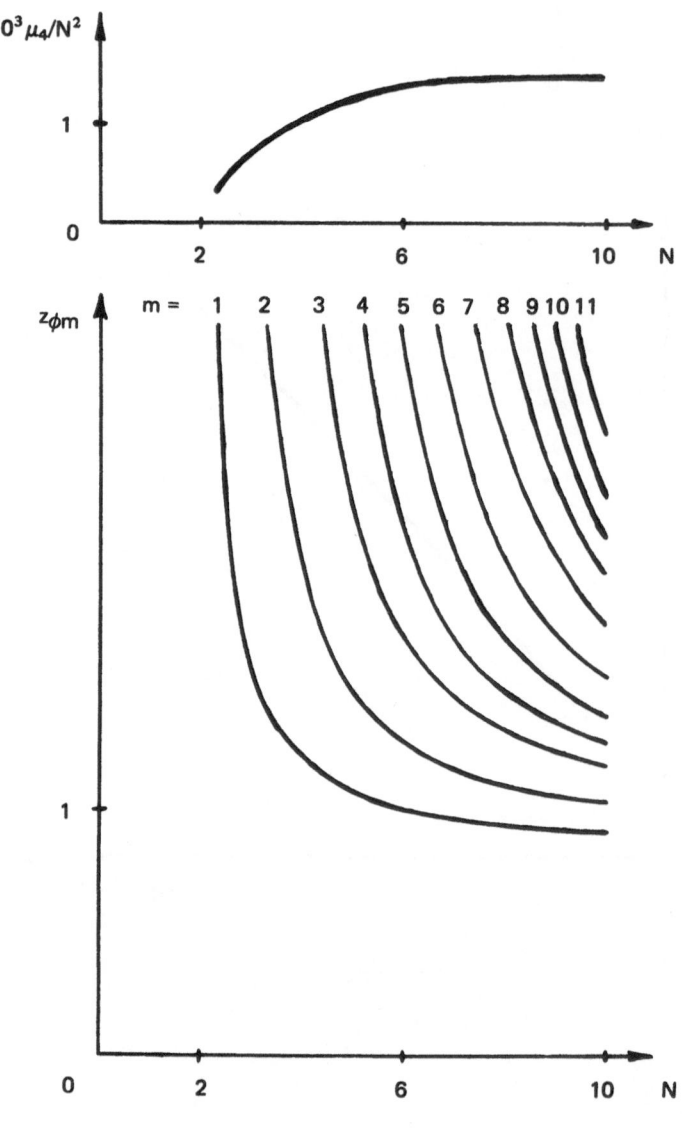

Figure 4

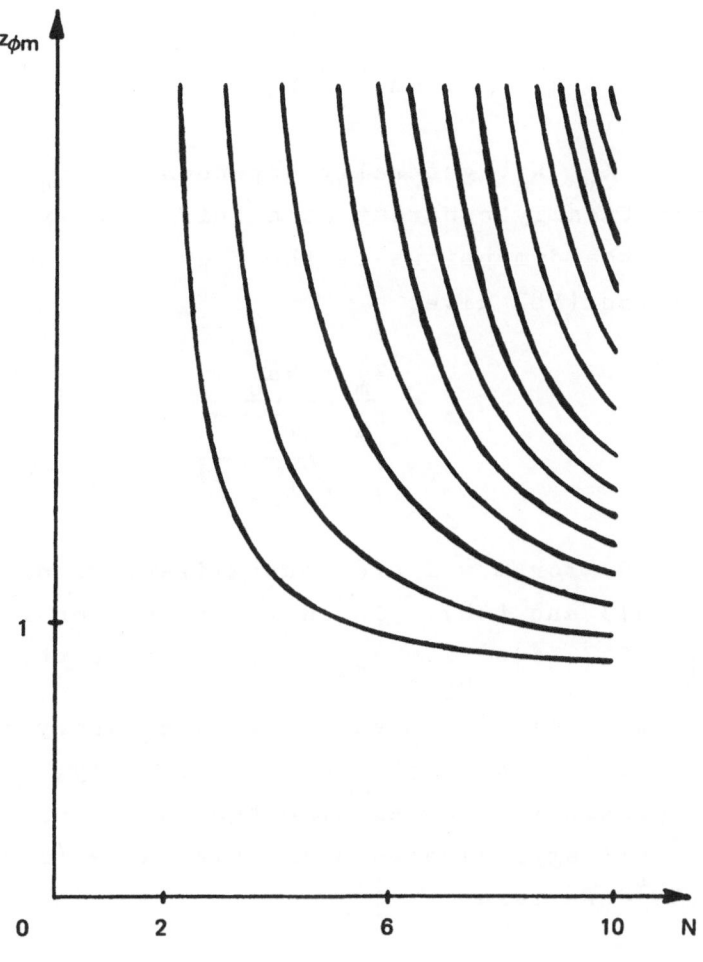

Figure 5

absorbed between the adjacent regions decreases and the positions of all foci become close to those determined by Fig. 5.

Thus, it is evident that foci positions $z_{\phi m}$ can be approximated by the analytical dependence

$$z_{\phi m} = \frac{X_m N}{N - N_m} \qquad (59a)$$

where values X_m, N_m, generally dependent on μ_4 and N, but are practically constant at a sufficiently low value of μ_4. For the dimensions of the foci positions the $\zeta_{\phi m} = z_{\phi m} \ell_x$ eq.(58) gives

$$\zeta_{\phi m} = \frac{X_m}{N_m} \frac{k \bar{a}_o^2}{\sqrt{\dfrac{P_o}{P_{cr}^{(m)}} - 1}} \qquad . \qquad (59b)$$

Equations (59) for $m = 1$ are generalizations of expressions (47) and (48). In fact, as is seen from Fig. 5, $\lim\limits_{\mu_4 \to o} X_1 \approx 0.7$, i.e. $\lim\limits_{\mu_4 \to o} z_{\phi 1} = z_1^*$. The quantities $P_{cr}^{(m)}$ and N_m included in eqs. (59) are easily found by means of eqs. (57) and (58), or more exactly, from Figs. 4 and 5. Consider as an example the curves of Fig. 4, which are well approximated over this range ($1 \leq m \leq 4$) if in eqs. (59) one takes

$$X_1 = 0.72, \ X_2 = 0.76, \ X_3 = 0.79, \ X_4 = 0.77$$

$$(60)$$

$$N_1 = 1.77, \ N_2 = 2.48, \ N_3 = 3.24, \ N_4 = 3.85 \quad ,$$

here values of N_m are close to the exact ones (though small differences arise from the approximations used.

In concluding this section we note that this picture of beam propagation in the medium also holds for the case of beams with differing "smooth" initial intensity distributions over a plane*

b) Stimulated Raman Scattering in Forward Direction

A comprehensive description of light beam propagation with allowance for SRS in a medium is generally very difficult since it is necessary to take into account δ-correlated sources of the primary scattered (Stokes) radiation that are distributed in the medium (see Lugovoi, 1968). Therefore, just consider an ideal situation where the energy is transferred from the main beam into the first Stokes component of SRS, which, in turn is initiated by a preset monochromatic "priming" beam incident, like the main beam, from the exterior on the boundary ($z = 0$) of the medium in question. If with respect to the first Stokes component beam one takes into account the Kerr effect, then independent focal regions might form in the latter due to its assumed coherence. In what follows we shall not be interested whether such situations are possible or not under real conditions**. So we shall take into account the Kerr effect for the

---

*The same description is correct for a beam prefocused by an ordinary lens (the boundary condition for this beam has the form (36)). A numerical solution of this problem was obtained by Dyshko et al (1969). A similar problem was considered by Talanov (1970). In the latter paper, however, the transformation of variables used in connection with this problem breaks down at $z = R$ (i.e. in the focal plane of a lens). Therefore, the picture of beam propagation for $z > R$ suggested in Talanov (1970) does not hold.

**See next page

main beam and not for the beam of the first Stokes
frequency.   The two beams are assumed to be axially
symmetric.   In most cases of practical interest one can
neglect the changing population of the ground vibrational
state of the molecules of the substance.* Under these
conditions the imaginary part of the dielectric permeability of the medium in the main beam field is posi-
tive and equal to $4\pi\Gamma|E_{-1}|^2$ and in the field of the first
Stokes component of SRS is negative and equal to $4\pi\Gamma|E|^2$,
where E is the complex amplitude corresponding to the
main beam (see (6)) and similarly E_{-1} is the complex
amplitude corresponding to the beam of the first Stokes
frequency, Γ is a coefficient (see Lugovoi, 1968).   Thus,
eq. (20) for E and the equation for E_{-1} in the same
approximation have the form

$$\frac{\partial^2 E}{\partial r^2} + \frac{1}{r}\frac{\partial E}{\partial r} + 2ik\frac{\partial E}{\partial z} + k^2(n_2|E|^2 + \frac{4\pi\Gamma}{\varepsilon_0}|E_{-1}|^2)E = 0 \ , \quad (61)$$

$$\frac{\partial^2 E_{-1}}{\partial r^2} + \frac{1}{r}\frac{\partial E_{-1}}{\partial r} + 2ik_{-1}\frac{\partial E_{-1}}{\partial z} \quad k_{-1}^2 \frac{4\pi\Gamma}{\varepsilon_0}|E|^2 E_{-1} = 0 \quad .$$

Here $k_{-1} = \frac{2\pi}{\lambda_{-1}}$; $\lambda_{-1} = \frac{2\pi c}{\omega_{-1}n_0}$; $\omega_{-1} = \omega-\omega_0$ is the first
Stokes frequency; ω_0 the frequency of vibrational transi-
tion of the substance associated with the SRS process.
     To be more exact, in analogy with eq. (23), consider
the system (61) under conditions

**There is an experimental work (Kundryavtseva, 1972)
  which supports the possibility.

---

*The role of excited vibrational states of molecules was
 analyzed in Chiao et al (1966) and Brewer and Townes
 (1967).

$$E\big|_{z=0} = E_o e^{\frac{-r^2}{2\bar{a}_o^2}}, \qquad E_{-1}\big|_{z=0} = E_{-10}\, e^{\frac{-r^2}{2\bar{a}_o^2}}, \qquad (62)$$

giving the Gaussian initial distribution of the two beams with the same distribution radius and their plane phase front in the initial cross-section. By introducing the notation of eq. (44) and writing also $H = k\bar{a}_o \sqrt{(\frac{4\pi\Gamma}{\varepsilon_o} E_o^2)}$, $Y = \frac{E_{-1}}{E_o}$, $\xi = \frac{k_{-1}}{k} = \frac{\omega_{-1}}{\omega}$, one derives the following system of equations for dimensionless variables X and Y:

$$\frac{\partial^2 X}{\partial r_1^2} + \frac{1}{r_1}\frac{\partial X}{\partial r_1} + 2iN\frac{\partial X}{\partial z_1} + (N^2|X|^2 + iH^2|X|^2)X = 0$$

$$\frac{\partial^2 Y}{\partial r_1^2} + \frac{1}{r_1}\frac{\partial Y}{\partial r_1} + 2iN\xi\frac{\partial Y}{\partial z_1} - i\xi^2 H^2|X|^2 Y = 0 \qquad (63)$$

with the boundary condition

$$X\big|_{z_1=0} = e^{-\frac{1}{2}r_1^2}, \qquad Y\big|_{z_1=0} = \alpha e^{-\frac{1}{2}r_1^2}, \qquad (64)$$

where $\alpha = E_{-10}/E_o$ is the ratio between the initial field strength on the priming beam axis and that of the main one. The numerical solution given in Dyshko et al (1971) for the values $\alpha = 10^{-4}$, $\xi = 0.9$ and for N and H ranging from 4 to 10 has shown that for every N there is a value $H = H_{cr}(N)$ such that at $H < H_{cr}$, $|X|^2$ as a function of z_1 increases infinitely on the beam axis while approaching the first focus*. For $H > H_{cr}$ the solution is completely defined, i.e. bounded (within eqs. (61)), over the whole range of $z > 0$. In this case

---

*See next page

if H is just a little larger than H_{cr} the solution is a
multifoci structure in which the positions of corresponding first foci differ only slightly from those of
Fig. 5. Further foci may not exist at all since in the
process of propagation a considerable quantity of energy
is transferred from the main to the Stokes beam. The
latter introduces absorption into the main beam. Thus,
intensity increases on the axis necessary in the formation of further foci is entirely suppressed by absorption.

Figure 6 is an example of $|X|^2$ vs. z_1 curves when
$N = 6$, $H = 1.04 \ H_{cr}$ ($H_{cr} \approx 7$). The three solid curves
in this figure show $|X|^2$ vs. z_1 on the beam axis ($r_1=0$)
and on two cylinders $r_1 = \frac{1}{12}$ and $r_1 = \frac{1}{6}$, with z_1 ranging
from 0.8 and 1.9 respectively. Each of the plots is
made to the same scale as Fig. 2. The dashed lines in
Fig. 6 schematically show foci formation and ring waves
leaving them (in our example there are three foci in the
main beam and all three foci lie at the indicated values
of z_1). The power $P_{\phi m}$ flowing through the central cross-
section of the mth focal region in this case for all the
three focal regions is practically the same,
$P_{\phi m} \approx 0.7 \ P_{cr}^{(1)}$ (with an absolute accuracy of
$0.01 \ P_{cr}^{(1)}$). Taking this into account and proceeding
from $|X_{\phi m}|^2$ as determined by the plot in Fig. 6 one can
easily determine cross-sections of the corresponding
focal regions. As is shown, these cross-sections rapidly

---

*Just as in Section 3, we do not touch here upon the
question of the existence of a mathematical singularity
in the solution, but assume that the rate of growth of
$|X|^2$ vs. z, while approaching the first focus, proves
to be so high that condition $\Lambda_{11} \gg \lambda$ (see eqs. (9))
necessary for initial eqs. (61) to be valid actually
breaks down.

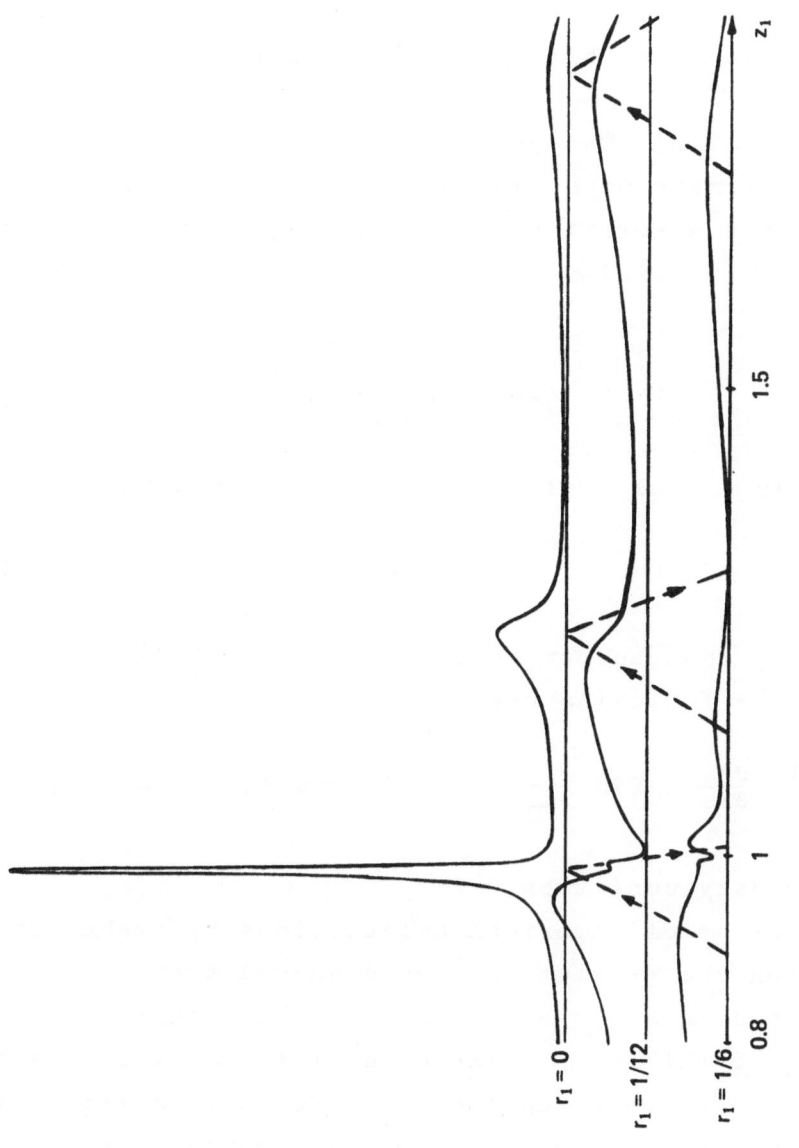

Figure 6

increase with the focus number due to sharp decrease of $|X_{\phi m}|^2$ with increasing m. For this reason, under experimental conditions one can register only one first focus if the receiver is not sensitive enough to register regions exhibiting far lower energy densities.

c)   Two-photon Absorption

If the main mechanism of absorption in a medium is a two-photon absorption, the imaginary part of the dielectric permeability of the medium is proportional to $|E|^2$:

$$\delta\epsilon'' = \epsilon_0 m_2 |E|^2 \qquad . \qquad (65)$$

Here m_2 is a real coefficient. By designating

$$\mu_2 = \frac{m_2}{n_2} N^2 \qquad (66)$$

and taking into account eqs. (22), (23), (24), (44) one derives the following equation for X:

$$\frac{\partial^2 X}{\partial r_1^2} + \frac{1}{r_1}\frac{\partial X}{\partial r_1} + 2iN\frac{\partial X}{\partial z_1} + (N^2|X|^2 + i\mu_2|X|^2)X = 0 \quad , \quad (67)$$

with boundary condition (46).

A series of numerical calculations by Dyshko et al (1971) for the problem concerned showed that in the plane N, μ_2 there is a curve $\mu_2 = \mu_{2cr}(N)$ such that at $\mu_2 < \mu_{2cr}$, $|X|^2$ vs. z_1 increases infinitely on the beam axis, while approaching the first focus (similar to the case when $H < H_{cr}$, see previous paragraph). At $\mu_2 > \mu_{2cr}$ the solution is perfectly well defined over the whole region $z_1 > 0$. In the latter case, if μ_2 is slightly larger than μ_{2cr} (for example $\mu_{2cr} < \mu_2 < 2\mu_{2cr}$)

the solution obtained is a multifoci structure. Its
specific feature is that the foci positions appreciably
differ from those of Fig. 5. This difference is due to
the fact that two-photon absorption beyond the foci
might be vanishingly small only at a sufficiently small
coefficient μ_2, whereas for $\mu_2 > \mu_{2cr}$ this absorption
will always be finite, i.e. the power absorbed between
neighboring focal regions (and between the initial plane
$z_1 = 0$ and the first focal region) will be finite. For
this reason, the total number of foci proves to be considerably smaller than in the absence of two-photon absorption in a medium. Nevertheless, the mechanism of
foci formation and the beam structure turns out to be
the same as in the two previous cases.

Fig. 7 shows three curves for $|X|^2$ vs. z_1 at $r_1=0$,
$\frac{1}{12}$, $\frac{1}{6}$; $N = 6$, $\mu_2 = 2.6$, respectively. Each of these
curves is drawn to the same scale as those of Fig. 2.
In the range $1 < z_1 < 2.1$ represented in Fig. 7 there
are first two foci. Comparison of Figures 2 and 7
shows that foci formations and the departure of ring
waves for the cases of two- or three-photon absorption
in a medium are qualitatively equal. The dependence of
relative beam power P/P_0 upon z_1 represented by the
upper curve of Fig. 7 shows that the power absorbed in
each focus is about $P_{cr}^{(1)}$. Besides, an appreciable
part of the beam power is absorbed between the initial
plane $z_1=0$ and the first focal region, and between
neighboring focal regions. The power $P_{\phi m}$ flowing
through the central cross-section of the m-th focal
region at m = 1.2 is equal to $P_{\phi 1} \approx 0.61\ P_{cr}^{(1)}$,
$P_{\phi 2} \approx 0.66\ P_{cr}^{(1)}$.

As for the positions of foci $z_{\phi m}$ on axis z_1 they
depend on N and μ_2. Fig. 8 shows a family of curves

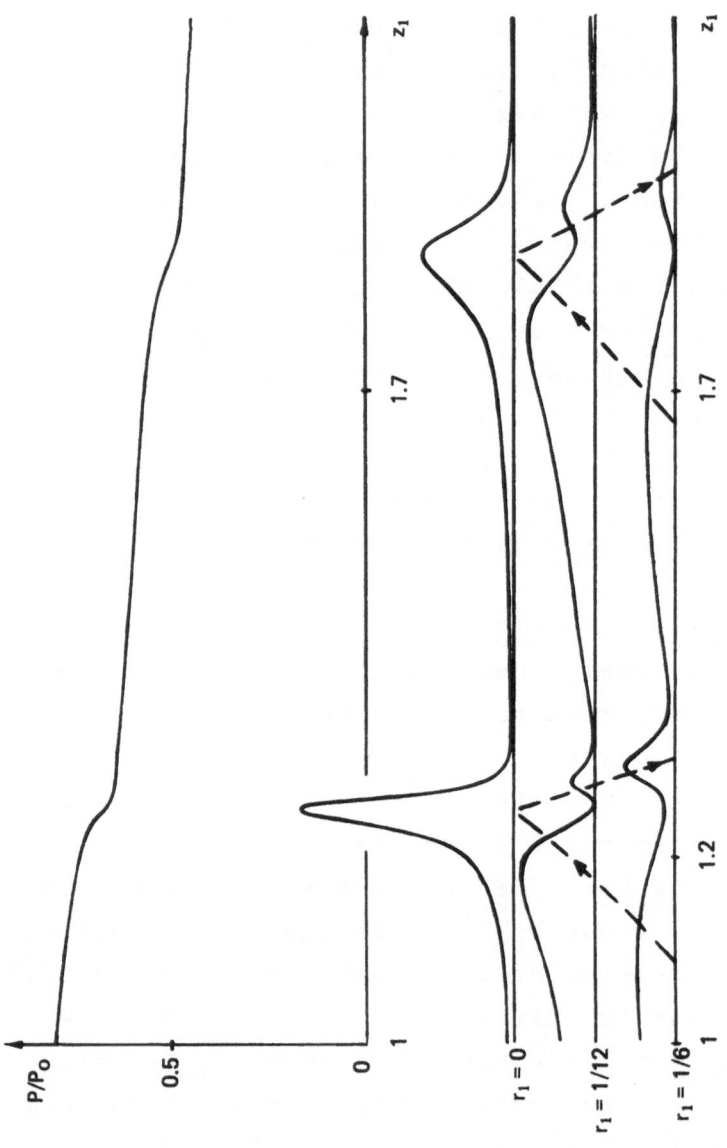

Figure 7

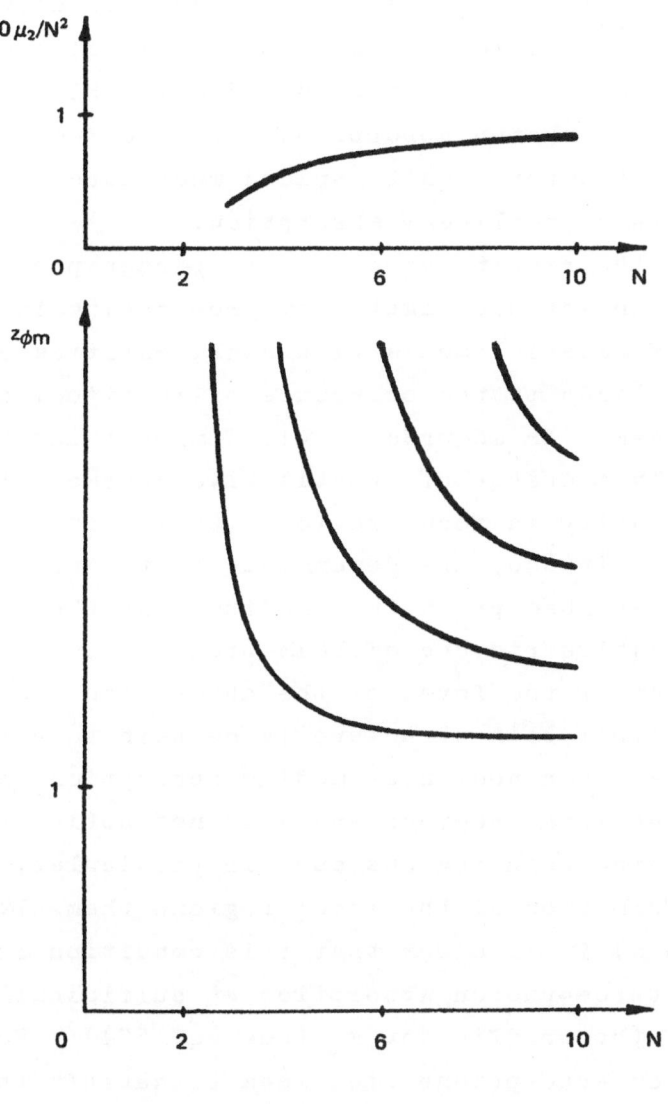

Figure 8

giving the dependence of $z_{\phi m}$ upon N at certain values of μ_2. Values of μ_2 for each N were chosen so that the value of $|X_{\phi 1}|^2$ was preserved and equal to about 110. The plot of corresponding μ_2/N^2 vs. N is also given in Fig. 8. Comparison of Figs. 4 and 8 (or Figs. 2,6,7) shows for two-photon absorption that the foci in a multifoci structure can be spaced much more than for other forms of nonlinear absorption.

Thus the results discussed in paragraphs a), b), c) show that an intense light beam propagating in a wide variety of material media with nonlinearities of the Kerr type leads to the occurence of multifoci structure for the beam. In accordance with Lugovoi and Prokhorov (1968) only a number of quantitative characteristics (energy density in focal regions, their size, number and relative positions) are determined by the concrete form of nonlinear absorption in a medium. At the same time the qualitative picture of beam propagation proves to be independent of the forms of the absorption considered.

Multifoci beam structure is certain to arise in those cases when nonlinear medium absorption shows up only in the focal regions and does not noticeably affect the remaining beam regions and, in particular, does not prevent formation of the focal regions themselves. From paragraph a) it is clear that this condition is satisfied for three-photon absorption at sufficiently low values of the coefficient m_4 (see eq. 52)). Four-photon, five-photon absorptions etc. seem to satisfy the same conditions for sufficiently small corresponding coefficients m_6, m_8, etc. The examples of other forms of nonlinear absorption considered in paragraphs b) and c) also show that multifoci beam structure can occur for high absorption beyond the focal regions. In the latter

case the total number of foci and their positions along
the beam axis strongly depend on the power absorbed
beyond the foci.

6. Foci Structure

The longitudinal structure of focal regions is
characterized by the dependence of $|X|^2$ on z_1 at $r_1 = 0$
in the vicinity of its maxima with respect to z_1. This
dependence is shown in Figs. 2,6,7, respectively. For
all the three forms of nonlinear absorption considered
above, these figures show that focal regions with
different numbers are similar to each other and the
dependence of $|X|^2$ on z_1 is asymmetric with regard to
maximum points of $|X|^2$ (the fall of the corresponding
curves is steeper than the rise). It is obvious that
the following approximation of $|X|^2$ is possible at $r_1 = 0$
in the vicinity of the m-th focal region:

$$|X|^2 = \begin{cases} \dfrac{|X_{\phi m}|^2}{[1+(\frac{z_1-z_{\phi m}}{\mu_m})^2]^\alpha} & , \quad z_1 \leq z_{\phi m} \\[3em] \dfrac{|X_{\phi m}|^2}{[1+(\frac{z_1-z_{\phi m}}{\mu_m})^2]^\beta} & , \quad z_1 \geq z_{\phi m} \end{cases} \quad (68)$$

where $\alpha \approx \frac{1}{2}$, and the value of β generally depends on the
form and magnitude of the nonlinear absorption and
usually is $\beta = 1$ to 2.

As regards the transverse structure of focal
regions (which is determined by the dependence of $|X|^2$

on r_1 in the cross-sections $z_1 = z_{\phi m}$), the corresponding curves for the case of a three-photon absorption (at $N = 6$, $\mu_4 = 0.05$) for the first three foci are given in Fig. 9. One can see that the transverse structure of different focal regions are similar too. This similarity is understandable if one takes into account that the formation mechanism of all foci is identical. The same structure of focal region cross-sections is preserved for the other forms of nonlinear absorption discussed above. In what follows $r_{\phi m}$ stands for the radial distribution over r_1 in the m-th focal region, i.e. the value of r_1 when $|X|^2 = \frac{1}{2} |X_{\phi m}|^2$. From Fig. 9 one can see that at $r_1 \leq r_{\phi m}$ the radial intensity distribution in the focal region is rather close to the Gaussian one:

$$|X|^2 = |X_{\phi m}|^2 \exp\left[-\left(\frac{r_1}{r_{\phi m}}\right)^2 \ln 2\right] \quad . \tag{69}$$

It is convenient to write down equations (68), (69) in the dimensional form:

$$|E|^2_{r=0} = \begin{cases} \dfrac{|E_{\phi m}|^2}{\left[1+\left(\dfrac{z-\zeta_{\phi m}}{\Lambda_{\phi m}}\right)^2\right]^{\alpha}} \quad , \quad z \leq \zeta_{\phi m} \\[40pt] \dfrac{|E_{\phi m}|^2}{\left[1+\left(\dfrac{z-\zeta_{\phi m}}{\Lambda_{\phi m}}\right)^2\right]^{\beta}} \quad , \quad z \geq \zeta_{\phi m} \end{cases} \tag{70}$$

$$|E|^2_{z} = \zeta_{\phi m} = |E_{\phi m}|^2 \exp\left(-\frac{4r^2}{d^2_{\phi m}} \ln 2\right) \quad , \quad r \leq \frac{d_{\phi m}}{2} \quad .$$

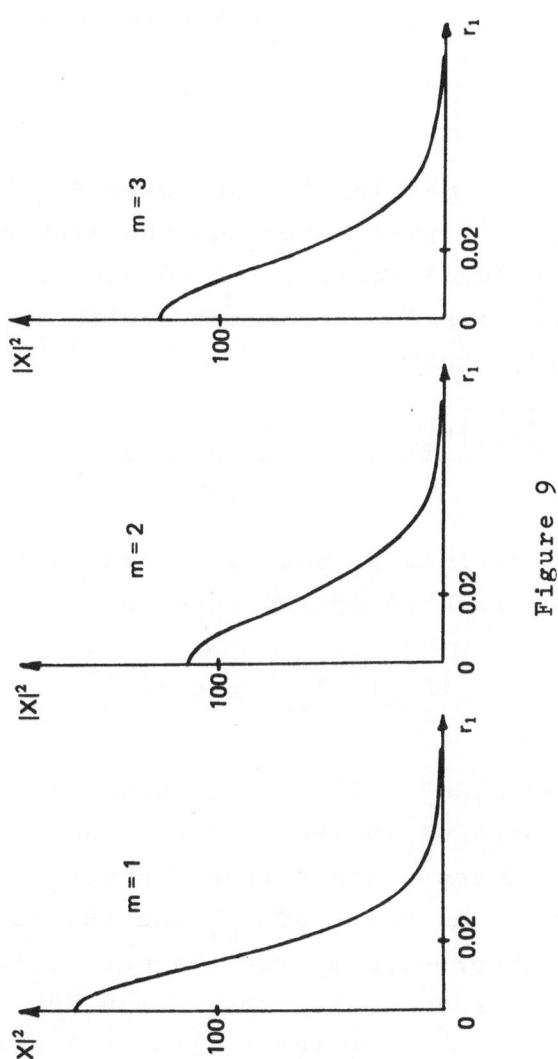

Figure 9

Here $E_{\phi m} = E_o X_{\phi m}$, $d_{\phi m} = 2 r_{\phi m} \bar{a}_o$ is the diameter of the focal region, $\Lambda_{\phi m} = \mu_m \ell_x$ is the scale of longitudinal changes of the field in the focal region ($\Lambda_{\phi m} \equiv \Lambda_{\parallel}$), associated with $\beta \sim 1$ to 2 and with the total length $\ell_{\phi m}$ of the focal region given by the relation

$$\ell_{\phi m} \approx 2.5 \, \Lambda_{\phi m} \qquad . \qquad (71)$$

According to Section 5, the power $P_{\phi m}$, which is close to $\frac{2}{3} P_{cr}^{(1)}$, passes through the central cross-section of any focal region, regardless of the conditions concerned and the form of nonlinear absorption. Because of (70), then

$$P_{\phi m} = \frac{c n_o d_{\phi m}^2 |E_{\phi m}|^2}{32 \ell n 2} = \frac{N^2 |X_{\phi m}|^2 r_{\phi m}^2}{N_1^2 \ell n 2} P_{cr}^{(1)} \qquad . \qquad (72)$$

Therefore the maximum intensity values $|E_{\phi m}|^2$ and diameters $d_{\phi m}$ are related by the equation

$$|E_{\phi m}|^2 \, d_{\phi m}^2 \approx 0.18 \, \frac{\lambda^2}{n_2} \qquad . \qquad (73)$$

Here one takes $N_1^2 \ell n 2 \approx 2.7$. The values of $d_{\phi m}$ depend on nonlinear absorption in the medium. The results represented in Figures 3 and 4 (see Section 5a) provide relations between the value of $d_{\phi 1}$ and the coefficient for three-photon absorption m_4 for the most interesting conditions. It is easy to see that for $\mu_4/N^2 \sim 2 \times 10^{-3}$ to 2×10^{-4} (i.e. $|E_{\phi 1}|^2$ from 100 to 1000 E_o^2) and $N \geq 4$ (i.e. $P_o \gtrsim 4 P_{cr}^{(1)}$) the energy density $|E_{\phi 1}|^2$ in the center of the first focus does not depend in practice on the initial beam power, and the corresponding value of $|E_{\phi 1}|^2$ is equal to

$$|E_{\phi 1}|^2 \approx 0.15 \frac{n_2}{m_4} \quad , \qquad (74a)$$

i.e. it is determined only by medium constants n_2 and m_4. Hence the diameter, $d_{\phi 1}$, of a focal region is also determined by these constants*

$$d_{\phi 1} \approx 1.1\lambda \frac{\sqrt{m_4}}{n_2} \quad . \qquad (74b)$$

Similar relations are valid for other foci. Here, note that in agreement with the above results, the measured foci diameter for a given substance are roughly independent of observation conditions and are equal to 5×10^{-4} cm for carbon bisulphide, 1.2×10^{-3} cm for nitrobenzene, 10^{-3} cm for toluene (Brewer et al, 1968; Loy and Shen, 1969). It is noteworthy that these values correspond to the ruby laser wavelength in the medium ($\lambda = 0.5 \times 10^{-4}$ cm) and greatly exceed the value λ.

Now we estimate the power absorbed in focal regions, say, for the case of three-photon absorption in a medium.

---

*Note that, according to (74b), in the focal region and, hence, generally in the medium, the inequality $\delta\varepsilon'' \ll \delta\varepsilon'$ holds. As was mentioned in Section 1 this allows us to ignore corrections to the real part of ε, which are due to the presence of $\delta\varepsilon''$. The direct numerical calculations of this problem including these corrections (when $\delta\varepsilon$ is wirtten in the form

$\delta\varepsilon = \frac{1}{2}\varepsilon_2 |E|^2 - \varepsilon_0 n_4 |E|^4 + i\varepsilon_0 m_4 |E|^4$, where $n_4 \lesssim m_4$) show

that the beam propagation picture is qualitatively the same. This picture is quantitatively changed, but very little. For example, the focal region diameters vary by not more than 15%.

According to Loy and Shen (1969), the ratio of the power
absorbed in a focus to that received by it, is about
$km_4|E|^2_{\phi m}\ell_{\phi m}$ or with reference to eqs. (73), (74a) is
about $.15Kn_2\ell_{\phi m}$. According to the results of Section 5,
this ratio should be unity and $\ell_{\phi m} \sim kd^2_{\phi m}$. Thus, the
relation between the focal region length and diameter is
the same as it is in case for focusing in a linear
medium.

More exact values of the numerical coefficient in
this relation can be found directly from Figures 2,6,7
and from the values of the power passing through the
central cross-section of the focal regions. From these
plots and for the parameters and types of a nonlinear
absorption corresponding to them we find respectively

$$a) \quad \ell_{\phi m} \approx 3kd^2_{\phi m} \quad , \tag{75a}$$

$$b) \quad \ell_{\phi m} \approx 2kd^2_{\phi m} \quad , \tag{75b}$$

$$c) \quad \ell_{\phi m} \approx 4kd^2_{\phi m} \quad . \tag{75c}$$

Equations (75) show that condition $\Lambda_{||} \gg \Lambda_{\perp}$, (or
$\ell_{\phi m} \gg d_{\phi m}$) which is necessary for the parabolic equation to apply (see (9)) is certain to be reached be-
cause the inequality $\Lambda_{\perp} \gg \lambda$ (or $d_{\phi m} \gg \lambda$) is fulfilled.

Now consider a numerical example for typical experimental conditions. The initial beam radius $\bar{a}_o$ is

---

*For example, from eq. (34) it follows that in case of
 focusing of a beam with a Gaussian intensity distribution in a linear medium this length is $\ell_\phi \approx 0.7kd^2_\phi$.
 For beams with other initial intensity distributions
 only the numerical coefficients are different in this
 equation.

taken to be 0.15 mm (Korobkin et al, 1970; Loy and Shen, 1970). The wavelength λ in a substance is taken to be 0.5×10^{-4} cm, which corresponds to the ruby laser frequency at refractive index $n_o \approx 1.5$. Let us consider the substance carbon bisulphide, where $n_2 \sim 0.5 \times 10^{-11}$ CGSE and hence, according to Garmire et al (1966) and eq. (40), $P_{cr}^{(1)} \sim 20$kw*. If the incident beam power is $P_o \sim 200$ kw ≈ 10 $P_{cr}^{(1)}$, according to (55) we obtain N = 6. By means of eqs. (49) and (60) (or directly from the curves of Figures 4,5,8) one finds the distances $\zeta_{\phi m}$ of several first foci from the input plane of the medium: $\zeta_{\phi m} \sim 5-10$cm. Then from eq. (72) it follows that

$$|X_{\phi m}|^2 \approx \frac{2N_1^2 \ln 2}{3N^2 r_{\phi m}^2} \approx \frac{7.2}{N^2} \left(\frac{a_o}{d_{\phi m}}\right)^2 .$$ Since for carbon bisul-

phide $d_{\phi 1} \approx 5 \times 10^{-4}$cm so $|X_{\phi 1}|^2 \approx 180$ (or equivalent by $|E_{\phi 1}|^2 \approx 180 \ E_o^2$). Thus, this example lies within the interval of parameters assumed in the numerical solutions and corresponding analytical approximation. The length of a focal region is found from eqs. (75) to be $\ell_{\phi 1} \sim 1$mm.

Now consider a somewhat different example. Let the initial beam radius $\bar{a}_o$ be 0.5 mm; and take the length of a cell of the substance studied to be $\ell = 10$ cm. Then according to eq. (75), the first focus will appear inside the cell only if the initial beam

---

*The value P_{cr} for carbon bisulphide was experimentally measured by Garmire et al (1966). Similarly P_{cr} was experimentally measured by Wang (1966) to be 55 kw for toluene and 19 kw for nitrobenzene, that is the coefficient n_2 for these substances is about 0.2×10^{-11} CGSE and 0.5×10^{-11} CGSE respectively.

power is more than 150 times the critical one. It is
obvious that at such an excess of the initial power over
a critical value the deviations from axial symmetry of
the initial distribution in the real beam may contain a
supercritical power, thereby forming independent multifoci structures in different areas of the beam. Under
these conditions, the initial radius and the initial
power for each such structure will be appreciably
smaller than 0.5 mm and 150 $P_{cr}^{(1)}$ respectively. This
picture of the beam separating into several independent
beams at extremely high supercritical powers has been
observed experimentally. The initial radius $\bar{a}_o$ for
every independent beam is about .1 mm (see Korobkin and
Serov, 1967), which makes the initial power of the order
of 10 $P_{cr}^{(1)}$. Note that the possibility of such separation of a beam with a high supercritical power was
theoretically predicted in Besgalov and Talanov (1966),
Zakharov et al (1971), and Dyshko et al (1969)*.

Now we estimate the longitudinal component E_z of
the electric field amplitude in the focal regions.
According to eqs. (13) and (70), one has:

$$|E_z| = \frac{1}{k}\left|\frac{\partial E}{\partial r}\cos\phi\right| \approx |E_{\phi m}\cos\phi| \frac{4r\ell n2}{kd_{\phi m}^2}\exp\left(-\frac{4r^2}{d_{\phi m}^2}\ell n2\right), (76)$$

where ϕ is the polar angle of a cylindrical system of
coordinates r, ϕ. According to eq. (76) the maximum

---

*From this point of view the theoretical consideration
(Zalcharov and Shabat, 1971) of two-dimensional beams
in the media in question is formal, since under real
conditions they separate into three-dimensional beams
(for example, approximately axially symmetrical ones).

of $|E_z|$ occurs at $\phi=0$, π and at $r=r_{max}$, where

$$|E_z|_{max} \approx 0.1 \frac{\lambda}{d_{\phi m}} |E_{\phi m}| \qquad . \qquad (77)$$

The amplitude of the transvere component of the electric field $|E_{\phi m}|$ is easily estimated by means of eq. (73). At $n_2 = 0.5 \times 10^{-11}$ CGSE, $d_{\phi m} = 5 \times 10^{-4}$ cm, $\lambda = 0.5 \times 10^{-4}$ cm one finds $|E_{\phi m}| \approx 1.9 \times 10^4$ CGSE and then by means of eq. (77) obtains $|E_z|_{max} \approx 1.9 \times 10^2$ CGSE.

In concluding this Section we note that in some efforts (see Gurevitch and Shvartsburg (1970) and Gurevitch et al (1971) light beam propagation in a media with Kerr nonlinearity is described in terms of geometrical optics approximations. However, the latter is not applicable to the description of multifoci structure. In fact, the first focus receives power $P_{\phi 1} \approx P_{cr}^{(1)}$, from which one finds $\ell_{H\Lambda} \sim \ell_d$ where $\ell_{H\Lambda} = \dfrac{d_{\phi 1}}{2\sqrt{(n_2|E_{\phi 1}|^2)}}$ is the characteristic length of the field change in the focal region associated with the substance nonlinearity (and determined by geometrical optics); $\ell_d \sim kd_{\phi 1}^2$ is the characteristic length of the diffractive divergence in a corresponding linear medium. Since $\ell_{H\Lambda}$ and ℓ_d are of the same order of magnitude, the beam behavior in a focal region is equally determined by geometrical-optical refraction due to medium nonlinearity and by diffraction. Therefore, the geometrical optics approximation fails in the first focal region and in all of the region beyond it.

IV.   PROPAGATION OF POWERFUL LIGHT PULSES IN A NONLINEAR MEDIUM.   THE THEORY OF MOVING FOCI

7.   Quasi-stationary Beams

To analyze the propagation of powerful light pulses in a nonlinear medium under quasistationary conditions, the time t according to eqs. (20) and (21) should be a parameter in the boundary condition we shall give as an example is

$$E\Big|_{z=0} = E_o(t)e^{-\dfrac{r^2}{2\bar{a}_o^2}} , \qquad (78)$$

i.e. we assume that at all t an incident beam has a Gaussian intensity distribution with constant radius $\bar{a}_o$ and a flat phase front. Function $E_o(t)$ in eq. (78) is determined by the laser pulse envelope, i.e. by the time dependence of the initial power P_o of the beam.

Now we analyze the solution obtained in the previous Chapter taking into account the change in the initial beam power with time. Since the longitudinal positions of all foci depend strongly on this power one can conclude at once that the foci will move along the beam axis following the laser pulse envelope (Lugovoi and Prokhorov, 1968). From (59b) one can derive time dependence of foci positions $\zeta_{\phi m}$:

$$\zeta_{\phi m}(t) = \frac{X_m}{N_m} \frac{k\bar{a}_o^2}{\sqrt{\dfrac{P_o(t)}{P_{cr}^{(m)}} - 1}} . \qquad (79)$$

Here X_m/N_m is a numerical coefficient (see (59)), and $P_o(t) = \dfrac{cn_o}{8} \bar{a}_o^2 E_o^2(t)$. The total number m of all foci at z > 0 and time t is determined by condition

$P_o(t) > P_{cr}^{(m)}$. If the maximum laser pulse power is just slightly above the critical value $P_{cr}^{(1)}$, one moving focus appears. Then further foci will be formed and they increase in number with increasing maximum initial power.

Under real conditions, a layer of medium always has finite thickness ℓ, depending, say, on the length of a cell containing the substance studied. The instants t_m at which each focal region passes through the output plane of a medium, as well as through any fixed plane inside it, will be different, and corresponding instants Δt_m characteristic of this passing are determined by the formula

$$\Delta t_m = \frac{\ell_{\phi m}}{|V_{\phi m}|} \quad , \tag{80}$$

where $V_{\phi m}$ are the corresponding velocities of movement

$$V_{\phi m} = \frac{d\zeta_{\phi m}}{dt}\Big|_{t=t_m} \quad . \tag{81}$$

In what follows we consider for example a pulse with an envelope, given by

$$P_o(t) = \frac{P_{max}}{[1+(\frac{1.3t}{\tau_u})^2]^2} \quad . \tag{82}$$

Here τ_u is the pulse width determined at the level $\frac{1}{2} P_{max}$; time is counted from the maximum power of an incident beam over cross section $z=0$ (pulses with other envelopes can be considered in the same manner and the results are very similar). Then, according to eq. (79), the foci positions $\zeta_{\phi m}(t)$ are

$$\zeta_{\phi m}(t) = \chi_m \ k\bar{a}_o^2 \ \frac{1+(\frac{1.3t}{\tau_u})^2}{N_{max}-N_m-N_m(\frac{1.3t}{\tau_u})^2} \quad , \quad (83)$$

$N_{max} = N_1 \ \sqrt{\frac{P_{max}(1)}{P_{cr}}}$. The instant t_m at which the mth focus passes through the plane $z = \ell$ is determined from the condition $\zeta_{\phi m} = \ell$ which gives

$$t_m = \mp \ \frac{\tau_u}{1.3} \ (\frac{N_{max}-N_m-\chi_m \ \frac{k\bar{a}_o^2}{\ell}}{N_m+\chi_m \ \frac{k\bar{a}_o^2}{\ell}})^{1/2} \quad . \quad (84)$$

It is evident that the condition of focus formation in the medium layer concerned ($0 < z < \ell$) is reduced to $N_{max} - N_m -\chi_m \ \frac{k\bar{a}_o^2}{2} \geq 0$. Under these circumstances, according to eqs. (83) and (84) the m-th focus first occurs in cross-section $z=\ell$ at the instant $t_m<0$ (the minus sign in formula (84)) and moves inside this layer. At $t=0$ the focus stops at point $z=\zeta_{min}$ where

$$\zeta_{min}^{(m)} = \frac{\chi_m k\bar{a}_o^2}{N_{max}-N_m} \quad (85)$$

and then moves towards boundary $z=\ell$. The instant t_m when the focus leaves the layer corresponds to the plus sign in formula (84).

We calculate the velocity of foci movement in the cross-section z. From (83) it is

$$v_{\phi m} = \frac{2.6}{\tau_u} \ \frac{\chi_m \ k\bar{a}_o^{-2} \ N_{max} \ \frac{1.3t}{\tau_u}}{[N_{max}-N_m-N_m(\frac{1.3t}{\tau_u})^2]^2} = \mp \ \frac{2.6}{\tau_u} \ \frac{z^2}{\chi_m \ k\bar{a}_o^{-2} \ N_{max}} \quad .(86)$$

As is seen, absolute values $|v_{\phi m}|$ increase monotonically with $|t|$ or z. They are maximum at the boundary $z=0$ of the layer considered. We estimate velocities typical of the foci movement $|v_{\phi m}|$ and characteristic times Δt_m during which they stay at a given point of the medium. Here, the beam parameters at maximum power are assumed to be the same as in the first numerical example of Section 6, i.e. one takes $N_{max} = 6$ (i.e. $P_{max} \approx 10 \, P_{cr}^{(1)}$), $\bar{a}_o = 0.15$ mm, $\lambda = 0.5 \times 10^{-4}$ cm, $\ell_{\phi m} = 0.1$ cm (i.e. $d_{\phi 1} \approx 5 \times 10^{-4}$ cm), $\ell = 10$ cm. Let the laser pulse width τ_u be 2×10^{-8} sec. Then for the distance $z = \ell = 10$ cm from the incident boundary plane of a medium, using eqs. (86) and (60) one finds $|v_{\phi 1}| \approx |v_{\phi 2}| \approx 1.2 \times 10^9$ cm/sec, $|v_{\phi 3}| \approx 0.9 \times 10^9$ cm/sec, where by means of (80) one finds $\Delta t_m \approx 10^{-10}$ sec. From these estimates it is clear that the characteristic times Δt_m, determining the temporal scale τ (see (9)) greatly exceed the time necessary for the Kerr reorientational effect to establish itself. For example, in our case of carbon bisulphide the corresponding value of τ_k is under 2×10^{-12} sec (N. Bloembergen, 1971). Therefore, condition (9d) required for the initial equation to be valid is certain to be reached even in the case of the orientational mechanism of the Kerr effect. In similar situations this condition proves to be fulfilled in other substances as well. For the electronic Kerr effect this condition is satisfied by a great margin. It is easy to see that condition (9c) is certain to be reached too. Now consider the last condition for the validity of equation (20), i.e. inequality (19). It may be rewritten as:

$$|v_{\phi m}| \ll v \quad , \qquad (87)$$

where v is the velocity of light in a medium (see eq. (14)). One can see that this relation in our example is valid also. However, for a shorter pulse width τ_u and longer cells, inequality (87) may break down. In the latter case one should proceed from (14) (see the following Section).

We note that expression (80) for the characteristic times Δt_m, during which a focal region stays at a given point of a medium, does hold provided the velocity of foci movement $v_{\phi m}$ is not noticeably changed during time Δt_m. This condition is known to break down at the "turning" points of focal regions, i.e. at t=0. At these points the characteristic times $\Delta t_m^{(o)}$ are found from condition $\ell_{\phi m}/2 = \zeta_{\phi m}(\Delta t_m^{(o)}/2) - \zeta_{\phi m}(0)$, which gives

$$\Delta t_m^{(o)} \approx \tau_u(N_{max}-N_m)\left(\frac{\ell_{\phi m}}{N_{max}\,\chi_m\,k\bar{a}_o^2}\right)^{1/2} \, . \qquad (88)$$

In the above numerical example, taking eq. (88) into consideration one obtains $\Delta t_m^{(o)} \approx 2 \times 10^{-9}$ sec. It is seen that the time $\Delta t_m^{(o)}$ during which a focus stays at its turning point is much longer than the corresponding value of Δt_m at the far boundary plane z=ℓ of medium. Therefore, processes occuring at high energy densities in substances and requiring some time for their development (say, breakdown in liquids) are expected to show up, first of all, at the foci turning points or near them (Lugovoi and Prokhorov, 1968). This seems to be the cause of the discrete character of optical breakdown in transparent dielectrics which was observed during laser pulse propagation by Korobkin and Serov (1967).

Now we estimate the total energy $W_m^{(o)}$ absorbed in a focal region during its turning time $\Delta t_m^{(o)}$ and the total energy W_m absorbed in this region during the time Δt_m when it passes the plane $z=\ell$. In general, we have $W_m^{(o)} \sim P_{cr}^{(1)} \Delta t_m^{(o)}$, $W_m \sim P_{cr}^{(1)} \Delta t_m$. At $P_{cr}^{(1)} = 20$ kw for these conditions we find $W_m^{(o)} \sim 4 \times 10^{-5}$ J, $W_m \sim 2 \times 10^{-6}$ J. Depending on the specific type of absorption, a fraction of this energy may be spent on medium heating. If, for example, the main type of absorption is SRS, then at $\frac{\omega_{-1}}{\omega} = 0.1 \sim .05$, part (90-95%) of this energy is transformed into the Stokes component energy and 5-10% dissipates in the medium. In what follows we assume, for example, that the medium is heated by 5% of the whole energy absorbed in the focal region. Then a fixed point of the medium which has been traversed by a focal region in the case of carbon bisulphide will be heated by $\Delta T^{(o)} \sim 100^{\circ}$ at the turning point and by $\Delta T \sim 5^{\circ}$ at $z=\ell$. For other substances (say, toluene and nitrobenzene), the corresponding values of $\Delta T^{(o)}$ and ΔT under similar conditions prove to be about an order of magnitude lower. Local heating produces a sound wave expanding the substance from the point of heating and, hence, changing by (Δn_ρ) its refractive index*. It is easy to see that at turning points of focal regions the calculated value of $|\Delta n_\rho|$ can noticeably exceed $\Delta n = \frac{1}{2} n_o n_2 |E_{\phi m}|^2$ found without considering medium heating.

---

*Note that since under typical conditions the foci movement is supersonic, it is obvious that the density perturbations in a medium are to be concentrated in a sonic cone. In the general case these perturbations are caused not only by heating but by electrostriction in the focal region as well.

This means that the multifoci structure may disappear
entirely while foci turn or approach the turning point.
At the same time if the foci move rapidly (say, at $z \approx \ell$)
the heating will not appreciably affect a multifoci
structure (or, at least, its first foci), ($|\Delta n_\rho| << \Delta n$)
because this heating is much weaker and the times Δt_m
are much shorter. Based on the observation of bright
spots at the end of the cell it was experimentally
established that the occurence of these spots is followed by sonic expansion of the substance (Butenin et
al., 1967).

Sometime a multifoci structure may disappear under
the influence of SRS*. In fact, the gain of the first
Stokes component over the length of a focal region say,
for three-photon absorption, namely $\exp \left(\frac{4\pi\Gamma}{\varepsilon_o} \frac{2\pi}{\lambda_{-1}} \right.$
$\left. |E_{\phi m}|^2 \ell_{\phi m} \right)$, or with reference to eqs. (73) and (75a),
$\exp \left(\frac{85\pi\Gamma}{\varepsilon} \frac{\omega_{-1}}{\omega} \frac{1}{n_o} \right)$, is calculated (on the basis of the
stationary theory) to be $\sim \exp 70$ for nitrobenzene (in
the calculations [see Lugovoi, 1968] the absolute cross-
section and line width of Raman scattering for these
substances were taken from Kato and Takuma [1971], and
Clements and Stoicheff [1968], and the value of n_2 for
all the substances was taken in accordance with the
footnote on page 38. Such large calculated gains indicate that under stationary conditions these foci (whose
sizes are governed by the inertialess type of absorption)
should not appear because first of all, the beam energy
would have been converted, due to amplification of

---

*In some cases the same effect can be introduced by
 stimulated Mandelstam-Brillouin scattering, stimulated
 scattering in the wing of Rayleigh line, etc.

spontaneously scattered light, into energy for the
first Stokes component of SRS*. Such a situation is not
excluded in the calculated foci turning points or near
them. At the same time, far away from these points,
with quickly moving foci, SRS generally becomes nonstationary (and hence, much less effective) and does
not prevent foci formation. Nonstationary behavior
of SRS is associated with the fact that at high gains
the setting time of the stationary regime greatly exceeds that of ordinary Raman scattering due to the
frequency dependence of the gain in SRS and to anomalous
substance dispersion near the first Stokes frequency
(Kroll, 1965, and Zeldovich, 1972). For example, the
setting time of ordinary Raman scattering in carbon
bisulphide is 2×10^{-11} sec (Clements and Stoicheff,
1968). At $\Delta t_m \sim 10^{-10}$ sec and a gain of about 800, SRS
is certain to be very nonstationary in the focal region.
Thus, on its way from the output plane of a medium to
the calculated turning point a focal region may disappear (or grow to a great extent) due to SRS. The

---

*According to the result of Section (5b), the gain at
 which a stationary (quasi-stationary) SRS begins
 noticeably to supress all foci (beginning at the first)
 is determined from the condition $H_{cr} = N$ (or which is
 the same $4\pi\Gamma/\varepsilon_o \approx n_2$). It is obvious that this value for
 the conditions accepted in (5b) is only $\sim$ exp 20, i.e.
 considerably lower than the above values. Note that for
 carbon bisulphide, the gain under stationary conditions
 in the focal region will be, in fact, lower than $\sim$ exp
 800, due to appreciable excitation of the upper vibrational state of molecules at SRS. However, the gain,
 taking into account excitation, is not essentially
 lower than that calculated without taking into account
 molecule excitation. So the indicated correction for
 the gain is not important in the situation considered.

time which it will occur seems to be determined from
the condition that absorption due to SRS in the focal
region becomes comparable with inertialess (say, three-
photon) absorption. It is also clear that at this
instant an ultrashort pulse of the SRS first Stokes
component will be excited [Korobkin et al.] (for more
detail see Section 9).

Note that multifoci structure disappearance during
foci turning in nitrobenzene and carbon bisulphide was
observed experimentally by Korobkin et al. (1970).
Disappearence of foci approaching the calculated turning
point in carbon bisulphide and toluene was observed
in Lay and Shen (1969). The effect of SRS and stimulated Mandelstam-Brillouin scattering on beam propagation from the input plane of a nonlinear medium to the
first focus was investigated by Maier et al. (1970).

8. Propagation of Short Pulses

Now we consider propagation of laser pulses sufficiently short so that condition (87) fails. In this
case one should proceed from eq. (14) with boundary
condition (21). Substitution of the variable $t = \xi + \frac{z}{V}$ which is well-known for equations of type (14) gives
the following equation for E:

$$\Delta_{\perp} E + 2ik \frac{\partial E}{\partial z} + n_2 k^2 |E|^2 E = 0 \quad , \qquad (89)$$

with boundary condition

$$E\big|_{z=0} = \phi(r_{\perp}, \xi) \quad . \qquad (90)$$

This problem is again a stationary one (here variable
ξ is a parameter only in the boundary condition).
Therefore, if a family of solutions for eqs. (89)-(90),

corresponding to various ξ is known and one takes for
this family $\xi = t - \frac{z}{v}$ one can obtain the solution of the
problem concerned, eqs. (14), (21). From this, one
immediately arrives at the following general conclusion:
a multifoci structure in a light beam whose maximum
power is greater than $P_{cr}^{(1)}$ occurs also in the case
when condition (87) is not fulfilled. Here the general
picture of the beam propagation is as follows. Just as
under the quasistationary conditions, the main specific
feature of this picture is moving foci on the beam
axis. In this case, at certain instants foci are formed
inside the medium layer concerned ($0 < z < \ell$), split
and move along the z axis both in the direction of incident beam propagation and in the opposite one.

a) Trajectories of Moving Foci*

For example, let the incident beam be determined by
boundary condition (78). According to the foregoing,
in order to obtain the foci positions $\xi_{\phi m}$ along the z
axis one should write $\xi = t - \frac{\zeta_{\phi m}}{v}$ instead of t in (79).
It provides the equation for value $\xi_{\phi m}$ at a given instant
t:

$$\zeta_{\phi m} = \frac{\chi_m}{N_m} \frac{k\bar{a}_o^2}{(\frac{p_o(t-\frac{\zeta_{\phi m}}{v})}{P_{cr}^{(m)}})^{1/2} - 1} . \quad (91)$$

However, it is most convenient to begin with the equivalent graphical method (to eq. (91)) of determining
the trajectory of moving foci. For this purpose, we
represent a system of curves $N = N(\frac{z}{k\bar{a}_o^2})$ with coordinates

---

*The discussion in this section is based largely on the
 work of Askaryan (1962); Abramov et al. (1969); Gustafson and Taran (1970).

$\frac{z}{k\bar{a}^2}$, N (see solid curves of Fig. 10) for absorption

(which gives the "reversed" dependence of $\frac{Nz}{k\bar{a}_o^2}$ upon N).

In the same coordinates we represent the function

$$N = \frac{1}{E_{cr}} \left| E_o \left(t - \frac{k\bar{a}_o^2}{v} \frac{z}{k\bar{a}_o^2} \right) \right| \text{ (see dashed curve in Fig. 10)}.$$

The values of $z/k\bar{a}_o^2$ corresponding to the intersection
points of solid and dashed curves indicate foci positions
(in the appropriate scale) along the beam axis at
instant t. If one considers that the dashed curve is
shifted by t at a velocity of $\frac{v}{k\bar{a}_o^2}$ along the $\frac{z}{k\bar{a}_o^2}$ axis
but preserves its shape, it is easy to elucidate the
picture of foci formation and time dependence of foci
positions. This picture is shown in Fig. 11. As can
be seen, each focus splits. Hence, two groups of foci
are formed. The velocities of the group moving in the
direction of the incident pulse propagation always exceed that of light, v, in a medium (they can be also
over the velocity of light in vacuum). This group of
foci is different than that arising under quasistationary
conditions. Each focus from the other group moves in
the same direction as it undergoes quasistationary
conditions: at first, it moves in the direction opposite to the incident beam, then it stops and moves
in the direction of the incident beam propagation.
Here, the results of quasistationary considerations
are obtained in the limit $\ell_\mu/\ell \to \infty$ where $\ell_\mu = \tau_u v$ is the
light train length corresponding to pulse width τ_u,
and ℓ is the width of layer of the medium concerned.

Now we consider an incident pulse with an envelope of shape (82) in more detail. Equation (91) gives

$$t = \frac{\xi_{\phi m}}{v} \mp \frac{\tau_u}{1.3} \left(\frac{N_{max} - N_m - \chi_m \frac{k\bar{a}_o^2}{\zeta_{\phi m}}}{N_m + \chi_m \frac{k\bar{a}_o^2}{\zeta_{\phi m}}} \right)^{1/2} , \qquad (92)$$

where, for the focus motion velocity $\upsilon_{\phi m} = \dfrac{d\zeta_{\phi m}}{dt}$, one finds the following expression

$$\upsilon_{\phi m} = \frac{v v_{\phi m}}{v + v_{\phi m}} , \qquad (93)$$

where

$$v_{\phi m} = \mp \frac{\xi_{\phi m}^2}{\chi_m k\bar{a}_o^2 N_{max}} \frac{2.6}{\tau_u} \left(N_{max} - N_m - \chi_m \frac{k\bar{a}_o^2}{\zeta_{\phi m}} \right)^{\frac{1}{2}} \left(N_m + \chi_m \frac{k\bar{a}_o^2}{\zeta_{\phi m}} \right)^{3/2}$$

.

is the corresponding value of focus movement velocity under quasistationary conditions (see 86)). For brevity, $v_{\phi m}$ will be referred to as the quasistationary velocity.* From Fig. 11 it is evident that the foci formation points are determined from condition $\dfrac{dt}{d\zeta_{\phi m}} = 0$ which, with reference to (93), leads to the equation

$$|v_{\phi m}| = v . \qquad (94)$$

The solution $\tilde{\zeta}_{\phi m}$ of this equation is easy to obtain for some limiting cases. For example, under condition

---

*It is not difficult to ascertain that relation (93) between the real velocity of focus motion $\upsilon_{\phi m}$ and the corresponding quasistationary value $v_{\phi m}$ is valid at any approximation of foci positions vs. the initial power $\zeta_{\phi m} = F_m(P_0)$ (see, for example (59)), in a corresponding stationary problem and at any shape of the laser pulse envelope $P_0(t)$ (see e.g. (82)). Under quasistationary conditions it is evident that $\upsilon_{\phi m} \approx v_{\phi m}$.

$$\tilde{\zeta}_{\phi m} - \zeta_{min}^{(m)} \ll k\bar{a}_o^2, \ \zeta_{min}^{(m)} \quad , \tag{95}$$

it has the form

$$\tilde{\zeta}_{\phi m} = \zeta_{min}^{(m)} - \frac{(N_{max}-N_m)^2(v\tau_u)^2}{(2.6)^2 N_{max}\chi_m k\bar{a}_o^2} \quad . \tag{96}$$

From (91) it is also evident that the foci turning points $z=\zeta_{min}^{(m)}$ are determined by the same equation as under quasistationary conditions

$$\zeta_{min}^{(m)} = \frac{\chi_m k\bar{a}_o^2}{N_{max}-N_m} \quad , \tag{97}$$

with the only exception being that the instants $t^{(m)}$ at which various foci turn are now different:

$$t^{(m)} = \frac{\zeta_{min}^{(m)}}{v} \quad . \tag{98}$$

It is noteworthy that, according to (93) and (86), for short pulses and a focus far enough from the boundary z=0 the value of its quasistationary velocity $v_{\phi m}$ can considerably exceed the velocity of light in a medium:

$$v_{\phi m} \gg v \quad . \tag{99}$$

Then from (93) it follows that its real velocity $\upsilon_{\phi m}$ is close to that of light v.

b) Structure of Moving Foci

Since the above substitution of variable $t=\xi + \frac{z}{v}$ in equation (14) does not contain the transverse co-ordinate $r_\perp$, the transverse structure of focal regions in our case is just that under quasistationary conditions. At the same time the longitudinal structure is generally changed. If one does not take into consideration the vicinity of foci formation points $\zeta_{\phi m}$

and their turning points $\zeta_{min}^{(m)}$, then these changes are reduced to the changes of the longitudinal scale, i.e. of focal region length. In this case, proceeding from expressions (70), where one should assume $\zeta_{\phi m} = F_m$ $[P_o(t-\frac{z}{v})]$ one finds in the vicinity of foci that

$$
|E|^2\big|_{z=0} =
\begin{cases}
\left\{\dfrac{|E_{\phi m}|^2}{1+\left[\dfrac{\Delta z - \upsilon_{\phi m}\Delta t}{(1-\frac{\upsilon_{\phi m}}{v})\Lambda_{\phi m}}\right]^2\right\}^\alpha \;, & \dfrac{\Delta z - \upsilon_{\phi m}\Delta t}{1-\frac{\upsilon_{\phi m}}{v}} < 0 \;, \\[6mm]
\left\{\dfrac{|E_{\phi m}|^2}{1+\left[\dfrac{\Delta z - \upsilon_{\phi m}\Delta t}{(1-\frac{\upsilon_{\phi m}}{v})\Lambda}\right]^2\right\}^\beta \;, & \dfrac{\Delta z - \upsilon_{\phi m}\Delta t}{1-\frac{\upsilon_{\phi m}}{v}} > 0 \;,
\end{cases}
\tag{100}
$$

where $\Delta t = t - t_o$, $\Delta z = z - z_o$, and t_o is the instant at which a focus passes cross-section z_o. From expressions (100) as well as from (70) and (69) it is evident that in the case concerned the focal region length $\ell_{\phi m}$ at the instant t_o is associated with the corresponding quasi-stationary value $\ell_{\phi m}$ (see (75)) by relation

$$
\tilde{\ell}_{\phi m} = \left| 1 - \frac{\upsilon_{\phi m}}{v} \right| \ell_{\phi m} \;.
\tag{101}
$$

Accordingly, the characteristic time $\Delta \tilde{t}_m$ that the focal region stays at given point z_o of the medium is equal to

$$
\Delta \tilde{t}_m = \frac{\tilde{\ell}_{\phi m}}{|\upsilon_{\phi m}|} = \left| 1 - \frac{\upsilon_{\phi m}}{v} \right| \frac{\ell_{\phi m}}{|\upsilon_{\phi m}|} \;.
\tag{102}
$$

Now consider the vicinity of the foci turning points. Proceeding from expressions (70) and using (82), and (98) one obtains, similarly to (100),

$$|E|^2|_{t=0} = \begin{cases} \dfrac{|E_{\phi m}|^2}{[1+(\frac{\Delta}{\Lambda_{\phi m}})^2]^\alpha} & , \quad \Delta < 0 \\[4em] \dfrac{|E_{\phi m}|^2}{[1+(\frac{\Delta}{\Lambda_{\phi m}})^2]^\beta} & , \quad \Delta > 0 \end{cases} \qquad (103)$$

where

$$\Delta = \Delta z - \chi_m k \bar{a}_o^2 \left(\frac{1.3}{\tau_u}\right)^2 \frac{N_{max}(\Delta t - \frac{\Delta z}{v})^2}{(N_{max}-N_n)^2} \quad ,$$

$$\Delta t = t - t^{(m)} \quad , \quad \Delta z = z - \zeta_{min}^{(m)} \quad .$$

$$(104)$$

As is shown in eqs. (103), (104), the characteristic time $\Delta \tilde{t}_m^{(0)}$ for the focal region to remain at a turning point is governed by the same expression as it is under quasistationary conditions,

$$\Delta \tilde{t}_m^{(o)} = \Delta t_m^{(o)} \approx \tau_u (N_{max}-N_m) \sqrt{\frac{\ell_{\phi m}}{\chi_m k \bar{a}_o^2 N_{max}}} \quad . \qquad (105)$$

As for the longitudinal structure of the focal regions at a fixed instant $t^{(m)}$, it is generally changed as follows. From eqs. (103), (104) one finds that the focal region structure and length remain the same as compared to the quasistationary conditions ($\tilde{\ell}_{\phi m}^{(0)} = \ell_{\phi m}^{(0)}$) only when

$$v\Delta t_m^{(o)} >> \ell_{\phi m} \quad . \qquad (106)$$

Proceeding from expressions (96), and (105) it is easy
to see that under this condition the turning point $\zeta_{min}^{(m)}$
and the point of formation $\zeta_{\phi m}$ of the mth focus are
rather well separated ($\zeta_{\phi m} - \zeta_{min}^{(m)} \gg \ell_{\phi m}$). In the
opposite case

$$v\Delta t_m^{(o)} \lesssim \ell_{\phi m} \ , \qquad (107)$$

these points coincide in the sense that $\zeta_{\phi m} - \zeta_{min}^{(m)} \lesssim \ell_{\phi m}$.
In this case expression (103) describes the vicinity
of the two points. From equations (103), under condition (107) it follows that at instant $t^{(m)}$, curve
$|E|^2_{\tau=0}$ has two maxima in this vicinity which correspond to the focus splitting after its formation. The
distance δ_m between these maxima is

$$\delta_m = \frac{(N_{max} - N_m)^2 (v\tau_u)^2}{(1.3)^2 N_{max} \chi_m k \bar{a}_o^2} \ , \qquad (108)$$

i.e. it is four times the distance $\tilde{\zeta}_{\phi m} - \zeta_{min}^{(m)}$ between the
focus formation and the turning points. At $v\Delta_m^{(o)} \ll \ell_{\phi m}$,
the minimum between the two maxima concerned is not
very sharp, i.e. at moment $t_m^{(o)}$ the focus is not yet
clearly split. In this case the total length of the
focal region is

$$\tilde{\ell}_{\phi m}^{(o)} = v\Delta t_m^{(o)} \qquad (109)$$

Now we consider a numerical example. The beam
parameters are taken to be the same as in the first
numerical example of Section 6, i.e. we assume $N_{max} = 6$, $\bar{a}_o = 0.15$ mm, $\lambda = 0.5 \times 10^{-4}$ cm, $\ell_{\phi m} = 0.1$ cm. Let
the laser pulse width τ_u be 3×10^{-11} sec and medium layer

$\ell = 10$ cm. For distance $z = \ell$ one obtains from (93) and (60) the following velocities of foci motion $\upsilon_{\phi 1} \approx \upsilon_{\phi 2} \approx v(1+2.6 \times 10^{-2})$, $\upsilon_{\phi 3} \approx v(1+3.3 \times 10^{-2})$. We see that in the cross-section under consideration the velocities of the foci motion are close to that of light in a medium. Under these conditions the longitudinal sizes $\tilde{\ell}_{\phi m}$ of the moving foci, according to (101), are quite small $\tilde{\ell}_{\phi m} \approx 3 \times 10^{-3}$ cm and times $\Delta \tilde{t}_m$ during which focal regions stay in the cross-section concerned, according to)102), are equal to $\Delta \tilde{t}_m \approx 1.5 \times 10^{-13}$ sec. The corresponding time $\Delta \tilde{t}_m^{(o)}$ and sizes $\tilde{\ell}_{\phi m}^{(o)}$ for the foci turning

points determined by relations (105), (107)-(109) are $\Delta t_m^{(0)} \approx 3 \times 10^{-12}$ sec, $\tilde{\ell}_{\phi m}^{(o)} \sim 0.1$ cm (here since $v \Delta t_m^{(o)} \approx 0.06$ cm $\sim \ell_{\phi m}$, in our example the foci formation points coincide with their turning points). It is interesting that in contrast to the quasistationary conditions the length of each focal region in our case changes with time in the process of light beam propagation in the medium. Note that the times considered $\Delta \tilde{t}_m^{(o)}$ and $\Delta \tilde{t}_m$ are so short that medium heating by the focal regions and its subsequent expansion during pulse propagation does not considerably change the refractive index change in the path of moving foci will show up only after the whole pulse passes*. Note also that for an electronic mechanism of the Kerr effect when $\tau_k \lesssim 10^{-15}$ sec,

---

*It is easy to see if one considers that the total energy $\tilde{W}_m$ absorbed in the volume of a focal region as it passes the fixed cross-section $z \neq \zeta_{min}^{(m)}$, $\tilde{\zeta}_{\phi m}$ is of the order of $P_{cr}^{(1)} \tilde{\ell}_{\phi m} \Delta \tilde{t}_m / \ell_{\phi m}$ and similarly, at $z = \zeta_{min}^{(m)}$, the corresponding energy $W_m^{(o)}$ is about $P_{cr}^{(1)} \tilde{\ell}_{\phi}^{(o)} \Delta \tilde{t}_m^{(o)} / \ell_{\phi m}$.

conditions (9) for the validity of equation (14) for
the case discussed are fulfilled.

9. Excitation of SRS Ultrashort Pulses

So far we have not specifically considered SRS in
the backward direction. However, Loy and Shen (1971)
have given attention to the fact that under some
conditions backward SRS can considerably affect certain
parts of foci trajectories. This effect shows up when
formation point $\tilde{\zeta}_{\phi m}$ of the corresponding focus is inside (or close to the boundary) of the medium layer
under consideration and well separated from the turning
points $\zeta_{min}^{(m)}$. Such conditions are realized for example,
under the following parameters of the incident beam and
medium: $\bar{a}_o$ = 0.15 mm, N_{max} = 3.5, λ = 0.5x10^{-4}cm, $\ell_{\phi m}$ =
0.1 cm, τ_u = 0.5x10^{-8}sec, and ℓ = 30 cm. Under these
conditions, from eqs. (85), and (60) it follows that in
the layer considered, there are only two foci (m = 1,2).
Here $\zeta_{min}^{(1)} \approx$ 12 cm, $\zeta_{min}^{(2)} \approx$ 21 cm. From (94) it also
follows that the formation point of the first focus
$\tilde{\zeta}_{\phi 1} \approx$ 26 cm is inside the layer concerned. At $(\tilde{\zeta}_{\phi 1}-$
$\zeta_{min}^{(1)})/\ell_{\phi 1} \approx$ 140 the formation point of this focus is
well separated from its turning point. As the focus
moves from a formation point to a turning point, its
velocity $\upsilon_{\phi 1}$ formally changes from ∞ to zero and at
$z = z_{1v} \approx$ 19 cm its value is equal to v, i.e. to the
velocity of light in a medium. Under these conditions
the light flux which has been initially scattered
spontaneously in the focal region in the backward direction propagates further in this region over path
Δz_{1v}, in general determined from the condition
$$\frac{v}{2} \frac{d^2 t}{d\zeta_{\phi m}^2}\bigg|\ \zeta_{\phi m} = z_{mv}\ \left(\frac{\Delta z_{mv}}{2}\right)^2 = \ell_{\phi m}$$ which with reference to

(2) gives

$$\Delta z_{mv} = \left\{ \frac{2\ell_{\phi m} z_{mv} (N_{max} - N_m - \chi_m \frac{k\bar{a}_o^2}{z_{mv}})(N_m + \chi_m \frac{k\bar{a}_o^2}{z_{mv}})}{N_m(N_{max} - N) + (\frac{1}{4}N_{max} - N_m)\chi_m \frac{k\bar{a}_o^2}{z_{mv}}} \right\}^{1/2} \tag{110}$$

From (110) in our example one finds $\Delta z_{1v} \simeq 1.5$ cm. As $\frac{\Delta z_{1v}}{\ell_{\phi 1}} \approx 8$, the total interactin time of the focal region with the previously scattered light flux has a sharp maximum for $z \simeq z_{1v}$ the energy of the focal region can be effectively converted into that of the first Stokes component of SRS. The length of the light pulse train of the corresponding first Stokes component seems to be about focal region length $\ell_{\phi m}$. So the width of the corresponding pulse due to its passage through a fixed medium plane will be about $\tilde{\ell}_{\phi m} v$. In our example $\tilde{\ell}_{\phi m}/v \approx 10^{-11}$ sec. An ultrashort SRS pulse excited in this way will propagate in the backward direction at the velocity of light in the medium. In the process of further interaction with an unperturbed part of the initial light beam this pulse will be enhanced and highly directed. A similar interaction process between two pulses in opposite directions is described in Maier et al. (1966, 1969). As regards the part of the initial light beam interacting with the ultrashort SRS pulse it can be appreciably suppressed, so that the foci no longer appear in it. In this case the whole section of the focus trajectory corresponding to condition $|v_{\phi m}| < v$ will not be realized because of backward SRS (in Fig. 11, this section is above the straight line $t = t_{mv} + \frac{z - z_{mv}}{v}$ where t_{mv} is the moment at which a focus is passing the cross-section z_{mv}).

Ultrashort pulses in backward SRS were already discovered experimentally by Maier et al. (1966). In the latter paper and in Maier et al. (1969) the authors paid attention to the correlation between the occurence of these pulses and light spots appearing on the cell and which contained the substance studied. The above mechanism of their excitation inside the cell was described in Loy and Shen (1971). This excitation mechanism for ultrashort SRS pulses was verified experimentally by observing the correlation between these pulses and the moving foci in a medium. Ultrashort light pulses (width about 10^{-11} sec) were observed experimentally in backward stimulated scattering in the wing of the Rayleigh line by Kyzylasov et al.(1969) and Kyzylasov and Stamnov (1969). Pulse excitation and their further formation in stimulated scattering in the wing of the Rayleigh line seems to be analogous to the corresponding process at SRS.

In concluding of this Section we note that if ultrashort pulses of SRS at $z = z_{mv}$ are not excited very effectively due to the nonstationary character of SRS, there is another possibility of this formation while focal regions are approaching their turning points. As was mentioned in Section 7, as a focus approaches its turning point the SRS process can at some moment be effective enough for the total energy of the focal region to be converted into that of the first Stokes component, thereby forming an ultrashort pulse. If such a conversion takes place near the turning point, i e. when $\upsilon_{\phi m} \ll v$, the first Stokes component pulse should appear equally both in the backward and forward directions. At a sufficiently short time interval of conversion the pulse width will be about $\tilde{\ell}_{\phi m}/v$.

However, if the interval of the focal region disappear-
ance is large compared to $\tilde{\ell}_{\phi m}/v$, it is the interval of
focal region disappearance that determines the pulse
width. We designate the instant of focal region disappearance by t_o and the cross-section z where it
occurs by z_o. The result of interaction between the
pulses discussed and the initial light beam is that
part of all foci trajectories which are below the
straight line $t=t_o + \frac{z-z_o}{v}$ (see Fig. 11) may not be
realized.[*]

Experimentally the ultrashort pulses of the first
Stokes component occuring inside the medium layer concerned and propagating in the forward (as well as in
the backward) directions were observed by Korobkin
et al.

10. Pulse Spectrum Broadening in a Nonlinear Medium

Above we were interested only in the intensity
distribution $|E|^2$ of a light beam in a medium. Now
consider the total value of complex amplitude $E = |E|e^{i\phi}$. This makes it possible to investigate spectral
characteristics of the beam which transversed the medium

---

[*]Note that this conclusion refers to sufficiently short
laser pulses only. In the opposite case of quasistationary beams, SRS pulses will leave a layer of a
nonlinear medium in a time much shorter than a laser
pulse width. This will provide a possibility of subsequent focusing of new portions of a laser pulse.
The rise time of the new part of a laser pulse seems
to correspond to the width of a SRS ultrashort pulse.
Thus, the time scale of the changing incident beam intensity can be equal to an ultrashort pulse width even
when an exciting laser is operated in a giant pulse
regime. The latter point can in particular considerably
affect the spectral properties of the moving foci
field (see Section 10).

layer concerned. Just as in the above discussion, an
incident beam is assumed to have a Gaussian intensity
distribution over a cross-section and a flat initial
phase front, i.e. the boundary condition at z=0 is
written in the form of (78). Here we proceed from
equation (14), i.e. we consider the short pulse case.
According to Section 8, one should first of all write
out the solution $E = E(r,z,E_o)$ of the corresponding
stationary problem, eqs. (21), (25). Then the solution
of the unstationary problem concerned will be determined by expression $E(r,z,t) = E[r,z,E_o(t-\frac{z}{v})]$. For
simplicity we shall restrict ourselves to considering
the field on the beam axis (i.e. at r = 0,z>0). From
eqs. (25,), (27), (30), (31) it follows that the
solution $E(0,z,E_o)$ of the stationary problem has the
form

$$E(0,z,E_o) = |E(0,z,E_o)|\exp[i\phi(0,z,E_o)] \quad , \quad (111)$$

where

$$\phi(0,z,E_o) =$$

$$\int_o^z \{\tfrac{1}{2}n_2|E(0,z',E_o)|^2-[ka(z',E_o)]^{-2}\}dz' \quad (112)$$

and $a(z',E_o) = \{\dfrac{\partial^2}{\partial r^2} \mid \dfrac{E(r,z',E_o)}{E(0,z',E_o)} \mid_{r=0}\}^{-1/2}$ is the

beam "radius" determined at r→0. Equations (111),
(112) define the total value of the complex amplitude
E on the beam axis if the axis intensity ($|E|^2$) and
radius a of the intensity distribution in the near-axis
region are known. Below we use expressions (111), (112)
to analyze the spectral properties of the beam.

a) Phase Modulation of Pulses

First consider the case

$$z \ll k\bar{a}_o^{-2}, \quad \frac{a}{\sqrt{(n_2 E_{o\ max}^2)}}, \qquad (113)$$

where $E_{o\ max}$ is the maximum value of E_o throughout a pulse. In this case the solution of the stationary problem, eqs. (21) and (25), gives for $|E|^2$ (see, e.g. (42), (43):

$$|E(r,z',E_o)|^2 \approx E_o^2 e^{-\frac{r^2}{\bar{a}_o^2}}, \qquad (114)$$

from which, according to (112),

$$\phi(0,z,E_o) \approx kz[\frac{1}{2}n_2 E_o^2 - (k\bar{a}_o)^{-2}] \qquad (115)$$

and for $E(r,z,t) = |E|e^{i\bar{\phi}}$, using eqs. (111), (115), one finds

$$|\bar{E}(0,z,t)| \approx |E_o(t-\frac{z}{v})| \qquad (116a)$$

$$\bar{\phi}(0,z,t) \approx \frac{1}{2} n_2 kz E_o^2(t-\frac{z}{v}) - \frac{z}{k\bar{a}^{-2}} \qquad (116b)$$

We can see that the envelope of a pulse propagating in a nonlinear medium under condition (113) is the same as in a linear medium. As for the phase $\bar{\phi}$, these two cases are quite different. According to (116b), in a nonlinear medium (with $n_2 \neq 0$) the phase $\bar{\phi}$ becomes time-dependent. It is modulated in accordance with the in-cident beam power changing in time over the initial cross-section. Phase modulation depth is proportional to z, i.e. the thickness of a layer of nonlinear medium traversed by the pulse. Phase modulation will change

the spectrum of a pulse propagation in a medium. This
change can be determined directly by calculating the
Fourier transforms of the function $Ee^{-i\omega_o t}$ (the exponential factor in this case is written down from eq.
(6)). For convenience, in this Section, ω_o will be
written instead of ω in Eq. (6). Such calculations for
pulses of various shapes were made in Treacy (1972).
Here for brevity we shall calculate only the instantaneous frequency $\omega(t)$ of the field oscillations in a given
cross-section Z. Because of eq. (6), we generally have
$\omega(t) = \omega_o + \Delta\omega(t)$ where

$$\Delta\omega(t) = -\frac{\partial\bar{\phi}}{\partial t} \qquad . \tag{117}$$

For example, for the envelope of the incident pulse
with a shape of (82), one derives from eqs. (116b) and
(117)

$$\Delta\omega(t)=2kzn_2 E_{o\ max}^2 \left(\frac{1.3}{\tau_u}\right)^2 (t-\frac{z}{v})\{1+[\frac{1.3(t-\frac{z}{v})}{\tau_u}]^2\}^{-3} \quad . \tag{118}$$

The corresponding curve $\Delta\omega(t)$ is represented in Fig. 12.
Function $\Delta\omega(t)$ is evidently an odd function of t with
respect to the point $\frac{z}{v}$ and ranges from $\Delta\omega_{max} \approx \frac{0.66}{\tau_u}$
$kzn_2 E_{o\ max}^2$ to $\Delta\omega_{min} = -\Delta\omega_{max}$. With respect to the same
point, the function $|E(0,z,t)|$ is even. Therefore, the
spectral distribution of the electric field oscillation
intensity on the beam axis is $I = |E(0,z,\omega)|^2|\sqrt{8\pi}$,
where

$$E(0,z,\omega) = \int_{-\infty}^{\infty} E(0,z,t)e^{-i(\omega_o-\omega)t} dt \tag{119}$$

is symmetrical about frequency ω_o and under the condition
$\Delta\omega_{max} \gg \frac{1}{\tau_u}$ its boundary frequencies appear to be
$\omega=\Delta\omega_{max}$, and $\Delta\omega_{min}$. Hence, the full width,

$\Delta\omega = \Delta\omega_{max} - \Delta\omega_{min}$, of the frequency distribution of intensity I is governed by equation [73].

$$\Delta\omega \approx \frac{1.3}{\tau_u} kzn_2 E_{o\ max}^2 \quad . \qquad (120)$$

If one considers that the width of the spectral distribution of the incident beam intensity is about $1/\tau_u$, formula (120) directly determines the extent of breadening ($\Delta\omega\tau_u$) of the spectrum while the beam propagates in a nonlinear medium. One can see that at phase modulation, the spectrum broadening is proportional to the path z covered by the pulse. It is also clear that the spectrum broadening is proportional to the velocity change in the refractive index of a substance with time, $(n_2 E_{o\ max}^2 / \tau_u)$, so it is of a Doppler character. The possibility of a Doppler spectrum broadening of a pulse which has passed through a layer of nonlinear medium was discussed in Zeldovich and Raisen (1966). In a later effort, (Shimizu, 1967) it was also pointed out that the same value of instant frequency $\omega(t)$ is realized for two different instants t (see (118)). If at these two times the phase $\overline{\phi}$ differs by $(2k+1)\pi$, where k is an integer, then distribution I has a minimum near the given frequency. Similarly, if the phases differ by $2k\pi$, the distribution of I has a minimum, i.e. the frequency distribution of intensity I has a structure with an oscillating dependence on ω and a great number of maxima and minima (Shimizu, 1967) (see Fig. 12). We estimate the number of these oscillations. According to (116b), the phase, $\overline{\phi}$, changes during time t from 0 to ∞ and is equal to $\Delta\overline{\phi} = -kzn_2 E_{o\ max}^2$. Therefore, under condition $\Delta\omega \gg \frac{1}{\tau_u}$ one has $|\Delta\overline{\phi}| \gg 1$. The total (for $-\infty < \omega < \infty$) number of oscillations (say minima) is of an order of $\frac{\Delta\phi}{2\pi} \gg 1$ and the "mean period" $\Delta\Omega$ of

these oscillations is equal to $\Delta\Omega = \frac{2\pi\Delta\omega}{|\Delta\phi|} \approx \frac{8}{\tau_u}$.

If the pulse spectrum broadening is large enough,
the dispersion of the linear part of refractive index
n_o may become important. The dispersion will disturb
the phase relations between frequency components; i.e.
phase modulation may lead to amplitude modulation. The
occurrence of beats between various frequency components
may result in further spectrum broadening which can be
described as it is done in the case of beats between
different components of multimode laser beams. For
example, frequencies ω_1 and $\omega_1 - \omega_2$ create an anti-
Stokes component, etc. This mechanism of laser beam
spectrum broadening is taken into account for $\omega_L + \omega_r$.

We note also that for the results described in
this Section to be valid condition (9) should be satisfied. However, if, for example, condition (9) fails,
the spectral distribution of intensity $I(\omega)$ becomes
asymmetrical with respect to ω_o (Cheung et al., 1968).
If conditions (9b), (9c) break down, the envelope of a
pulse propagating in a medium may also undergo considerable changes (DeMartini et al. (1967)). The latter
fact will lead to an asymmetry in the spectrum of $I(\omega)$
with respect to ω_o and broadening mainly (by energy)
occurs in the Stokes region (Gustafson et al., 1969).

b) Field Spectrum of Moving Foci

Now consider field $E(t)$ spectrum over the fixed
plant $z=z_o$, $(z_o \neq \tilde{\zeta}_{\phi m}$, $\zeta_{min}^{(m)})$ due to a moving focal region
passing through this plane (in this case condition (113)
fails but conditions (9) are assumed to hold). Now the
solution of the appropriate stationary problem for

$|E|^2_{r=0}$ is determined by expressions (70). The value
$a(z',E_o)$, also absorbed in eq. (112), can be estimated
as follows. Power $KP^{(1)}_{cr}$ passes over the cross-section
z in the vicinity of a focal region (where coefficient
K is about unity and decreases with increasing z; say,
at $\zeta_{\phi m}-z<<\ell_{\phi m}$ one has $K \approx 1$ and at $z=\zeta_{\phi m}$; $K \approx \frac{2}{3}$).
Taking into account that the transverse intensity dis-
tribution in the vicinity of a focus is close to
Gaussian (see (69)), one derives $\frac{1}{2}n_2|E(0,z',E_o|^2 =$
$2K(z')[Ka(z',E_o)]^{-2}$. To expose the main specific features of the field spectrum of moving foci we suppose
for simplicity that $K(z')\equiv 1$ and in equations (70) take
$\alpha=\beta=1$, i.e. we restrict ourselves to the symmetrical
(with respect to point $\zeta_{\phi m}$) approximation of the de-
pendence of $|E|^2$ in the vicinity of a focus. Then
equations (111), (112) give:

$$\overline{E}(0,z_o,t) = \frac{|E_{\phi m}|\Delta t_x \exp(i\chi_o)}{\sqrt{1+(\frac{z-\zeta_{\phi m}(E_o)}{\Lambda_{\phi m}})^2}} \qquad (121)$$

$$\exp[\frac{i}{4}k\Lambda_{\phi m}n_2|E_{\phi m}|^2 \arctan(\frac{z-\zeta_{\phi m}(E_o)}{\Lambda_{\phi m}})],$$

where χ_o is a constant and the value of $\overline{E}$ near the
instant t_o when the focal region passes through cross-
section z is

$$\overline{E}(0,z_o,t) = \frac{|E_{\phi m}|\Delta t_x \exp(i\chi_o)}{\sqrt{(\Delta t_x)^2+(t-t_o)^2}} [\frac{\Delta t_x+i(t-t_o)}{\Delta t_x-i(t-t_o)}]^{\alpha/2} \qquad (122)$$

where

$$\Delta t_x = \frac{\Lambda_{\phi m}}{|v_{\phi m}|}, \quad \alpha = \frac{(-1)^s}{4} k \Lambda_{\phi m} n_2 |E_{\phi m}|^2, \quad (s=0,1) \tag{123}$$

$v_{\phi m}$ is the quasi-stationary velocity of the focus corresponding to cross-section of z concerned; s=0 at $v_{\phi m}$ <0, s=1 at $v_{\phi m}$> 0.

The Fourier transformation (119) of function (122) is determined by the expression (Lugovoi and Prokhorov, 1970).

$$E(o,z_o,\omega) = \begin{cases} \dfrac{F \ \exp[i(\omega-\omega_o)t_o]}{\sqrt{\omega_o - \omega} \ \Gamma\left(\dfrac{1+\alpha}{2}\right)} \ W_{\frac{\alpha}{2},0}[2\Delta t_x(\omega_o-\omega)], & (\omega<\omega_o) \\[4mm] \dfrac{F \ \exp[i(\omega-\omega_o)t_o]}{\sqrt{\omega - \omega_o} \ \Gamma\left(\dfrac{1-\alpha}{2}\right)} \ W_{-\frac{\alpha}{2},0}[2\Delta t_x(\omega-\omega_o)], & (\omega>\omega_o) \end{cases} \tag{124}$$

where $F = \pi E_{\phi m} \sqrt{2\Delta t_x} \exp(i\chi_o)$, $\Gamma(z)$ is the gamma function, $W_{\lambda,\mu}(z)$ is the Whittaker function, and α in this expression may easily be estimated by means of eqs. (71), (75), (73). For example, for the case of three-photon absorption in a medium one has $|\alpha|\approx2$. According to (124), at $|\alpha|=2$ the spectral intensity distribution is asymmetrical with respect to the point $\omega=\omega_o$. With $v_{\phi m}<0$ a greater part of the energy is distributed at $\omega<\omega_o$ (i.e. the major spectrum broadening goes into the Stokes region). In the opposite case (when $v_{\phi m}>0$), more energy is distributed at $\omega>\omega_o$, i.e. in the anti-Stokes region. Here in the both cases concerned the total width of spectrum $\Delta\omega$ of a moving focus is governed by relation*,

$$\Delta\omega \approx \frac{\omega_o}{2} \ \frac{|v_{\phi m}|}{c} \ \Delta n_{\phi m}, \tag{125a}$$

---

*It is easy to see that this expression determines the full range of instantaneous frequence $\omega(t)$ at point z_o.

where $\Delta n_{\phi m} = \frac{1}{2} n_0 n_2 |E_{\phi m}|^2$ is the increase in the refractive index at the point of the maximum energy density in the focal region. Expression (125a), with reference to (73), may be also represented in the form:

$$\Delta \omega \approx 0.3 \ |v_{\phi m}| \ \frac{\lambda}{d^2_{\phi m}} \ . \qquad (125b)$$

We note that $v_{\phi m}$ in (125) of the quasistationary velocity of the focus coincides with the real value of its velocity $\upsilon_{\phi m}$ under quasistationary conditions only. In the general case of short pulses, according to (93), this value is expressed in terms of $\upsilon_{\phi m}$ as follows:

$$v_{\phi m} = \frac{v \upsilon_{\phi m}}{v - \upsilon_{\phi m}} \ . \qquad (125c)$$

Expression (125a) was established in Lugovoi and Prokhorov (1970) and it is equivalent, taking into account equation (125b), to the corresponding expression obtained in Shen and Loy (1971).

From eqs. (123)-(125) it follows that in the portion of a focus trajectory determined by condition $-\infty < \upsilon_{\phi m} < 0$, the spectrum mostly broadens into the Stokes region. For $0 < \upsilon_{\phi m} < v$, the main spectrum broadening goes into the anti-Stokes region and, finally for $\upsilon_{\phi m} > v$ the spectrum broadens, as was pointed out by Shen and Loy (1971), largely into the Stokes region. From eqs. (125) it is evident that under quasistationary conditions the spectrum width $\Delta \omega$ is proportional to the velocity of focal region movement $|\upsilon_{\phi m}|$ and therefore, at constant values of $|E_{\phi m}|^2$ this width will increase monotonically with the distance from the turning point of the focal region to the plane z_o*. For short pulses, the spectrum width $\Delta\omega$ due to a moving focus, will also increase with the distance z_o from the initial boundary plane of the medium because $\dfrac{\vartheta_{\phi m} - v}{v}$ approaches zero as $z_o \to \infty$. Of course, under real conditions this increase will take place as long as $|E_{\phi m}|^2$ remains unchanged. Generally, it should be kept in mind that in the corresponding stationary problem for the first (and similarly any other) focus, one has $N \to N_1$ as $z_o \to \infty$ and at the same time a fixed value of the nonlinear (say three-photon) absorption coefficient m_4, while $|E_{\phi m}|^2$, beginning with $N < 4$, falls appreciably according to the upper plot of Fig. 4. Thus $|E_{\phi m}|^2$ is practically constant only over a certain range of values z_o while a rather great distance z_o between it and the initial boundary plane of medium, it rapidly falls. The latter fact means that the foci disappears**.

In experiments (Chiao et al., 1966; Loy and Shen, 1970) which used giant pulses it was established that the bright spots on the end of the cell disappear if the cell is about 50-100 cm long.

Now we estimate the spectrum width $\Delta\omega$ of moving foci in a number of cases. In the numerical example discussed in Section 7 (at $\tau_u = 2 \times 10^{-8}$ sec) one obtains

---

*These results are not in agreement with the theoretical conclusions of Denarier-Roberge and Taran where the spectrum width $\Delta\omega$ to a moving focal region was assumed to be maximum when the distance between the turning point and the output plane of the medium is equal to about the longitudinal size $\ell_{\phi m}$ of the focal region.

**In contrast to the case of foci disappearence due to substance expansion or simulated scattering considered in Sections 7 and 9, here they disappear due to an inertialess and spatially local nonlinear absorption.

$\Delta\omega\approx0.3$ cm^{-1}. This value is in agreement with the one
measured by Loy and Shen (1969) under similar conditions
(the value of $\Delta\omega$ measured in this paper does not exceed
1 cm^{-1}). In the numerical example presented in Section
(8b) ($\tau_u=3\times10^{-11}$sec) one finds $\Delta\omega\approx200$ cm^{-1}. If in this
example the laser pulse width is assumed to be 3×10^{-12}
sec. then rigorously speaking, one arrives at the
relation $\tilde{\ell}_{\phi m} \approx 3\times10^{-4}$cm $\sim d_{\phi m}$, i.e. one of the conditions
(9b) fails. However this is unlikely to appreciably
affect $\Delta\omega$. Under this assumption one finds $\Delta\omega\approx2000$ cm^{-1}.
Such values are in agreement with measurements made by
Polloni et al., (1969), Cubeddu et al.,(1971), Cubeddu
and Zaraga (1971) when the laser pulse width is about
$10^{-11} \sim 3\times10^{-12}$sec. Finally, at $d_{\phi m} = 10^{-4}$cm (this is
the cross-section of bright spots in a number of glasses)
and $\tau_u = 3\times10^{-12}$sec from (125), one obtains $\Delta\omega\approx5\times10^{4}$cm^{-1}.
This value exceeds the laser frequency itself $\omega_o\approx$
14000cm^{-1}, i.e. the incident pulse spectrum "super-
broadens". Spectrum superbroadening was experimentally
observed by Bondarenko et al. (1970) in glasses with a
laser pulse width equal to about 3×10^{-12}sec.

It is noteworthy that at such high values of $\Delta\omega$
as 10^3 to 10^4cm^{-1} the detailed description of the field
spectrum of moving focal regions may necessitate consideration of dispersion in the linear part of the
refractive index of substance. However, this problem
has not yet been considered.

Note also that spectrum interference of two (or
more) focal regions may lead to a quasistationary
structure in the total (resultant) frequency intensity
distribution. The period $\Delta\Omega$ of this structure will be
$\Delta\Omega\approx\frac{\pi}{\Delta t}$ where $\Delta t=t_m-t_n$ is the time interval between focal
regions passing through given section z_o. Depending

on conditions, both spectral distributions of a solid
type (Bolshov and Venkin, 1968; Bubnov, 1969) and distributions with a quasiperiodic structure (Shimizu,
1967; Cheung et al., 1968; Bolshov and Venkin, 1968;
Bubnov, 1969) have been experimentally observed.

V. CONCLUSION

In the above discussion we have investigated the
specific features of intensive light beam propagation
in a media with Kerr nonlinearity, provided conditions
(9), which are of special practical interest, are fulfilled. One of the consequences of the theory developed
is that under typical conditions the foci diameters are
practically independent of the incident beam power and
of the corresponding laser pulse width. This very
situation has been experimentally observed both for giant
and ultrashort (picosecond) laser pulses. The latter
point provides a reason for supposing that relations
(9) are satisfied under real conditions not only for
giant pulses but for picosecond ones as well and, hence,
that the Kerr effect, just like nonlinear absorption
in a medium, is determined by a sufficiently fast
mechanism.

1. Generalization of the Theory

Now we discuss the question of light beam propagation in a medium, provided conditions (9) are not fulfilled. Note, first of all, that recently the attention
of some authors (see, for example, Bloembergen (1971)
and Loy and Shen, (1970) has been attracted to the case
where condition (9d) is not valid, i.e. the Kerr effect
is rather inertial. This may be the case, say, for
picosecond laser pulses if the main mechanism of the

medium nonlinearity is the orientational Kerr effect.
Generally, two possibilities should be distinguished.
First, suppose that the inertial Kerr effect does not
influence focus formation and takes place only in the
focal region itself. Here, the focal region sizes will
depend on the finite setting time of the Kerr effect.
Since this process is more sensitive to changes in the
imaginary part of the refractive index than to the real
one, a direct mechanism of limiting intensity in a focus
may be nonlinear absorption associated with the delayed response of the medium polarization with respect
to the electric field. Therefore under such conditions
one may expect a multifoci structure of the light beam
similar to that discussed in Section 5. Second, suppose that the Kerr effect is rather inertial in all
beam regions in a medium. The solution of the corresponding problem can be reached by numerical methods
(attempts at its analytical solution made elsewhere are
based on an assumption that the beam preserves the
Gaussian form in a medium and therefore are not sufficiently reliable [compare Section 2]). The results
of the numerical solution of this problem are given in
Fleck and Kelley (1969), though no comprehensive solution
establishing a complete beam propagation picture is
available at present. The authors draw the conclusion
the the light propagation is not self-trapped in a
waveguide. It is noteworthy that from the theoretical
point of view for the two possibilities the diameters
of filaments and the trajectory of the moving focus
will essentially depend on detailed conditions of
observation (beam power, laser pulse width and shape)
which, however, does not at present seem to be in
agreement with experimental data and which seems to

indicate that the Kerr effect is not inertial under
the conditions involved. Therefore, we shall not now
dwell on this class of problems.

Now consider the case where nonlinear absorption
in a medium is very low, such that conditions (9a-9c)
are not satisfied in a focal region. According to the
above, these inequalities under typical conditions
have only one independent (small) parameter, for
example, $\frac{\lambda}{\Lambda_{\perp}}$, and their breakdown means that $\Lambda_{\perp} \sim \lambda$. In
this case the beam propagation picture is described
directly by the use of the Maxwell equations within
which a wave reflected from the collapse pointz* (see
Sections 3,5) in the backward direction or at great
angles to the beam axis should be formed. As to the
wave which passed in the forward direction through
the collapse point, this reflection seems to be equivalent to nonlinear absorption (taking place only at a
transverse diameter of an order of λ). Therefore, one
may expect that the light beam formation will have a
multifoci structure similar to that discussed in
Section 5 even when there is no real nonlinear absorption in a medium.

Under some special conditions (say, in the case of
appreciable material ionization in the focal region)
the main mechanism limiting intensity in this focus
may be weakening nonlinearity of the real part of n,
the refractive index due to the appearance of free
electrons. Strictly speaking, under these conditions
the dependence of the real part of the refractive index
on the light intensity ($|E|^2$) may be represented only
in the form of a functional. This problem is difficult
to compute (bearing in mind the memory and speed of
existing computers) even for the case when the

parabolic equation is applicable. Therefore, we may
consider a rough model represented by a medium with
refractive index n depending on $|E|^2$ with a suitable
functional form, say, eq. (17), i.e. as a saturable
Kerr nonlinearity (certainly, with artificially lowered
value of $|E_s|^2$). Numerical solutions (Dyshko et al.,
to be published) based on the parabolic equation show
that the model of the saturable Kerr nonlinearity
even without nonlinear absorption in a medium at
$P>P_{cr}^{(1)}$, $|E_{\phi 1}|^2 \geq 100$ (compare Section 6) also provides a
multifoci structure of the light beam. Here the structure
of the foci themselves can be described by expressions
(67)-(69) if one takes $\beta \tilde{} \alpha$.

2. Experimental Results

Now we consider briefly the experimental results on
intensive laser beam propagation in a medium. The theory
of moving foci formulated in Lugovoi and Prokhorov
(1968) was followed by a number of experimental works
which cleared up what was actually observed--moving
foci or waveguide filaments (about waveguid filaments
see "Introduction"). It has been established by a number of workers (Loy and Shen, 1969, 1970, 1971; Korobkin
et al., 1970; Lipatov et al., 1970) that in giant
laser pulses only moving foci were observed. The photographs of moving foci trajectories in Korobkin et al.,
(1970) were taken by a high-speed sweep of the processes investigated. A multifoci structure stationary in
time was observed by Lipatov et al. (1970). Loy and
Shen (1970) registered foci moving with superliminary
speed (the corresponding theory is given in Section 8a).
Later these authors (Loy and Shen, 1971) established a
good quantitative agreement between the calculated and

observed distance between a moving focus and an ultrashort pulse of a backward SRS for various instants of
time over a wide range of incident light pulse powers.
The diameter of focal region in a benzene solution was
investigated and, on the basis of the corresponding
theoretical and experimental values, the main mechanism
limiting the energy density in the focal region was
shown to be forward SRS (just as in Section 5c). The
observed values of spectrum broadening were found to
be a quantitative agreement with corresponding theoretical calculations (see expressions (125)) for various
cell lengths and incident light pulse powers.

At the same time an attempt was made to discover
a waveguide filament gave no positive results. The
earlier statement (Askaryan et al., 1971) about observations of a waveguide filament is erroneous since the
length (~10cm) of the filament observed by these
authors is not larger than, say, $\ell_\phi = 0.7kd^2 = 1.4\frac{\pi d^2}{\lambda}$ (see
the footnote on page 47) of the focal region of the
Gaussian beam in a linear medium corresponding to the
diameter of a "filament" d = 1 to 15×10^{-2}cm ($\lambda \approx 0.5 \times 10^{-4}$cm)
measured in Askaryan et al., (1971).

In picosecond laser pulses according to the theory
(see Section 8) there should also arise a structure,
at least, for a sufficiently rapid (say, electronic)
mechanism for the Kerr effect. Note also that a typical
size of the bright foci observed in the cell is
$d \sim 10^{-3}$cm and the length ℓ_ϕ of a focal region in a linear
medium, even for a Gaussian beam, is 10^{-1}cm, i.e. is
about the whole length of the light pulse train for a
typical pulse width $\tau_u \sim 3.10^{-12}$sec. Therefore, it is
clear that waveguide filaments cannot be formed under
such conditions.

We briefly consider the experimental investigations of picosecond laser pulse propagation in a medium. Brewer and Lee (1968) were the first to register bright spots at the end of the cell during these pulse propagations. Furthermore, Cebeddu et al., (1971), on the basis of spectrum broadening observations, conclude that a propagating light train has a thin core with a high energy density and intensity smoothly distributed along the longitudinal coordinate, and about as long as the whole light train. A similar picture on a basis of a model was described earlier by Akhmanov et al., (1966, 1967). However, in our opinion it is possible to explain the results represented in [80] by the multifoci structure formed. In a subsequent publication, Cebeddu and Zaroya (1971) conclude that a moving focus is formed under these conditions. Recently, Korobkin et al., (1972), while observing picosecond pulses propagating in liquid dielectrics the end of the cell have recorded bright points whose positions well coincide with the predicted (see eq. (97)) foci turning points of a multifoci structure. The discrete character of this picture is explained as follows: the times during which the focal region stays at the turning points are maximum and, therefore, most of the effect was obtained just at these points (compare Section 7). In Korobkin et al., (1972) it was concluded that a multifoci structure was formed. Zverev et al., (to be published) recorded discrete breakdowns in solid dielectrics after picosecond laser pulse propagation through them. The positions of breakdowns correspond to calculated positions of foci turning points of the multifoci structure.

Thus for giant laser pulses, the theory of multifoci structure and moving foci has total experimental

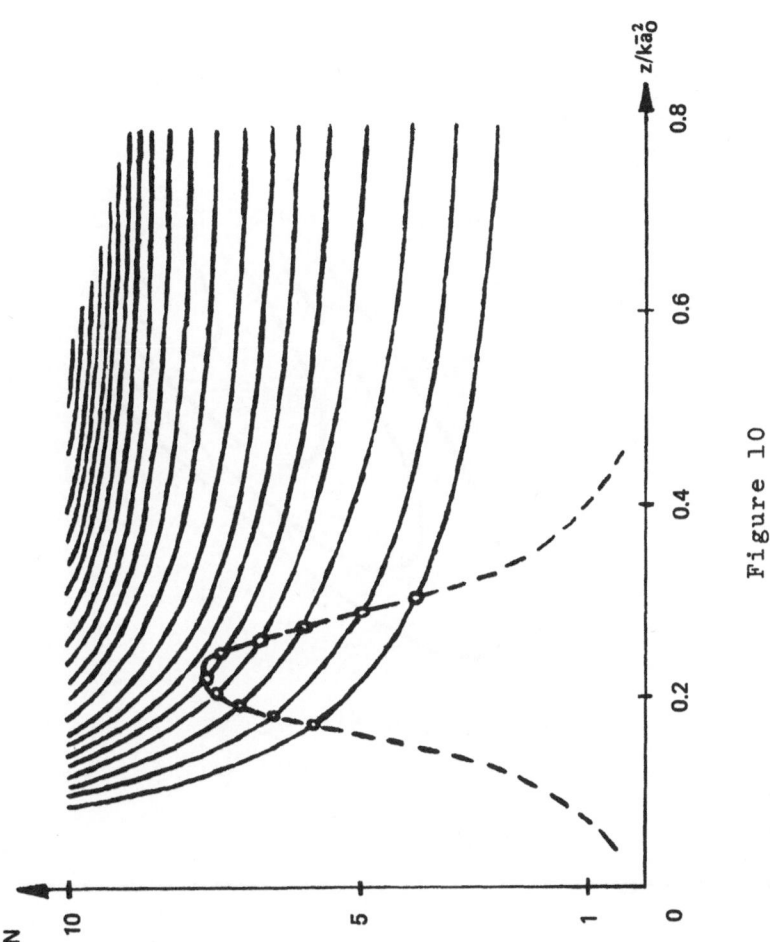

Figure 10

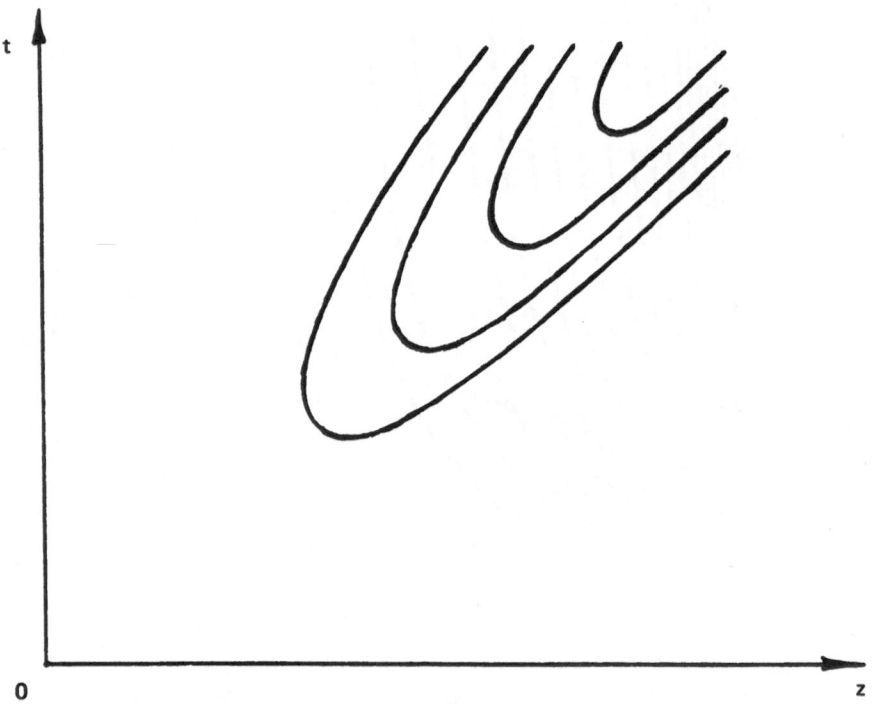

Figure 11

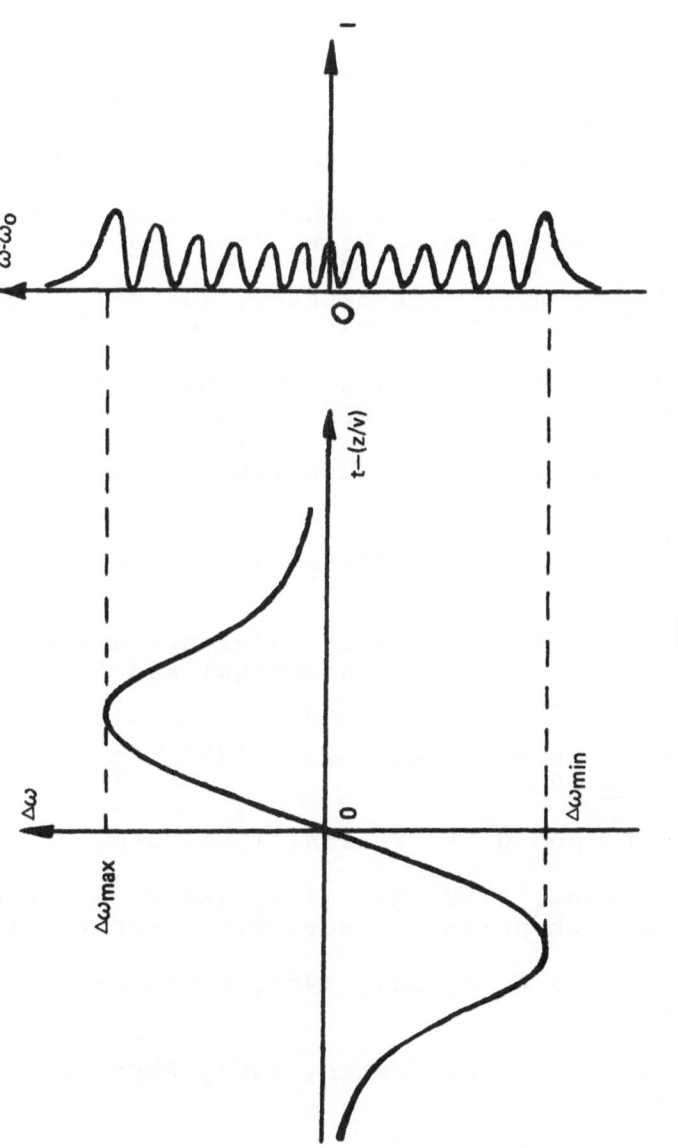

Figure 12

confirmation. For picosecond laser pulses the first
experimental data available is explained on the basis
of this theory.

REFERENCES

Abramov, A. A., V. N. Lugovoi, and A. M. Prokhorov,
    1969, Zh. Eksperim i. Teor. Fiz. Letters, 9, 675.

Akhmanov, S. A., A. P. Sukhorukhov, and R. V. Khokhlov,
    1966, Ah. Eksperim i. Teor. Fiz., 50, 1537.

Akhmanov, S. A., A. P. Sikhorukhov, and R. V. Khokhlov,
    1967, Usp. Fiz. Nauk, 93, 19.

Askaryan, G. A., Kh. A. Dyanov, and M. M. Mukhamdzhanov,
    Zh. Eksperim i. Teor. Fiz. Letters, 1971, 14, 452.

Askaryan, G. A., 1962, Zh. Eksperim i. Teor. Fiz., 42,
    1567.

Bespalov, V. I. and V. I. Talanov, 1966, Zh. Eksperim i.
    Teor. Fiz. Letters, 3, 471.

Bloembergen, N., August, 1971, paper presented at the
    Esfahan Symposium on Fundamental and Applied Laser
    Physics, Iran.

Bloembergen, N. and P. Lallemand, 1966, Phys. Rev.
    Lett., 16, 81.

Bolshov, M. A. and G. V. Venkin, 1968, JPS, 9, 1050.

Bondarenko, N. G., I. V. Eriomina, and V. I. Talanov,
    1970, Zh. Eksperim i. Teor. Fiz. Letters, 12, 125.

Brewer, R. G. and C. H. Lee, 1968, Phys. Rev. Letters,
    21, 267.

Brewer, R. G. and C. H. Townes, 1967, Phys. Rev. Letters,
    18, 196.

Brewer, R. G., J. R. Lifsitz, E. Garmire, R. Y. Chiao,
    and C. H. Townes, 1968, Phys. Rev., 166, 326.

Brewer, R. G. and J. R. Lifsitz, 1966, Phys. Rev. Let-
    ters, 23, 79.

Bubnov, M. M., 1969, Diploma Work, M.G.U.

Butenin, A. V., V. V. Korobkin, A. A. Malyutin, and
    M. Ja. Schelev, 1967, Zh. Eksperim i. Teor. Fiz.,
    6, 687.

Butylkin, V. S., A. E. Kaplan, and J. G. Kronopulo,
    1969, IZVESTYA VUZOV, Radiophysica, 12, 1792.

Cheung, A. V., D. M. Rank, R. Y. Chiao, and C. H.
    Townes, 1968, Phys. Rev. Letters, 20, 786.

Chiao, R. Y., E. Garmire, and C. H. Townes, 1964, Phys.
    Rev. Letters, 13, 479.

Chiao, R. Y., M. A. Krinsky, C. H. Townes, H. A. Smith,
    and E. Garmire, 1966, IEEE J. of Quant. Electron.,
    QE-2, 467.

Clements, W. R. L. and B. P. Stoicheff, 1968, Appl.
    Phys. Letters, 12, 246.

Cubeddu, R., R. Polloni, G. A. Sacchi, O. Svelto, F.
    Zaraga, 1971, Phys. Rev. Letters, 26, 1009.

Cubeddu, R., and F. Zaraga, 1971, Opt. Communications,
    3, 310.

Dawes, E. L. and J. H. Marburger, 1969, Phys. Rev.,
    179, 862.

Denarier-Roberge, M. M., and J. P. Taran, 1969, Appl.
    Phys. Letters, 14, 205.

DeMartini, F., C. H. Townes, T. K. Gustafson, and P. L.
    Kelley, 1967, Phys. Rev., 164, 312.

Duguay, M. A., W. Hansen, and S. L. Shapiro, 1970, IEEE
    J. Quant. Electron., 6, 725.

Dyshko, A. L., V. N. Lugovoi, and A. M. Prkhorov, 1967,
    Zh. Eksperim i. Teor. Fiz. Letters, 6, 725.

Dyshko, A. L., V. N. Lugovoi, and A. M. Prokhorov, 1969,
    Dokl. Akad. Nauk, 188, 792.

Dyshko, A. L., V. N. Lugovoi, and A. M. Prokhorov, 1971,
    Zh. Eksperim i. Teor. Fiz., 61, 2305.

Dyshko, A. L., V. N. Lugovoi, and A. M. Prokhorov, (to
    be published),

Fabelinskii, I. L. and V. S. Starunov, 1967, Appl. Opt.,
    6, 1793.

Fleck, T. A. and P. L. Kelley, 1969, Appl. Phys. Let-
    ters, 15, 313.

Garmire, E., R. V. Chiao, and C. H. Townes, 1966, Phys.
    Rev. Letters, 16, 347.

Goldberg, V. N. and V. I. Talanov, 1967, R. E. Erm,
    IZVESTIYA MUSSO, Radiophysica, 40, 674.

Gurevitch, A. V. and A. B. Shvartsburg, 1970, Zh.
    Eksperim i. Teor. Fiz., 58, 2012.

Gurevich, A. V., L. V. Pariiskaya, and A. B. Shvarts-
    burg, 1971, Sh. Eksperim i. Teor. Fiz., 61, 1979.

Gustafson, T. K. and C. H. Townes, 1972, Influence of
    Steric and Compressibility on Nonlinear Response
    to Laser Pulses and the Diameters of Self-Trapped
    Filaments, preprint.

Gustafson, T. K., and J-P. E. Taran, 1970, IEEE J.
    Quant, Electron., QE-5, 381.

Gustafson, T. K., P. L. Kelley, R. Y. Chiao, and R. G.
    Brewer, 1968, Appl. Phys. Letters, 12, 165.

Gustafson, T. K., T. P. Taran, H. A. Haus, T. R. Lif-
    sitz, and P. L. Kelley, 1969, Phys. Rev., 177,
    306.

Hellwarth, R. W., A. Owyoung, and N. George, 1971, Phys.
    Rev., A., 4, 2342.

Hellwarth, R. W., 1966, Phys. Rev., 152, 156.

Kato, Y. and H. Takuma, 1971, JOSA, 61, 347.

Kelley, P. L., 1965, Phys. Rev. Letters, 15, 1005.

Kerr, E. L., 1971, Phys. Rev. A., 4, 1195.

Korobkin, V. V. and R. V. Serov, 1967, Zh. Eksperim i.
    Teor. Fiz. Letters, 6, 642.

Korobkin, V. V., A. M. Prokhorov, R. V. Serov, and M.
    Ja. Schelev, 1970, Zh. Eksperim i. Teor. Fiz.
    Letters, 11, 153.

Korobkin, V. V., V. N. Lugovoi, A. M. Prokhorov, and
    R. V. Serov, Zh. Eksperim i. Teor Fiz. Letters,
    (to be published).

Korobkin, V. V., 1972, UFN, 107, 512; V. V. Korobkin,
    V. A. Korshunov, and A. A. Malutin (to be published).

Kroll, N., 1965, J. Appl. Phys., 36, 34.

Kudryavtseva, A. D., A. M. Sokolovskaya, M. M. Suschinsky,
    1972, Quantovaya Electronica, 7, 73.

Kyzylasov, J. I., V. S. Stamnov, and I. L. Fabelinsky,
    1969, Zh. Eksperim i. Teor. Fiz. Letters, 9, 383.

Kyzylasov, J. I. and V. S. Stamnov, 1969, Zh. Eksperim
    i. Teor. Fiz. Letters, 9, 648.

Lallemand, P. and N. Bloembergen, 1965, Phys. Rev.
    Letters, 15, 1010.

Lamb, G. L., Jr., 1971, Rev. Mod. Physics, 43, Part 1,
    99.

Lipatov, N. I., A. A. Manenkov, and A. M. Prokhorov,
    1970, Zh. Eksperim i. Teor. Fiz. Letters, 11, 444.

Loy, M M. T. and Y. R. Shen, 1969, Phys. Rev. Letters,
    22, 994.

Loy, M. M. T. and Y. R. Shen, 1970, Phys. Rev. Letters,
    25, 1333.

Loy, M. M. T. and Y. R. Shen, 1971, Appl. Phys. Letters,
    19, 285.

Lugovoi, V. N., and A. M. Prokhorov, 1968, Zh. Eksperim
    i. Teor. Fiz. Letters, 7, 153.

Lugovoi, V. N. and A. M. Prokhorov, 1970, Zh. Eksperim
    i. Teor. Fiz. Letters, 12, 478.

Lugovoi, V. N., and I. I. Sobelman, 1970, Zh. Eksperim
    i. Teor. Fiz., 58, 1283.

Lugovoi, V. N., Dokl. Akad. Nauk, 1967, 176, 58.

Lugovoi, V. N., 1968, Vvendenie v. Teoriyu vynuzhdennogo
    kombinationnogo rasseyaniva (Introduction to the
    Theory of Stimulated Raman Scattering), Nauka.

Maier, M., C. Wondl, and W. Kaiser, 1970, Phys. Rev.
    Letters, 24, 352.

Maier, M., W. Kaiser, and A. Giordmaine, 1966, Phys.
    Rev. Letters, 17, 1275.

Maier, M., W. Kaiser and A. Giordmaine, 1969, Phys.
    Rev., 177, 580.

Marburger, J. H. and E. Dawes, 1968, Phys. Rev. Letters,
    21, 556.

Polloni, R., C. A. Sacchi, and O. Svelto, 1969, Phys.
    Rev. Letters, 23, 690.

Raizer, Yu. P., 1967, Zh. Eksperim i. Teor. Fiz., 52,
    470.

Shen, Y. R. and M. M. Loy, 1971, Phys. Rev., A3, 2099.

Shen, Y. R., and Y. J. Shaman, 1967, Phys. Rev., 163,
    224.

Shen, Y. R. and Y. J. Shaman, 1965, Phys. Rev. Letters,
    15, 1008.

Shen, Y. R., 1966, Phys. Letters, 20, 378.

Shimizu, F., 1967, Phys. Rev. Letters, 19, 1097.

Szöke, A., 1964, Bull. Amer. Phys. Soc., 9, 490.

Talanov, V. I., Zh. Eksperim i. Teor. Fiz. Letters,
    1965, 2, 218.

Talanov, V. I., Zh. Eksperim i. Teor. Fiz. Letters,
    1970, II, 303.

Talanov, V. I., 1964, IVESTYA VUZOV, Radiophysica, 7,
    564.

Treacy, E. B., 1972, Measurement and Interpretation of
    Dynamic Spectrograms of Picosecond Light Pulses,
    preprint.

Volkov, T. F., 1958, Plasma Physics and the Problem of Controlled Fusion, V. III, 336.

Wang, C. C., 1966, Phys. Rev. Letters, 16, 344.

Zakharov, V. E., and A. B. Shabat, 1971, Exact Theory of Two-dimensional Self-focusing and One-dimensional Auto-modulation of Waves in Nonlinear Media, preprint I JaF AN SSR.

Zakharov, V. E., V. V. Sobolev, and V. S. Synakh, 1971, Zh. Eksperim i. Teor. Fiz., 60, 136.

Zeldovich, Ja. B. and Yu. P. Raizer, 1966, Zh. Eksperim i. Teor. Fiz. Letters, 3, 137.

Zeldovich, Ja. B., 1972, Zh. Eksperim i. Teor. Fiz. Letters, 15, 226.

Zverev, G. M., V. S. Naumov, V. A. Pashkov (to be published).

PARTICIPANTS

Laser Fusion and Lasers

R. Andrews
U. S. Naval Research Lab.
Washington, D. C.

John Apel
University of Miami

Uri Bernstein
Center for Theoretical
  Studies
University of Miami

Kenneth Billman
NASA/Ames Research Center
Moffett Field, California

John S. Blakemore
Florida Atlantic Univ.

Jean Louis Bobin
Commissariat A L'Energie
  Atomique, France

Stephen Bodner
U. S. Naval Research Lab.
Washington, D. C.

Ronald Bousek
University of Arizona

Keith Boyer
Los Alamos Scientific Lab.
University of California

Arthur Broyles
University of Florida

C. D. Cantrell
Los Alamos Scientific Lab.
University of California

George Chapline
Lawrence Livermore Lab.
University of California

Mikael Ciftan
Army Research Office
Durham, North Carolina

C. B. Collins
University of Texas, Dallas

Ralph Cooper
Los Alamos Scientific Lab.
University of California

K. Das Gupta
Texas Technical University

J. D. Daugherty
AVCO Everett Res. Lab.
Everett, Massachusetts

Raymond Elton
U. S. Naval Research Lab.
Washington, D. C.

J. Forsyth
University of Rochester

John Garrison
Lawrence Livermore Lab.
University of California

Damon Giovanielli
Los Alamos Scientific Lab.
University of California

Arthur Guenther
Air Force Weapons Lab.
Kirtland Air Force Base

Dale Henderson
Los Alamos Scientific Lab.
University of California

Robert Hofstadter
Stanford University

Heinrich Hora
Australian National Univ.
Canberra, Australia

Charles Hendricks
Lawrence Livermore Lab.
University of California

Joseph Hubbard
Center for Theoretical
  Studies
University of Miami

John L. Hughes
Australian National Univ.
Canberra, Australia

Reed Jensen
Los Alamos Scientific Lab.
University of California

O. Dean Judd
Los Alamos Scientific Lab.
University of California

Arthur Kantrowitz
AVCO-Everett Res. Lab.
Everett, Massachusetts

W. L. Kruer
Lawrence Livermore Lab.
University of California

Willis E. Lamb, Jr.
University of Arizona

Benjamin Lax
Massachusetts Institute
  of Technology

Melvin Lax
City College of the City
  University of New York

Leslie Levine
U. S. Naval Research Lab.
Washington, D. C.

John Lindl
Lawrence Livermore Lab.
University of California

William Louisell
University of Southern
  California

Moshe Lubin
University of Rochester

Philip Mallozzi
Battelle Memorial Institute

Robert Malone
Los Alamos Scientific Lab.
University of California

James Maniscalto
Lawrence Livermore Lab.
University of California

P. L. Mascheroni
University of Texas, Austin

Robert McCarthy
University of Arizona

R. McCorkle
I.B.M. Corporation
T. J. Watson Research Ctr.

William McCrory
Los Alamos Scientific Lab.
University of California

William McKnight
University of Alabama

William Mead
Lawrence Livermore Lab.
University of California

Laurence Mittag
Center for Theoretical Stds.
University of Miami

David Mosher
U. S. Naval Research Lab.
Washington, D. C.

John Nuckolls
Lawrence Livermore Lab.
University of California

Erol Oktay
Atomic Energy Commission
Washington, D. C.

Lars Onsager
Center for Theoretical
  Studies
University of Miami

Harry Robertson
University of Miami

Helmut Schwarz
Universidade de Brasilia

Marlan Scully
University of Arizona

John F. Seely
University of
  Southern California

Robert Shnidman
U.S.A. Ballistic Res. Lab.
Aberdeen Proving Grounds

George Soukup
Center for Theoretical
  Studies
University of Miami

Ian Spalding
UKAEA Research Group
England

Edward Teller
Lawrence Livermore Lab.
University of California

F. Winterberg
University of Nevada

E. A. Witalis
Research Institute of the
  Swedish National Defense
Stockholm, Sweden

Jack Wong
Lawrence Livermore Lab.
University of California

Lowell Wood
Lawrence Livermore Lab.
University of California

Gerold Yonas
Sandia Laboratory
Albuquerque, New Mexico

PARTICIPANTS

High Energy

Stephen L. Adler
Institute of Advanced
  Studies

Barry C. Barish
California Institute of
  Technology

Bruce Barnett
University of Maryland

Itzhak Bars
University of California
  at Berkeley

Sidney Bludman
University of Pennsylvania

Laurie Brown
Northwestern University

Peter A. Carruthers
Los Alamos Scientific Lab.
University of California

Demetrios Christodoulou
International Centre for
  Theoretical Physics
Trieste, Italy

W. John Cocke
Steward Observatory
University of Arizona

Fred Cooper
Belfer Graduate School
Yeshiva University

Joseph R. Cox
Florida Atlantic University

Richard Dalitz
Oxford University

P. A. M. Dirac
Florida State University

Paul Fishbane
University of Virginia

Daniel Freedman
State University of New
  York at Stony Brook

Howard M. Georgi III
Harvard University

Frederick Gilman
Stanford Linear
  Accelerator Center
Stanford University

Sheldon Glashow
Harvard University

Ahmad Ali Golestaneh
Mount Union College

O. W. Greenberg
University of Maryland

Gerald Guralnik
Brown University

Leopold Halpern
Florida State University

Roman Jackiw
Massachusetts Institute
  of Technology

Arthur M. Jaffee
Harvard University

Hans A. Kastrup
Institut für Theoretische
  Physik der Rheinisch-
  Westfälischen
  Technischen Hochschule
Aachen, Germany

Nicholas Kemmer
University of Edinburgh

Abraham Klein
University of Pennsylvania

Behram Kursunoglu
Center for Theoretical
   Studies
University of Miami

Christian LeMonnier
   de Gouville
Center for Theoretical
   Studies
University of Miami

Harvey Lynch
Stanford Linear
   Accelerator Center
Stanford University

K. T. Manhanthappa
University of Colorado

Sydney Meshkov
National Bureau of
   Standards
Washington, D. C.

Stephen Mintz
Florida International Univ.

R. Mohapatra
City College of the City
   University of New York

Robert J. Oakes
Northwestern University

Reinhard Oehme
University of Chicago

Heinz R. Pagels
Rockefeller University

Leonard Parker
University of Wisconsin
   at Milwaukee

Jogesh Pati
University of Maryland

Arnold Perlmutter
Center for Theoretical Stds.
University of Miami

Donald Pettengill
Center for Theoretical Stds.
University of Miami

Paul Roman
Boston University

Igor Saavedra
Universidad de Chile

Abdus Salam
International Centre for
   Theoretical Physics
Trieste, Italy

John H. Schwarz
California Institute
   of Technology

Charles Sommerfield
Yale University

Karl Strauch
Harvard University

E. C. G. Sudarshan
University of Texas, Austin

Leonard Susskind
Yeshiva University

Ivan Todorov
Institute for Advanced
   Studies

Hiroomi Umezawa
University of Wisconsin
   at Milwaukee

Kameshwar Wali
Syracuse University

Ming-Yang Wang
Center for Theoretical Stds.
University of Miami

Fredrik Zachariasen
California Institute
   of Technology

Anthony Zee
Princeton University

SUBJECT INDEX